A Modern Course on the Theory of Equations

A Modern Course on the Theory of Equations 2nd edition

David Dobbs, University of Tennessee
and Robert Hanks

with an appendix on microcomputer programming
by Michael Weinstein

Polygonal Publishing House
Washington, New Jersey

interior artwork by Jose Garcia
cover art by Marc Grabler

Library of Congress Cataloging-in-Publication Data
Dobbs, David E.
A modern course on the theory of equations / David E. Dobbs and Robert Hanks ; with an appendix by Michael Weinstein. -- 2nd ed.
p. cm.
Includes bibliographical references and index.
ISBN 0-936428-14-7
1. Equations, Theory of. I. Hanks, Robert. II. Title.
QA211.D63 1992
512.9'42--dc20 92-7178
CIP

manufactured in the United States of America
by Thomson-Shore
printed on acid-free paper

Polygonal Publishing House
Box 357
Washington, NJ 07882

Contents

Introduction: a Word to the Reader

This text is intended to provide a modern introduction to the theory of polynomial equations. We hope that it will be used by a varied clientele, in keeping with the variety of reasons for studying such equations. First, it is natural to study polynomials because they are already familiar: a polynomial, such as $2x^2 - 7x + (3/8)$, is built from comprehensible constants (in this case, 2, -7, and 3/8) and the "variable" x by means of the fundamental arithmetic processes of addition, subtraction, and multiplication. Moreover, most readers already know from high-school that the important game to be played with polynomials concerns finding their roots: for the above example, the quadratic formula computes the roots to be $(7 + \sqrt{46})/4$ and $(7 - \sqrt{46})/4$. Second, it is a happy circumstance that concrete objects such as polynomials are actually useful. Some applications are practical, as in the analysis of solutions of differential equations governing a projectile's motion. Theoretical applications include Weierstrass's theorem that, over suitable domains, arbitrary continuous functions can be uniformly approximated by real polynomials.

The "modern" nature of this book arises in two ways. First, we develop the theory of polynomials using a thoroughly algebraic approach, which we believe is welcomed by the "new math" generations of students. Second, thanks to the availability of relatively inexpensive hand-held calculators, we have devoted an entire chapter (6) to the implementation of several nontrivial root-approximation methods. In accordance with today's undergraduate curricular emphasis on applied mathematics and numerical methods, it should be noted that, by covering only the unstarred sections listed in the table of contents, one may fashion a one-quarter course in which up to one-third of the time is devoted to Chapter 6. Indeed, this text originated as a project, suggested by the University of Tennessee's Mathematics Education Committee, in response to the needs of Mathematics 3720, a one-quarter course populated primarily by (future and current) secondary-school mathematics teachers. We believe that, in its present form, the text more than meets the needs of any such teacher-training course. It is, moreover, sufficiently detailed, with careful introductions for each section and chapter, to be utilized for self-study, independently of any formal instruction. In addition, as explained below, a year's course based on the entire text presents a viable alternative to the usual junior/senior-level courses on abstract algebra.

Because we anticipate a varied audience for this book, due to the variety of its possible uses, we have endeavored to keep its prerequisites to a minimum. A year's course in calculus will certainly guarantee the necessary background. Specifically, the reader is assumed to be acquainted with high-school trigonometry, as well as high-school algebra at the level of the binomial theorem (concerning the expansion of $(a + b)^n$) and the fundamental theorem of arithmetic (stating that each positive integer exceeding 1 is expressible, uniquely apart from the order of factors, as the product of finitely many positive prime integers). For Chapters 5 and 6, the reader should recall calculus's intermediate-value theorem for continuous functions and the geometric interpretation of derivatives. He/she will also need an inexpensive hand-held scientific calculator; the precise functions required on such an instrument are described in the introduction to Chapter 6.

In writing this book, we have rejected any alleged dichotomy between theory and applications. While, by and large, Chapters 1 and 2 may be said to cover the algebraic theory, with special cases treated in Chapters 3–5 and numerical methods handled in Chapter 6, the text is really more integrated than such a facile description might suggest. For example, Chapter 6's root-approximation methods of Horner, Graeffe, and Bernoulli are previewed in Chapter 2, thereby reinforcing the interdependence of theory and applications. Similarly, the first step of Chapter 5's root-separation technique (due to Sturm) is actually carried out in Chapter 2's algorithm for identifying the multiplicities of roots. We expect that the integrated aspect of the text will allow flexibility in its coverage. In particular, the reader may proceed from Section 4 of Chapter 2 to the beginning of any subsequent chapter. Skimming is facilitated by the nature of the examples and comments in the text's first section: these are met and verified, respectively, several times in the book, using increasingly deep parts of the subject. We have also tried to encourage the reader to keep his/her feet wet, that is, to attempt homework before completing any section, by making the ordering of exercises at the end of each section parallel the order of the text's discussion. Also in the interest of moving the reader quickly through the preliminaries, we recommend that he/she not allow the first section's "new math" axiomatics to impede progress. From a *first* reading of that section, one needs only to learn the difference between a domain and a field.

We next turn to unusual aspects of this text on the theory of equations. One of these is its possible use in an undergraduate year-long course on "concrete" abstract algebra. To be sure, many of the typical concerns of abstract algebra are present in the unstarred sections. For example, with respect to ring theory, our approach emphasizes the role of the homomorphism properties of evaluation maps and complex conjugation. In the same vein, the notions of factor ring and reduction modulo a prime are used to prove a theorem of Schur which extends Euclid's theorem on the infinitude of primes. Although group theory's role in the unstarred sections is not explicit, our discussions of roots of unity and symmetric polynomials prepare the way for later study of that subject. While linear algebra, specifically the dimension theory for vector spaces, is not a prerequisite for the text, minor uses of it contribute to a strong, modern unit on field theory and Galois theory. (See the following starred sections—Section 6 of Chapter 2 and Sections 3 and 5 of Chapter 3. As with all starred sections, they point to deeper results, and the student/instructor should be flexible in covering them. Portions of the unit on field theory may well prove challenging to beginning graduate students. In particular, note that Section 5 of Chapter 3 uses some clearly stated results from group theory.)

Rather than more abstract algebra of the above type, a year-long course would also cover those starred sections which were not mentioned above. By and large, these sections prove hard classical theorems in the theory of equations, such as Gauss's fundamental theorem of algebra; the fundamental theorem about symmetric polynomials (proved algorithmically); the irreducibility results of Gauss and Eisenstein; the algorithm of Kronecker for factoring rational polynomials; alternate root-separation techniques of Vincent and Budan; estimates of error for the root-approximation methods in Chapter 6; and Newton's formulas—the backbone of Bernoulli iteration—via formal power series, with an application to the combinatorial symbol ${}_nH_r$.

There are several other unusual aspects of this book. The foundational material in Chapter 1 is couched in terms of polynomials in one variable over a domain, rather than over (the more customary, but less general, case of) a field. Two technical advantages are the later rigorous treatment of (possibly symmetric) polynomials in several variables; and the first-named author's recent algorithm for verifying trigonometric identities. (For the latter, see Section 5 of Chapter 1.) The reader may also note that our approach to Cardan's cubic formula is, technically, rather unusual. In addition, to accommodate the needs of readers who are interested in roots but not in the more general questions involving factorization, we have devised a weak version of Eisenstein's criterion, for inclusion in (the unstarred) Section 1 of Chapter 4. The traditional (strong) version of Eisenstein's criterion and Gauss's lemma are proved in (the starred) Section 2 of Chapter 4 in an unusually elegant way that utilizes the algebraic preliminaries established in the first two chapters.

To keep the book down to a manageable size, we have refrained from relating much of the fascinating history associated with the development of the theory of equations. We do, however, say something about the lives of Cardan, Tartaglia, Abel, and Galois. The interested reader (student *or* teacher) is encouraged to augment such tidbits and to discover others through independent study.

As the reader's studies will (we hope) not end with the reading of this book, we have included an epilogue, *Quo Vadis*?, to aid him/her in selecting appropriate subsequent texts and courses of study. In conjunction with the epilogue, an annotated list of references is given, both to aid in those subsequent studies and to provide references which elaborate upon the text's treatment at key points.

The reader may note slight differences in style between Chapter 6 and the remainder of this book. We hope that these variations won't detract from the book's unity or usefulness. To some extent, the nature of the material suggested these differences, but the principal explanation for them arises from the book's dual authorship: Hanks had primary responsibility for the writing of Chapter 6, while Dobbs wrote the rest. However, our interaction has involved each of us in the entire project, and we would both appreciate receiving any reader's comments that might improve later editions' exposition.

We owe expressions of thanks to our colleagues on the Mathematics Education Committee, especially Donald J. Dessart and Henry Frandsen, for encouraging our pursuit of the project which has resulted in this book. The secretarial support for the manuscript's typing was splendid, as was the subsequent cooperation of Michael Weinstein at Polygonal Publishing House. The assistance of our students who served as guinea pigs in Mathematics 3720 for this book's first two drafts is also gratefully noted. In addition, Dobbs wishes to acknowledge the influence of Nathan Mendelsohn who, nearly twenty years ago, taught him the theory of equations and much more. Our most profound gratitude is, of course, to our wives, Elaine Dobbs and Jacquelyn Hanks, for their patient and understanding support while keeping the respective home fires burning.

Introduction to the Second Edition

We are glad that the response to the first edition has been so favorable that this second edition has seen the light of day. The book's rationale remains unchanged: please read "Introduction: a word to the reader," if you have not already done so. In addition to correcting typographical errors in the text and answers to selected problems, we have made one substantial addition: an appendix, written by Michael Weinstein of Polygonal Publishing House, gives programs implementing a variety of numerical approximation schemes for roots. Just as hand-held calculators had become commonplace at the time of the first edition, many students and instructors now have access to personal computers and have requested a revision with such supplementary material. In this way, we hope that the second edition of this concrete introduction to abstract algebra will continue to merit the term "modern" in its title.

1

Fields and Polynomials

In the introduction, several reasons were suggested for the study of polynomial equations, and this chapter begins such a study. Section 3 revisits the algebraic operations on polynomials, which will be familiar to most readers from their high school studies, culminating with the fundamental algorithm of Euclid for computing greatest common divisors. The optional sections 4 and 5 give immediate dividends of the preceding material by proving facts which most readers will have accepted on faith in earlier courses in calculus and trigonometry. Of course, such an effective algebraic arsenal can only be assembled after some axiomatics. Accordingly, Section 1 lays bare the nature of some algebraic systems in which the coefficients and roots of polynomials reside—veterans of "new math" should not be upset by such matters. Section 2 forms a healthy concrete counterpart, through the analysis of the algebra of the complex number system; and foreshadows the treatment of polynomials, by a trigonometric description of n-th roots.

1. Integral Domains and Fields

The reader is certainly acquainted with properties of the binary operations of addition and multiplication on the sets **Z**, **Q**, and **R**. (As usual, **Z** is the set of integers $0, 1, -1, 2, -2, 3, -3, \ldots$; **Q** is the set of rational numbers, that is, fractions a/b, with $a \in \mathbf{Z}$ and $0 \neq b \in \mathbf{Z}$; and **R** is the set of real numbers, that is, sums of integers and (possibly nonrepeating) decimal expansions. We have the inclusions $\mathbf{Z} \subset \mathbf{Q} \subset \mathbf{R}$, and the operations of addition and multiplication on the larger algebraic system **R** restrict in the obvious sense, to the corresponding operations on the smaller sets, **Q** and **Z**.) Specifically, it is beneficial to collect *those* of the properties which will serve to define the realms over which the abstract study of polynomials most fruitfully proceeds, and we now proceed to do this.

Consider a set S (for example, **Z**, **Q**, or **R**), together with two binary operations called *addition* and *multiplication* (denoted in the usual way) subject to the following conditions:

(i) Closure: for each $a \in S$ and $b \in S$, both the *sum* $a + b$ and the *product* ab are elements of S;

(ii) Commutative laws: for each $a \in S$ and $b \in S$, $a + b = b + a$ and $ab = ba$;

(iii) Associative laws: for each $a \in S$, $b \in S$, and $c \in S$, $(a + b) + c = a + (b + c)$ and $(ab)c = a(bc)$;

(iv) Neutral elements: there exist unequal elements 0 and 1 in S such that $a + 0 = a$ and $a1 = a$ for each $a \in S$;

(v) Additive inverses: for each $a \in S$, there exists an element denoted $-a$ in S (and called the *additive inverse* of a) such that $a + (-a) = 0$, where 0 is as in (iv);

(vi) Cancellation law of multiplication: whenever $ab = ac$ for elements a, b, c of S and $a \neq 0$, then $b = c$;

(vii) Distributive law: for each $a \in S$, $b \in S$ and $c \in S$, $a(b + c) = ab + ac$.

An algebraic system S (fortified with an addition and multiplication) which satisfies the conditions (i)–(vii) will be called an *integral domain*, or simply a *domain*. (The older terminology was "domain of integrity," as it was felt that integrity is exhibited by a in (vi), in that multiplication by (nonzero) a cannot cause an equation such as $ab = ac$ except in the presence of the equation $b = c$.) Note that we have rigged matters so that **Z**, **Q**, and **R** are each *examples of domains.* Before giving further examples, we pause to draw some elementary consequences from (i)–(vii).

Some consequences are truly immediate. For example, (ii) and (iv) combine to show that $0 + a = a$ and $1a = a$ for each $a \in S$; similarly, (ii) and (vii) yield $(a + b)c = ac + bc$ for all elements a, b, and c of S. A slightly deeper upshot is the "anything times zero is zero" law of arithmetic; that is, $a0 = 0$ for each $a \in S$. Its proof begins with the impeccable equation $0 + 0 = 0$, passes to $a(0 + 0) = a0$, uses (vii) to conclude $a0 + a0 = a0$, adds $-(a0)$ to both sides to get $(a0 + a0) + (-(a0)) = 0$, and rearranges the left-hand side as $a0 + (a0 + (-(a0)))$, from which the assertion is clear. Note that so "simple" a proof made appeals—sometimes tacitly—to five of the seven properties (i)–(vii): only (ii) and (vi) were not used! It should now be clear why 0 and 1 were assumed unequal in(iv); indeed, if $0 = 1$, then S would consist merely of the single element 0 (a most uninteresting situation), for any $a \in S$ would satisfy $a = a1 = a0 = 0$.

Not only are 0 and 1 *distinct* in a domain, they are *uniquely determined,* too. To see this, say for the case of 0, suppose that $b \in S$ also satisfies $a = a + b$ for each $a \in S$. Then (taking $a = 0$) we have $0 = 0 + b = b$, that is, $0 = b$. One may also prove that additive inverses are uniquely determined, for if $a + c = a + (-a)$, then adding $(-a)$ to (the left of) both sides gives, thanks to (ii) and (iii), the equation $0 + c = 0 + (-a)$, whence $c = -a$. The familiar arithmetic law $a = -(-a)$ now follows for elements a in *any* integral domain, since a has the property demanded of the (uniquely determined) element $-(-a)$, namely that $-a + a = 0$. Similarly, we may infer the law $(-a)b = -(ab) = a(-b)$. Indeed, $(-a)b$ is *an* (and hence *the*) additive inverse for ab because $ab + (-a)b = (a + (-a))b = 0b = b0 = 0$; a like argument yields $a(-b) = -(ab)$.

Addition satisfies a cancellation law in the sense of (vi); in other words, given $a + b = a + c$ in some domain, one may conclude that $b = c$. For a proof, add $-a$ to the left of both sides and simplify, a trick used in each of the preceding two paragraphs. The same trick shows that (polynomial) equations of the form $x + b = a$ have unique solutions in a domain S. The solution $x = a + (-b)$ is usually denoted $a - b$ and serves to introduce the binary operation of *subtraction* on S. (Question: which, if any, of the analogues of (i)–(vi) does subtraction satisfy?) It is a simple matter to deduce the distributive laws $a(b - c) = ab - ac$ and $(a - b)c = ac - bc$. (Do it.) As a result, the "integrity" condition, (vi), is readily shown to be equivalent to

(vi)′ Triviality of zerodivisors: if $a \in S$ and $b \in S$ satisfy $ab = 0$, then either $a = 0$ or $b = 0$ (or both).

(Prove, in the context of (i)–(v) and (vii), that (vi) is equivalent to (vi)′.)

The reader will have noticed that the chief ingredient in proving the cancellation law of addition was condition (v) having to do with the existence of additive inverses. It is reasonable to ask why, instead of postulating the cancellation law of multiplication (vi), did we not list an axiom about existence of "multiplicative inverses." (By a *multiplicative inverse* of an element a of some domain S is meant some $b \in S$ satisfying $ab = 1$; such b is typically denoted a^{-1}.) Certainly, multiplicative inverses do exist in abundance: for instance, $(-1)^{-1} = -1$ in **Z**, $(\frac{7}{2})^{-1} = \frac{2}{7}$ in **Q**, and $\pi^{-1} = 1 \div \pi$ in **R**. However, 0^{-1} does not exist in *any* domain, since the equation $0b = 1$ would entail the forbidden equation $0 = 1$. Steering clear of 0^{-1}, we are thus led to considering

(viii) Multiplicative inverses of nonzero elements: for each $a \in S$ such that $a \neq 0$, there exists an element denoted a^{-1} in S such that $aa^{-1} = 1$.

Condition (viii) represents something qualitatively new in our list, inasmuch as **Q** and **R** each satisfy (viii), but **Z** does not. (Why not? What is 2^{-1}?) A condition bringing about so important a distinction merits special terminology, which we next introduce. An algebraic system satisfying (i)—(viii) is called a *field*. **Q** and **R** are *examples of fields*. **Z** is an example of a *domain which is not a field*. Of course, any field is a domain.

Fields are ideally suited for the operation of division. If a and b are elements of a field S such that $a \neq 0$, then the equation $ax = b$ has the unique solution $x = ba^{-1}$, usually denoted b/a or $b \div a$, in S; the induced binary operation on the set of nonzero elements of S is called *division*. (Thus we see that the arithmetic maxim, "Don't divide by 0," is due to the nonexistence of 0^{-1}). The notations b/a and $b \div a$ may be used in any domain S in which a^{-1} exists, although the binary operation of division is then defined on the set of elements of S which have multiplicative inverses. (Question: which, if any, of the analogues of (i)—(viii) must division satisfy?) The reader has doubtlessly divided in **Q** for quite some time using the "invert and multiply" rule, as in $(2/7)/(3/14) = (2/7)(14/3) = 4/3$, the first step of which is just a special case of $b/a = ba^{-1}$. Similarly, the rule "$0 \div a = 0$ if $a \neq 0$" follows from $a0 = 0$.

Before listing some more examples of fields, we shall make two additional comments. First, for any domain S, there is essentially only one way to construct a field F such that $S \subset F$, the binary operations of addition and multiplication on F restrict to the corresponding operations on S, and each element of F may be expressed in the form ba^{-1} for some elements $b \in S$ and (nonzero) $a \in S$. As above, ba^{-1} may be viewed as either $b \div a$ or b/a, and so it is customary to call F the *field of fractions* (or the *quotient field*) of S, denoted $F = \text{qf}(S)$. The most familiar example of a quotient field arises from $S = \mathbf{Z}$, in which case $F = \mathbf{Q}$. (For a more esoteric example, see the formal Laurent series in the optional final section of

Chapter 2.) The diligent reader will be able to verify that qf(S) may be constructed, together with the needed binary operations, as follows. The elements of qf(S) are the equivalence classes obtained from the equivalence relation given by: $(b_1, a_1) \sim (b_2, a_2)$ if and only if $b_1 a_2 = b_2 a_1$, where b_1, a_1, b_2, a_2 are members of S and a_1 and a_2 are nonzero. (To motivate the definition, think of (b_i, a_i) as the fraction b_i / a_i, and recall the "cross-multiplying" trick.) Addition and multiplication on qf(S) are defined by aping the usual rules for operating with fractions in **Q**. For the sake of brevity, we omit the details and verification that qf(S), thus constructed, has the desired properties.

The second comment will reduce the details involved in checking that a given structure is, indeed, a field. Specifically, if S satisfies conditions (i)—(v), (vii), and (viii), then S automatically satisfies (vi), and so S is a field. The proof that (vi) is a consequence of the other axioms is simplicity itself: given $ab = ac$ and $a \neq 0$, multiply by a^{-1} (whose existence is guaranteed by (viii)), and use commutativity and associativity to conclude that $b = 1b = (a^{-1}a)b = (a^{-1}a)c = 1c = c$, as required.

Our first application of the preceding comment will be to verify that $\mathbf{Q}(\sqrt{2})$, defined as $\{a + b\sqrt{2} \in \mathbf{R} : a \in \mathbf{Q} \text{ and } b \in \mathbf{Q}\}$, is a field. (As usual, $\sqrt{2}$ is taken to be *the* positive real number satisfying $(\sqrt{2})^2 = 2$. Similarly, $\sqrt{3}$, $\sqrt{6}$, and $\sqrt[3]{2}$ are positive, according to the usual conventions of high-school algebra. Of course, $\sqrt[3]{-8}$ is negative. We shall avoid notation such as $\sqrt[4]{-8}$, but the substantive issues broached by suggesting such symbols will be treated in the next section.) To show that $\mathbf{Q}(\sqrt{2})$ is a *subfield* of **R** (which we *do* know *is* a field), it is enough to verify conditions (i), (iv), (v), and (viii) for $\mathbf{Q}(\sqrt{2})$, as the other conditions (commutative laws, etc.) will then be inherited from **R**. Of these, all but (viii) are easy to check. For (i), note that $(a + b\sqrt{2}) + (c + d\sqrt{2}) = (a + c) + (b + d)\sqrt{2}$ and $(a + b\sqrt{2})(c + d\sqrt{2}) = (ac + 2bd) + (ad + bc)\sqrt{2}$; to verify (iv) and (v), observe $0 = 0 + 0\sqrt{2}$, $1 = 1 + 0\sqrt{2}$, and $-(a + b\sqrt{2}) = (-a) + (-b)\sqrt{2}$. The final detail, checking (viii), involves a pesky fact, namely that $\sqrt{2}$ is irrational. We trust that many readers will have seen Pythagoras' proof of this fact, using the fundamental theorem of arithmetic. More generally, we will assume $\sqrt{n}$ is irrational whenever n is a *squarefree* integer (exceeding 1). The reader may prove this by adapting Pythagoras' methods or he/she may accept it without proof, as we *shall* establish a more general result in the first section of Chapter 4. Now, to verify (viii), consider some nonzero element $e = a + b\sqrt{2}$, with $a \in \mathbf{Q}$ and $b \in \mathbf{Q}$. It is enough to prove that e^{-1}, which is guaranteed to exist in **R**, actually lies in $\mathbf{Q}(\sqrt{2})$. The computation *seems* a straightforward enough rationalization in **R**:

$$e^{-1} = \frac{1}{a + b\sqrt{2}} = \frac{a - b\sqrt{2}}{(a + b\sqrt{2})(a - b\sqrt{2})} = \left(\frac{a}{a^2 - 2b^2}\right) + \left(\frac{-b}{a^2 - 2b^2}\right)\sqrt{2}.$$

The detail overlooked involves the division by $a - b\sqrt{2}$: this is fine provided that $a - b\sqrt{2}$ is nonzero, and that is where the pesky fact comes in. If $a - b\sqrt{2}$ were 0, then b would have to be 0, for otherwise $\sqrt{2} = ab^{-1}$ would be rational, contrary to fact; thus, if $a - b\sqrt{2} = 0$, then $a = b\sqrt{2} = 0\sqrt{2} = 0$, whence $e = 0 + 0\sqrt{2} = 0$, another contradiction. Accordingly, $a - b\sqrt{2}$ is *not* 0, and $\mathbf{Q}(\sqrt{2})$ *is* a field. By the same rationalization technique, one may show for *each* squarefree integer $n > 1$ that $\mathbf{Q}(\sqrt{n}) = \{a + b\sqrt{n} : a \in \mathbf{Q} \text{ and } b \in \mathbf{Q}\}$ is a field.

If $\{F_1, F_2, \ldots\}$ is a set of fields, each contained in a given field F, then the intersection $\bigcap F_i$ is also a subfield of F. (To prove this, show that the intersection inherits (i)–(viii).) In particular, $\mathbf{Q}(\sqrt{2}) \cap \mathbf{Q}(\sqrt{3})$ is a field: we claim that it is merely $\mathbf{Q}$. Indeed, if $r = a + b\sqrt{2} = c + d\sqrt{3}$ for some rational numbers a, b, c, and d, then $b\sqrt{2} = e + d\sqrt{3}$ where $e = c - a$, and squaring both sides leads to $2b^2 = e^2 + 3d^2 + 2ed\sqrt{3}$. Arguing as in the preceding paragraph, we infer from the irrationality of $\sqrt{3}$ that $2b^2 = e^2 + 3d^2$ and $2ed = 0$. As $2 \neq 0$ (more about that soon!), (vi)′ shows that either $d = 0$ or $e = 0$. If $d = 0$, then $r = c \in \mathbf{Q}$, as desired. Now, if $e = 0$, then $2b^2 = 3d^2$. The conclusion that $r \in \mathbf{Q}$ follows in this case, too, although more pesky number theory intervenes first. Indeed, $3d^2$ must be an even number (since $2b^2$ is visibly even), and so d is even (it can't be odd, after all!), say $d = 2f$, with $f \in \mathbf{Z}$. Then $2b^2 = 12f^2$ and, by (vi), $b^2 = 6f^2$; as $\sqrt{6}$ is irrational, we conclude $f^2 = 0$, whence $b^2 = 0$ and $b = 0$, so that $r = a \in \mathbf{Q}$. Hence, $\mathbf{Q}(\sqrt{2}) \cap \mathbf{Q}(\sqrt{3}) \subset \mathbf{Q}$; as the reverse inclusion is evident, the above claim has been proved. More generally, one may prove that if n and m are distinct squarefree integers exceeding 1, then $\mathbf{Q}(\sqrt{n}) \cap \mathbf{Q}(\sqrt{m}) = \mathbf{Q}$.

We are now ready to verify that $\mathbf{Q}(\sqrt{2}, \sqrt{3}) = \{a + b\sqrt{2} + c\sqrt{3} + d\sqrt{6} \in \mathbf{R} : a, b, c, d \text{ elements of } \mathbf{Q}\}$ is a subfield of $\mathbf{R}$. As above, it is just a matter of verifying (i), (iv), (v), and (viii). Since $\sqrt{2}\sqrt{3} = \sqrt{6}$, we have $\sqrt{6}\sqrt{2} = 2\sqrt{3}$ and $\sqrt{6}\sqrt{3} = 3\sqrt{2}$, which easily leads to closure under multiplication. The other details in checking (i), (iv), and (v) are even easier, and so we are left with (viii). Consider a nonzero element $e = a + b\sqrt{2} + c\sqrt{3} + d\sqrt{6}$ of $\mathbf{Q}(\sqrt{2}, \sqrt{3})$; let $e^* = a + b\sqrt{2} - c\sqrt{3} - d\sqrt{6}$. If $e^* = 0$, then $a + b\sqrt{2} = (c + d\sqrt{2})\sqrt{3}$ and, since we proved in the preceding paragraph that $\sqrt{3} \notin \mathbf{Q}(\sqrt{2})$, it follows that $c + d\sqrt{2} = 0 = a + b\sqrt{2}$, whence $a = b = c = d = 0$, contradicting $e \neq 0$. Consequently, $e^* \neq 0$ and rationalization gives $e^{-1} = e^*/ee^*$. However, $ee^* \in \mathbf{Q}(\sqrt{2}\,)$ (expand and check), and so $(ee^*)^{-1}$ belongs to $\mathbf{Q}(\sqrt{2}\,)$ which is contained in $\mathbf{Q}(\sqrt{2}\,,\sqrt{3}\,)$, whence e^{-1} is (a product of two elements) in $\mathbf{Q}(\sqrt{2}\,,\sqrt{3}\,)$, completing the verification.

Another example of a subfield of $\mathbf{R}$ is

$$\mathbf{Q}(\sqrt[3]{2}\,) = \left\{ a + b\sqrt[3]{2} + c\sqrt[3]{4}\ : a, b, c \text{ elements of } \mathbf{Q} \right\}.$$

As usual, verification is routine, except for (viii). Construction of multiplicative inverses is excessively tedious here (possible, though, by Cramer's rule, not by rationalization), so we omit the details for this case, with the promise that an algorithm will be developed in Section 3 for a more general situation.

It is important to note that the few axioms that characterize fields are sufficiently general to permit very unfamiliar examples of fields. To illustrate this, consider distinct symbols 0 and 1 (or a and b, if you prefer) and subject $S = \{0, 1\}$ to binary operations of addition and multiplication given in the following tables:

+	0	1
0	0	1
1	1	0

and

·	0	1
0	0	0
1	0	1

An enjoyable analysis of cases reveals that S, though *finite*, is a field.

The preceding construction may be generalized. Let n be any positive integer exceeding 1. (In the preceding example, n is 2.) Let $S = \{0, 1, 2, 3, \ldots, n-1\}$ be the set of the first n nonnegative integers. Operations of addition and multiplication are induced on S by the usual operations on $\mathbf{Z}$, closure being guaranteed by passage to the remainder upon division by n. For example, if $n = 5$, then $1 + 3 = 4$, $1 + 4 = 0$, and $3 \cdot 4 = 2$. When thus equipped with binary operations, S is called the mathematical system of *integers modulo n,* and is denoted $\mathbf{Z}_n$. We observed above that $\mathbf{Z}_2$ is a field. It is easy to check that $\mathbf{Z}_3$ is a field (check this). However, $\mathbf{Z}_4$ is not a field: indeed, $\mathbf{Z}_4$ is not even a domain, since (vi)′ is violated by the equation $2 \cdot 2 = 0$ in $\mathbf{Z}_4$. If $\mathbf{Z}_n$'s construction is new to the reader, he/she will discover its origin and a possible application to "clock arithmetic" by contemplating a clock (nondigital!) whose face looks like the one shown in Figure 1.

FIGURE 1

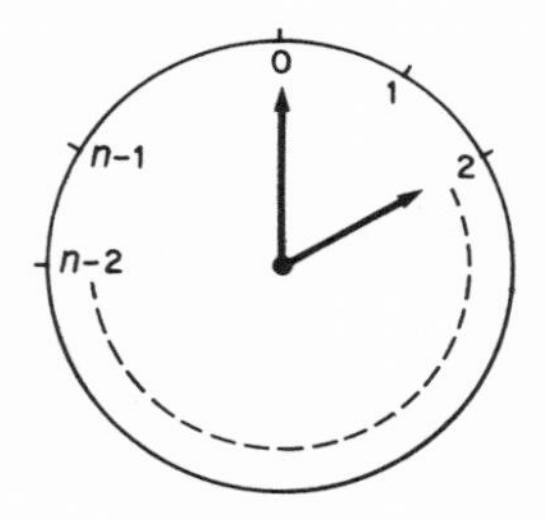

Two questions arise naturally, as a result of the preceding examples. For which n is $\mathbf{Z}_n$ a field? For which n is $\mathbf{Z}_n$ a domain? Somewhat surprisingly, these questions have the same answer: for *prime* n. (Recall that an integer $n > 1$ is called *prime* in case n cannot be expressed as a product of integers exceeding 1. Examples of prime numbers include 2, 3, 5, 7, and 23; 4 and 6 are examples of numbers which are not prime. A famous result of Euclid is that there are infinitely many prime integers—we give an extension of this at the close of Section 1 of Chapter 2.) To show that n is prime whenever $\mathbf{Z}_n$ is a domain, we need the notion of *characteristic*. If S is a domain, we shall say that S has characteristic n and write $\text{char}(S) = n$ in case n is the least positive integer with the property that the sum of n 1s is 0 in S; if no such positive n exists, the domain S is said to have *characteristic zero*, and we write $\text{char}(S) = 0$ in this case. For example, $\text{char}(\mathbf{Z}_2) = 2$; $\text{char}(\mathbf{Z}_3) = 3$; and $\mathbf{Z}$, $\mathbf{Q}$, $\mathbf{Q}\sqrt{2}$, and $\mathbf{R}$ all have characteristic 0. A fundamental result is that, if S is a domain of characteristic n and $n \neq 0$, then n is a prime. (The indirect proof is amusing: if $n = rd$ for integers $r > 1$ and $d > 1$, then the "generalized distributive law" gives

$$0 = \underbrace{1 + 1 + \ldots + 1}_{n \text{ summands}} = \underbrace{(1 + \ldots + 1)}_{r \text{ summands}} \cdot \underbrace{(1 + \ldots + 1)}_{d \text{ summands}};$$

hence, (vi)′ implies that one of the factors on the right-hand side is 0, so that either $\text{char}(S) \leq r$ or $\text{char}(S) \leq d$, a contradiction since $n > r$ and $n > d$.) Since a domain of type $\mathbf{Z}_n$ clearly has characteristic n, we infer that whenever $\mathbf{Z}_n$ is a domain, n is prime. It will be proved at the end of Section 3 that, conversely, $\mathbf{Z}_p$ is a field whenever p is prime.

Despite the preceding remarks (and the fact that 4 is not a prime), a field with exactly four elements *does* exist. While an elaborate discussion of the theory and construction of finite fields is beyond the scope of this chapter, we *can*, without explanation, provide tables for this example:

+	0	1	a	b
0	0	1	a	b
1	1	0	b	a
a	a	b	0	1
b	b	a	1	0

and

•	0	1	a	b
0	0	0	0	0
1	0	1	a	b
a	0	a	b	1
b	0	b	1	a

Verify that $\{0, 1, a, b\}$ is indeed a four-element field!

The next remark explains some of the theoretical importance of the mathematical systems $\mathbf{Z}_n$. If S is any integral domain, then S contains as a *subdomain* a mathematical system which is a copy of either the integers $\mathbf{Z}$ or one of the fields $\mathbf{Z}_p$ (corresponding to a prime integer p). Indeed, the copy in question is the intersection of the set of all subdomains of S; that is, the smallest subdomain D of S. If $\text{char}(S) = p > 0$, then

$$D = \{0, 1, 1+1, \ldots, \underbrace{1+1+\ldots+1}_{p-1 \text{ summands}}\},$$

which is evidently a copy of $\mathbf{Z}_p$; if char$(S) = 0$, then $D = \{0, 1, -1, 1 + 1, -1 + (-1), \ldots\}$, a copy of $\mathbf{Z}$.

With the above examples in mind, the reader should be well prepared for the later material, since the nature of a polynomial's roots depends upon the field in which one seeks for roots. The upshot of the optional material in Section 6 of Chapter 2 and Section 6 of Chapter 3 is that any polynomial over any field has "enough" roots in "suitable" fields. (For the meaning of "enough," skim Sections 1 and 3 of Chapter 2. The "suitable" constructions are of two kinds: one is given in the next section; the other proceeds from polynomial algebra via prime polynomials (discussed in Section 3) much as the construction of $\mathbf{Z}_p$ proceeds from the algebra of $\mathbf{Z}$ via prime integers p.)

To close this section, we comment further on domains and fields. We have seen that $\mathbf{Z}$ is a domain which is not a field; other examples of domains which are not fields are given in the exercises below. As all these examples are based on infinite sets, the reader may suspect that there are no finite examples. Certainly, the above information about $\mathbf{Z}_n$ supports such a conjecture and, in fact, the suspicion *is* right. We next prove that *any finite domain S is a field.* If $S = \{0, r_1, r_2, \ldots, r_n\}$ consists of 0 and n nonzero elements, we show that a typical nonzero element r (say, $r = r_1$) has a multiplicative inverse as follows. By the cancellation law of multiplication, the n elements $rr_1, rr_2, rr_3, \ldots, rr_n$ are distinct, and so one of them, say rr_i, must be the element $1 \in S$. Then $r_i = r^{-1}$, (viii) holds, and S is a field, as claimed.

To close with some historical notes, the reader should know of the existence of mathematical structures called *division rings* (or *skew fields*). These are systems satisfying all the properties of fields except, possibly, commutativity of multiplication. The most familiar example of a division ring (other than a field) is $\mathbf{H}$, the set of quaternions (denoted in honor of their discoverer, Hamilton), with operations akin to those of ordinary three-dimensional vectors, multiplication corresponding to cross-product of vectors. As cross-product is not commutative, $\mathbf{H}$ is not a field. In the spirit of the last paragraph, we note that $\mathbf{H}$ is infinite; in fact, it is known that any finite division ring must be a field. This *algebraic* fact implies that any finite projective plane which is Desarguesian (that is, satisfies the conclusion of Desargues's theorem) must also be Pappian (that is, satisfies the conclusion of Pappus's theorem). No purely *geometric* proof of this result is known! Of course, any Pappian projective plane must be Desarguesian (Hessenberg's theorem), and non-Desarguesian projective planes *do* exist (Moulton's example).

EXERCISES 1.1

1. (a) Let S be the mathematical system of positive integers, with the usual definitions of addition and multiplication inherited from $\mathbf{Z}$. Prove that S is not a domain.
(b) Repeat (a), with "positive" replaced by "nonnegative."
2. (a) Let S be the set of rational numbers of the form a/b, where a is an integer and b is an odd integer. Prove that S is a domain and that S is not a field. Identify the quotient field of S.
(b) Repeat (a), with the condition on denominators b changed to: b is a power of 2, that is, $b = 2^n$ for some nonnegative integer n. (*Hint*: to show the domain is not a field, use the fundamental theorem of arithmetic.)
3. (a) Let S be a domain. Consider a subdomain T of S, that is, a subset T of S satisfying conditions (i), (iv), and (v). Prove that T is, in fact, a domain.
(b) Let S and T be as in (a). Prove that the elements 0 and 1 in T are equal to the corresponding neutral elements in S. State and prove an analogous result about additive inverses.
4. (a) Consider an element a of some domain S. Prove that a has at most one multiplicative inverse in S. Conclude that, if a^{-1} exists, then $(a^{-1})^{-1} = a$.
(b) Let T be a subdomain of a domain S. Let $a \in T$ be such that a has a multiplicative inverse, b, in T and also a multiplicative inverse, c, in S. Prove that $b = c$.
(c) Give an example of a domain S, a subdomain T, and an element $a \in T$ such that a has a multiplicative inverse in S but a has no multiplicative inverse in T.
(d) If a and b are elements of a field F and $a \neq 0$, prove that there is exactly one element $r \in F$ such that $ar + b = 0$.
(e) Prove that a domain S is a field if and only if, whenever a and b are elements of S and $a \neq 0$, there exists $r \in S$ such that $ar + b = 0$.
5. (a) State and prove the generalized associative and distributive laws for an arbitrary domain S.
(b) By using all the axioms for domains except commutativity of addition (possibly including a strengthened version of (iv)), infer commutativity of addition, by expanding $(a + b)(1 + 1)$ in two different ways.
6. (a) If T is a subdomain of a domain S, prove that $\text{char}(T) = \text{char}(S)$.
(b) If F is a field, prove that $\text{char}(F) = 0$ if and only if F contains as a subdomain a mathematical system which is a copy of $\mathbf{Q}$.
(c) Prove that a domain S has characteristic 2 if and only if $(a + b)^2 = a^2 + b^2$ whenever $a \in S$ and $b \in S$. State and prove a similar result for characteristic 3.
7. Verify in detail that $\mathbf{Q}(\sqrt{3})$ is a field. Same for $\mathbf{Q}(\sqrt{5})$.
8. In $\mathbf{Z}_{12}$, compute $2 + 9$, $3 + 10$, $2 \cdot 5$, $3 \cdot 7$, and $3 \cdot 4$. What should $11 - 3$, $3 - 11$, $3 \div 7$, and $6 \div 5$ mean? Why is $3 \div 4$ senseless in $\mathbf{Z}_{12}$?

9. Prove that $\mathbf{Q}(\sqrt{2})$ is the *subfield of* **R** *generated* by $\sqrt{2}$, that is, the smallest subfield of **R** which contains $\sqrt{2}$. Identify the subfield of **R** generated by $\sqrt{2}$ and $\sqrt{3}$. Generalize.
10. Prove that there is no field containing precisely six elements. (*Hint*: eliminate the possible characteristics.)

2. Complex Numbers and the Polynomials $x^n - a$

This section introduces some of the properties of the mathematical system **C** of complex numbers. In a sense which is made explicit in the fundamental theorem of algebra in Section 6 of Chapter 3, **C** constitutes the final stop on the itinerary of structures such as **Z**, **Q**, and **R**. The fundamental theorem will lead to the basic information about roots of polynomials over **R** studied in Section 1 of Chapter 5. More immediate practical applications of the theory of **C** are familiar in mathematics and engineering, in the study of areas such as differential equations and electricity. Despite these very real and genuine uses, **C** was for a time regarded suspiciously because it contains an element i with behavior not present in **R**: indeed, i^2 equals the *negative* number, -1. (Accordingly, the notation i was chosen to denote *imaginary* behavior.) However, there is no avoiding **C**: for instance, the quadratic formula, when applied to quadratics such as $2x^2 + 2x + 1$ with negative discriminant, forces us to pass from **R** to **C**. This section brings about that passage. The reader will need access to a hand-held calculator (or tables) to handle n-th roots (of positive real numbers) and the trigonometric functions.

As a set, **C** may be regarded as the set of points in the Euclidean plane. By analytic geometry, **C** is just as reasonably viewed as the set of ordered

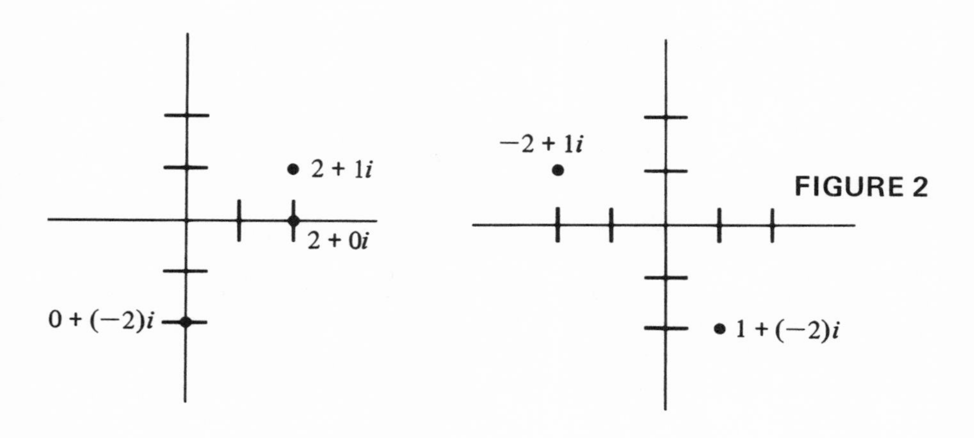

FIGURE 2

pairs of real numbers, the complex number (a, b) typically being written as $a + bi$. Graphically, one may represent complex numbers via analytic geometry (with the so-called *Argand diagrams*) as shown in Figure 2. According to the definition of ordered pairs, $a + bi = c + di$ (for real numbers a, b, c, and d) if and only if both $a = c$ and $b = d$. The binary operation of addition in $\mathbf{C}$ is defined precisely like vector addition in the plane: $(a + bi) + (c + di)$ is, by definition, $(a + c) + (b + d)i$. To define multiplication in $\mathbf{C}$, the requirements of commutativity and distributivity, together with $i^2 = -1$, lead to defining the product $(a + bi) \cdot (c + di)$ to be $(ac - bd) + (ad + bc)i$.

When real numbers a are viewed as complex numbers $a + 0i$, the mathematical system $\mathbf{R}$ becomes a subsystem of $\mathbf{C}$ (that is, addition and multiplication in $\mathbf{C}$ restrict to the corresponding operations in $\mathbf{R}$). If one writes bi for the *pure imaginary* number $0 + bi$, then the complex number $a + bi$ is indeed the sum of the *real complex* number a and the pure imaginary number bi. Of course, i denotes $1i$ and $-i$ is $(-1)i$.

From these pages' viewpoint, the fundamental consequence of the above structure is that $\mathbf{C}$ *is a field* (with subfield $\mathbf{R}$). By checking coordinates of ordered pairs, most of the verification is routine. For example, the elements 0 and 1 in $\mathbf{R}$ serve as appropriate neutral elements in $\mathbf{C}$; additive inverses are given by $-(a + bi) = (-a) + (-b)i$. To verify (viii), the existence of multiplicative inverses of a nonzero $e = a + bi$, "rationalize" as in the computation for $\mathbf{Q}(\sqrt{2})$ in Section 1: e^{-1} ought to be

$$\frac{a - bi}{(a + bi)(a - bi)},$$

which is $(a - bi)/(a^2 + b^2)$. It is easy to check that

$$f = \frac{a}{a^2 + b^2} + \frac{-b}{a^2 + b^2} i$$

satisfies $ef = 1$, so that $f = e^{-1}$.

The technique used above is effective in the practical matter of dividing in $\mathbf{C}$: for example,

$$\frac{2 - 3i}{-2 + 7i} = \frac{(2 - 3i)(-2 - 7i)}{(-2 + 7i)(-2 - 7i)}$$
$$= \frac{-25 - 8i}{53} = \frac{-25}{53} + \frac{-8}{53} i.$$

The reader will have noted that the verification of (viii) succeeded because $a^2 + b^2 > 0$ whenever $a + bi \neq 0$. Indeed, $a^2 + b^2$ has a geometric meaning which is evident from the Argand diagram: using the distance theorem (or, more basically, the theorem of Pythagoras about right triangles), we see that $a^2 + b^2$ is the square of the distance from $a + bi$ to the origin. (This is true in all cases, even if $a + bi = 0$.) To extend the meaning of absolute value on **R**, we define the *absolute value* (or *modulus*) of $a + bi$ by $|a + bi| = \sqrt{a^2 + b^2}$. In other words, if $z \in \mathbf{C}$, then $|z|$ is the distance from z to 0. For example, $|-2 + 3i| = \sqrt{4 + 9} = \sqrt{13}$, $|i/2| = 1/2$, and $|-3/2| = 3/2$. One important property of absolute value is its multiplicativity: $|z_1 z_2| = |z_1| \cdot |z_2|$ for all complex numbers z_1 and z_2. Its proof, or rather the proof of the equivalent fact that $|z_1 z_2|^2 = |z_1|^2 \cdot |z_2|^2$, amounts to the straightforward check that $(ac - bd)^2 + (ad + bc)^2 = (a^2 + b^2) \cdot (c^2 + d^2)$.

Students of trigonometry (or of polar coordinates) will realize that a complex number $z = a + bi$ may be totally described in terms of its absolute value $|z|$ and the angle whose radian measure θ satisfies $a = |z| \cos\theta$ and $b = |z| \sin\theta$. Customarily, θ satisfies $0 \leqslant \theta < 2\pi$ (since 2π radians = 360°), and we write $\theta = \arg(z)$, the *argument* of z. For example, $\arg(2) = 0$, $\arg(-2 + i) \doteq 168° \, 28' \doteq 0.94\pi$ (radians), and $\arg(-i/2) = 3\pi/2$; $\arg(0)$ is unspecified and usually innocuous. The Argand diagram, with all the attendant data, for the complex number $z = -2 + 2i$ is shown in Figure 3. The fundamental fact about arg is that $\arg(z_1 z_2) = \arg(z_1) + \arg(z_2) +$ an integral multiple of 2π. This will be proved at the end of the next paragraph.

For any real θ, it is customary to write $\text{cis}(\theta) = \cos\theta + (\sin\theta)i$. (For the reader who has studied complex calculus: $\text{cis}(\theta) = e^{i\theta}$.) Thus, any nonzero complex number $z = a + bi$ is expressible in *polar form* as $z = |z|[(a/|z|) + (b/|z|)i] = |z|[\cos(\arg z) + \sin(\arg z)i]$; that is, $z = |z| \,\text{cis}(\arg z)$. Since $|z_1 z_2| = |z_1| \cdot |z_2|$ by our above comments, the fundamental fact about arg asserted in the preceding paragraph will follow from

$$\text{cis}(\theta_1) \cdot \text{cis}(\theta_2) = \text{cis}(\theta_1 + \theta_2),$$

which is itself immediate from the familiar trigonometric expansion laws $\cos(\theta_1)\cos(\theta_2) - \sin(\theta_1)\sin(\theta_2) = \cos(\theta_1 + \theta_2)$ and $\cos(\theta_1)\sin(\theta_2) + \sin(\theta_1) \cdot \cos(\theta_2) = \sin(\theta_1 + \theta_2)$.

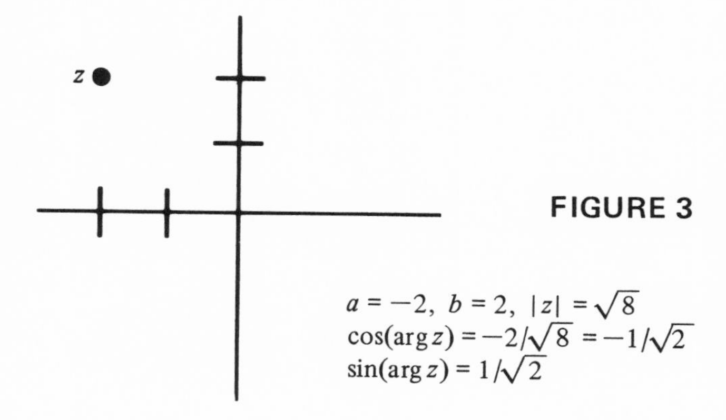

FIGURE 3

The fundamental fact about arg is usually summarized as $\arg(z_1z_2) = \arg(z_1) + \arg(z_2)$, a harmless pretense since (angles with) measures differing by an integral multiple of 2π have the same sines, cosines, etc. Thus, $\arg(z^2) = 2\arg(z)$, $\arg(z^3) = \arg(z^2z) = \arg(z^2) + \arg(z) = 2\arg(z) + \arg(z) = 3\arg(z), \ldots,$ and $\arg(z^n) = n\arg(z)$ in general for any positive integer n, a result known as *De Moivre's theorem*. (In fact, the assertion holds for all $n \in \mathbf{Z}$. For $n = 0$, it is a matter of observing that $\arg(1) = 0$. For $n < 0$, note that $\arg(z^n) = \arg[(z^{-1})^{-n}] = -n\arg(z^{-1})$, and so the task reduces to proving $\arg(z^{-1}) = -\arg(z)$. Prove it.)

Two typical uses of polar form and De Moivre's theorem are given next. The first concerns the computation of powers. For example, to calculate $(1+i)^4$, begin by finding the polar form of $z = 1 + i$. Now, $|z| = \sqrt{1+1} = \sqrt{2}$, $\cos(\arg z) = 1/\sqrt{2} = \sin(\arg z)$, and so $\arg z = \pi/4$. Then, $(1+i)^4 = z^4 = (\sqrt{2}\ \mathrm{cis}(\pi/4))^4 = (\sqrt{2})^4 \mathrm{cis}(4\pi/4) = 4\,\mathrm{cis}(\pi) = 4(-1 + 0i) = -4$. The second use expresses $\cos(n\theta)$ and $\sin(n\theta)$ in terms of $\sin\theta$ and $\cos\theta$. We illustrate the technique for $n = 3$. Since

$$\cos(3\theta) + i\sin(3\theta) = \mathrm{cis}(3\theta) = (\mathrm{cis}\,\theta)^3 = (\cos\theta + i\sin\theta)^3$$

$$= \cos^3\theta + 3i\cos^2\theta\sin\theta - 3\cos\theta\sin^2\theta - i\sin^3\theta,$$

the definition of equality for complex numbers gives, upon equating corresponding components, the identities $\cos(3\theta) = \cos^3\theta - 3\cos\theta\sin^2\theta = \cos^3\theta - 3\cos\theta(1 - \cos^2\theta) = -3\cos\theta + 4\cos^3\theta$ (which will pop up again in Section 3 of Chapter 3) and $\sin(3\theta) = 3\cos^2\theta\sin\theta - \sin^3\theta = 3(1 - \sin^2\theta)\cdot\sin\theta - \sin^3\theta = 3\sin\theta - 4\sin^3\theta$.

If the reader looks back at this section's introduction, he/she will probably be anxious to apply the quadratic formula in **C** and, in particular, to find square roots of complex numbers. In fact, the preceding material on polar forms and De Moivre's theorem leads to the computation of *all* n-th roots in **C**. Here's how.

As usual, by an n-th *root* of a complex number z is meant a complex number w such that $w^n = z$. Since **C** is a domain, $z = 0$ forces $w = 0$ (assuming, as we do, that n is a positive integer); that is, 0 is the only n-th root of 0. On the other hand, we shall see that if $z \neq 0$, then there are precisely n n-th roots of z: in case z has the polar form $z = r\,\mathrm{cis}\,\theta$ (with $0 < r = |z|$ and $\theta = \arg(z)$), then the n-th roots of z are the numbers $\sqrt[n]{r}\ \mathrm{cis}((\theta/n) + k(2\pi/n))$, where k assumes the n values $0, 1, 2, \ldots, n-1$, and as usual $\sqrt[n]{r}$ denotes the (uniquely determined, real) positive n-th root of r. Before proving this assertion, let's consider several applications of it.

First, to find the two square roots of $z = 4$, take $r = 4$, $\theta = 0$, and $n = 2$, and obtain from the formula $\sqrt{4}\ \mathrm{cis}(k\pi)$, for $k = 0, 1$. Thus, $2\,\mathrm{cis}(0) = 2$ and $2\,\mathrm{cis}(\pi) = -2$ are *the square roots* of 4. But $\sqrt{4}$ continues to denote only 2, not -2. Similarly, to compute the square root of i, take $n = 2$,

$r = 1$, and $\theta = \pi/2$; the formula produces $\text{cis}(\pi/4)$ and $\text{cis}(5\pi/4)$, that is, $(1/\sqrt{2}) + (i/\sqrt{2})$ and $(-1/\sqrt{2}) - (i/\sqrt{2})$. *Cubic* roots present no greater obstacle: for instance, the three cubic roots of -1 are found (using $n = 3$, $r = 1$, and $\theta = \pi$) to be $\text{cis}((\pi/3) + k(2\pi/3))$ for $k = 0, 1, 2$. (Express these values in the form $a + bi$, and find decimal expansions approximating a and b via tables or a calculator.) A final example, for arbitrary n, is to find the n-th roots of $1 + i$: here $r = \sqrt{2}$ and $\theta = \pi/4$, and the formula produces the n values $\sqrt[2n]{2}\ \text{cis}((\pi/4n) + k(2\pi/n))$ for $k = 0, 1, 2, \ldots, n-1$.

It will be helpful later to identify the n-th roots of 1: these are the numbers $\text{cis}(k2\pi/n)$, for $k = 0, 1, 2, \ldots, n-1$. If we let $\omega = \text{cis}(2\pi/n)$, then De Moivre's theorem shows that the n-th roots of 1 are just $1, \omega, \omega^2, \ldots,$ and ω^{n-1}. (How are these points situated, relative to one another, on an Argand diagram?) In particular, if $n = 3$, notice that the cubic roots of 1 are 1, ω, and ω^2, where $\omega = \text{cis}(2\pi/3) = (-1/2) + (\sqrt{3}/2)i$.

Are we finally prepared to solve for the roots of the quadratic $2x^2 + 2x + 1$ mentioned earlier? Yes, provided that the reader has verified that the two square roots of -1 are i and $-i$. (If you haven't, do so via the n-th root formula before proceeding. Also show that the square roots of -4 are $2i$ and $-2i$.) Now, by the quadratic formula, the given quadratic has roots $(-2 \pm \sqrt{4-8})/4$, that is, $(-2 \pm \sqrt{-4})/4$. The expression "$\pm\sqrt{-4}$" pertains to the two square roots of -4, that is, to $2i$ and $-2i$, and so the required roots are $(-1 + i)/2$ and $(-1 - i)/2$. In Chapter 3, the reader will encounter formulas to solve harder problems, for example equations such as $x^3 - 2x^2 + ix + 6 = 0$; as may be anticipated, this section's work involving trigonometry and the quadratic formula will play a crucial role in applying those new formulas.

It's time to keep our promise to prove the n-th root formula. Recall that $0 \neq z = r\,\text{cis}(\theta)$ is given. The question concerns which $w = p\,\text{cis}(\psi)$ satisfy $w^n = z$, for a given positive integer n. Certainly, the values of w cited in the formula *do* have the required property, for if $w_k = \sqrt[n]{r}\ \text{cis}((\theta/n) + k(2\pi/n))$, then $|(w_k)^n| = |w_k|^n = (\sqrt[n]{r})^n = r = |z|$ and $\arg((w_k)^n) = n\arg(w_k) = n((\theta/n) + k(2\pi/n)) = \theta + k2\pi = \arg(z) + k2\pi$. Moreover, the w_k's form n distinct values (isn't this clear geometrically from an Argand diagram?), since $w_k = w_m$ for $0 \leq k < m \leq n-1$ would force $((\theta/n) + (m2\pi/n)) - ((\theta/n) + (k2\pi/n)) = (m-k)2\pi/n$ to be $t2\pi$ for some positive integer t, whence $n > m - k = nt \geq n$, an absurdity.

The proof is only half over. (It *would* be complete if we could appeal now to the results in Section 1 of Chapter 2.) We still have to show that no other values of p and ψ are possible. First, to dispatch p, note that $w^n = z$ gives the equality of positive real numbers, $p^n = |w|^n = |w^n| = |z| = r$, whence $p = \sqrt[n]{r}$ as desired. Then $w^n = z$, that is, $p^n\text{cis}(n\psi) = r\,\text{cis}\,\theta$, becomes $\text{cis}(n\psi) = \text{cis}\,\theta$ after cancelling r (which *is* nonzero), whence $n\psi$

$= \theta + m2\pi$, for some $m \in \mathbf{Z}$. Dividing m by n in $\mathbf{Z}$ leads to a quotient $q \in \mathbf{Z}$ and a remainder $k \in \mathbf{Z}$ with $0 \leq k \leq n - 1$; that is, $m = qn + k$. Hence, $\psi = (\theta/n) + (m/n)2\pi = (\theta/n) + (q + (k/n))2\pi = \arg(w_k) + q2\pi$, and so $w = \rho\, \text{cis}(\psi) = \sqrt[n]{r}\ \text{cis}(\arg(w_k)) = w_k$, to complete the proof.

The final topic in this section will find use in Chapters 5 and 6. It is the subject of *conjugation* and may be motivated in two ways. The first of these concerns the "rationalization" trick employed above in the proof that **C** is a field; it shows that passage from $a + bi$ to $a - bi$ often has computational merit. Secondly, such passage has the pleasant geometric interpretation as reflection through the horizontal x-axis in the Argand diagram. Accordingly, we devise the following terminology. If $z = a + bi$ is any complex number (with, as usual, a and b real), the *conjugate* of z is the complex number $z^* = a - bi$. (Some authors write $\bar{z}$ instead of z^*, but we shall put the overhead-bar symbol to different use in Section 1 of Chapter 2.) For example, $(2 - 7i)^* = 2 + 7i$, $(-3)^* = -3$, and $(9i)^* = -9i$. The basic properties of the conjugation operation, which are readily verified from the appropriate definitions, are $(z^*)^* = z$, $(z_1 + z_2)^* = (z_1)^* + (z_2)^*$, and $(z_1z_2)^* = (z_1)^* \cdot (z_2)^*$.

Finally, we infer a key fact which will help in classifying roots in Section 1 of Chapter 5. To wit: if $z = a + bi$ (as always, with a and b real) and $a_0(a + bi)^n + a_1(a + bi)^{n-1} + \ldots + a_{n-1}(a + bi) + a_n = 0$ for certain *real* numbers $a_0, a_1, \ldots, a_{n-1}, a_n$, then $a_0(a - bi)^n + a_1(a - bi)^{n-1} + \ldots + a_{n-1}(a - bi) + a_n = 0$. The proof of this fact would be complicated if one worked directly from the definitions, but is simple using the concept of conjugation. Indeed, the preceding remarks show that the left-hand side which is to be *shown* to be 0 is merely the conjugate of the left-hand side which is *known* to be 0, and so the proof boils down to the triviality that $0^* = 0$.

EXERCISES 1.2

1. Simplify and express each of the following complex numbers in the standard form $a + bi$: $-(\frac{1}{2} - 7i)$, $(\frac{3}{4} + \pi i) + ((9/2) - 3i)$, $(\pi - 6i) - (-2 + 4i)$, $-2(-3 + 6i)$, $((3/2) + 6i)\cdot((7/4) - 3i)$, $(3 + 6i)^{-1}$, $(-2 + 7i)/(3 + 6i)$, and $(1 - i)^6$.
2. For each of the eight complex numbers z in question 1, draw an Argand diagram, compute the absolute value $|z|$ using a calculator with a square-root key, find the argument $\arg(z)$ in radian measure (with trigonometric tables/calculator), and hence write down the polar form of z.
3. (a) Several of the text's arguments assumed that $\text{cis}(\theta_1) = \text{cis}(\theta_2)$ forces $\theta_1 - \theta_2$ to be of the form $n2\pi$, with $n \in \mathbf{Z}$. Using trigonometry, explain why this is so. (Notice that "cis" is needed: it cannot be replaced by a single function such as sine.)

(b) The text also supposed that $|\operatorname{cis}\theta| = 1$ for each real θ. What familiar trigonometric fact is involved in proving this?

4. If z is any nonzero complex number, find formulas for $|z^{-1}|$ and $\arg(z^{-1})$ in terms of $|z|$ and $\arg(z)$. (*Hint*: see the parenthetical remark after the statement of De Moivre's theorem.) Hence, infer formulas for $|z_1/z_2|$ and $\arg(z_1/z_2)$ for any nonzero complex numbers z_1 and z_2. Using polar form, find an alternate construction of z^{-1} for any nonzero $z \in \mathbf{C}$, and thus obtain a new proof that $\mathbf{C}$ is a field.
5. Obtain formulas expressing $\cos(4\theta)$ and $\sin(4\theta)$ in terms of $\cos\theta$ and $\sin\theta$. Generalize.
6. (a) Using the polar form and De Moivre's theorem (and tables/calculator), find approximate values of a and b such that $(1 - i)^9 = a + bi$. Compute the actual values of a and b by expanding $(1 - i)^9$, say by the binomial theorem, and simplifying. Compare the two answers.
(b) Find all the fifth roots of 1, the 19-th roots of $1 - i$, and the sixth roots of -2. (Approximate via tables/calculator as needed.)
7. Find all complex numbers x such that
(a) $2x^2 - 3x - 9 = 0$.
(b) $4x^2 + 4x - i = 0$.
(c) $4x^2 + 4x + 3 = 0$.
(d) $x^4 + ix^3 - 3x^2 - 3ix = 0$. (*Hint*: factor and peek at Chapter 2.)
(e) $2x^4 + 2x^2 + 1 = 0$.
8. Let $\omega = (-1/2) + (\sqrt{3}/2)i$, the cubic root of 1 considered above. Prove:
(a) $1 + \omega + \omega^2 = 0$. (*Hint*: $\omega^3 - 1 = (\omega - 1)(\omega^2 + \omega + 1)$.)
(b) $(\omega^3)^{-1} + (\omega^2)^{-1} + \omega^{-1} = 0$. (*Hint*: common denominator.)
(c) $\omega^{-1} + \omega = -1$.
9. Let z be a nonzero complex number and n a positive integer. Let $w_1, \ldots, w_n$ be the n-th roots of z. Prove that $w_1 \cdot w_2 \cdot w_3 \cdot \ldots \cdot w_n = (-1)^{n-1}z$. (*Hint*: $1 + 2 + \ldots + (n - 1) = (n - 1)n/2$.) If $z = 1$, prove that $w_1 + w_2 + \ldots + w_n = 0$. (*Hint*: see 8(a).) Generalize for real z.
10. (a) Find the conjugates of $-1 + i, -1 - i, 1 + i$, and $1 - i$.
(b) What can you say about z if $z^* = z$? If $z^* = -z$? If $z^* = 0$?
(c) Find formulas for $|z^*|$ and $\arg(z^*)$ in terms of $|z|$ and $\arg(z)$.
(d) If z_1 and z_2 have the same conjugates, prove that $z_1 = z_2$.
(e) Find real a, b and complex z_1, z_2 such that $z_1(a + bi) + z_2 = 0 \neq z_1(a - bi) + z_2$.
11. What should be meant by "the *subdomain* of $\mathbf{C}$ *generated by* $\sqrt{2}$ and i"? Identify its elements.

3. The Algebra of Polynomials, the Division Algorithm, and Greatest Common Divisors

In this section we introduce polynomials, this book's principal object of study. The reader has surely encountered these objects before—in high-school algebra, if not earlier—and he/she probably knows that the study of a polynomial is intimately concerned with the polynomial's roots. With uncharacteristic patience, we shall delay the discussion of roots until Chapter 2, choosing to concentrate at present on algebraic preliminaries which will be put to use throughout the following chapters. The final portion of this section contains the initial applications of those preliminaries.

First things first: what *is* a polynomial? Having fixed a domain S for the ensuing discussion, we define a *polynomial over S in the variable* (or *indeterminate*) x to be an expression

$$f(x) = a_0x^n + a_1x^{n-1} + a_2x^{n-2} + \dots + a_n,$$

in short, $f(x) = \sum_{i=0}^{n} a_i x^{n-i}$, arising from elements $a_0, a_1, a_2, \dots, a_n$ of S. The suspicions about the meaning of the complex number i apply equally well to this (traditional) definition involving the mysterious x. To dispel them, just as in the case of **C**, we may banish x by choosing to identify $f(x)$ with the $(n + 1)$-tuple $(a_0, a_1, a_2, \dots, a_n)$, which could just as well be denoted f now. The notions of equality and addition for tuples are unassailable—they happen componentwise—and so we now know what polynomials *are* (that is, "expressions of the form . . . " are just certain tuples) and how to add them. For example, $2x^2 + 3x + 1 = (1 + 1)x^2 + 3x + 1 \neq 2x^2 + 2x + 1$ and $(2x + 3) + (9x^2 + 4x + 6) = 9x^2 + 6x + 9$.

Thus far, there have been no surprises: add by combining like terms. To define multiplication of polynomials, use the preceding section for motivation, as follows. Just as **R** was viewed as a subsystem of **C**, view S as imbedded in the set of polynomials by identifying $a_0 \in S$ with the 1-tuple (a_0); as i was identified with the planar point $(0, 1)$ in **C**, let x stand for the polynomial $(1, 0)$, that is, $1x^1 + 0$. Now, the requirements of commutativity, associativity, and distributivity, together with the usual "laws of exponents," force the definition of multiplication: if $f(x) = \sum_{i=0}^{n} a_i x^{n-i}$ and $g(x) = \sum_{i=0}^{m} b_i x^{m-i}$, then $f(x)g(x) = \sum_{i=0}^{n+m} c_i x^{n+m-i}$, where $c_i = \sum_k a_{n-k} b_{k+i-n}$, the index k running over those values for which a_{n-k} and b_{k+i-n} are defined. Although this definition was forced, the plot has thickened to the point of congealing, and an example is in order. If $f(x) = 2x^3 + 0x^2 + (-1)x + (-7)$ and $g(x) = 1x^2 + 0x + (-2)$, the product formula permits the ignoring of coefficients which equal 0 (why?) and, after sensibly rewriting $f(x)$ as $2x^3 - x - 7$ and $g(x)$ as $x^2 - 2$, we obtain $f(x)g(x) = 2x^5 + 0x^4 + (-4 - 1)x^3 + (-7)x^2 + 2x + 14 = 2x^5 - 5x^3 - 7x^2 + 2x + 14$, just as in high school! The point is, addition, subtraction (how?), and multiplication of polynomials will be performed precisely as in the reader's earlier education. As usual, the "new math" symbolism merely alters (and hopefully heightens) one's understanding of the facts of mathematical life, but those facts themselves are not changed by the new symbolism.

To emphasize the way in which calculations are to be carried out in practice, multiplication in **C** of "expressions of the form $a + bi$" and multiplication of polynomials over S (other expressions of another form) each proceed routinely accordingly to the "laws" studied in Section 1, the sole difference being that $i \cdot i$ $(= i^2)$ simplifies to -1, while $x \cdot x$ is x^2, which cannot be simplified further.

Some notation is overdue. Let $S[x]$ denote the set of polynomials over S in the variable x. It is clear that S is a subsystem of $S[x]$, in the sense that addition and multiplication in $S[x]$ restrict to the corresponding operations in S. Moreover, just as $a + bi$ turned out to be the sum of a and bi in **C**, it is readily shown that $f(x) = a_0x^n + a_1x^{n-1} + \ldots + a_n \in S[x]$ is the sum of the $n + 1$ quantities $a_0x^n, a_1x^{n-1}, \ldots, a_n$ in $S[x]$.

Another word about notation. We shall retain the "x" in denoting a typical polynomial $f(x)$, so as to facilitate "substitution for x." This process, to be termed *evaluation* in Section 1 of Chapter 2, is at the heart of our subject, and is impeded by a notation such as "f" which does not have an x for which you can substitute. Readers who have studied techniques of integration in calculus will recall a similar convention in integral notation, making $\int xe^{x^2}dx$ more tractable than its *alter ego* $\int xe^{x^2}$.

Now that S has been shown to be a subsystem of $S[x]$, it is fair to ask: what *sort* of system is $S[x]$? We next give the answer: $S[x]$, *the system of polynomials over* S in x, *is a domain*. For the proof, only two items need to be checked, the others being easy exercises. First, one must establish commutativity of multiplication. This will follow from commutativity of multiplication in S. Indeed, if $f(x) = \sum a_ix^{n-i}$ and $g(x) = \sum b_ix^{m-i}$, then $f(x)g(x) = \sum c_ix^{n+m-i}$, where the earlier formula for the coefficient c_i may be rewritten as $c_i = \sum_v a_vb_{i-v}$, the index v running over those values for which both a_ν and $b_{i-\nu}$ are defined. Commutativity of multiplication on S gives $c_i = \sum b_{i-\nu}a_\nu = \sum_\mu b_\mu a_{i-\mu}$, with index μ taking all values for which b_μ and $a_{i-\mu}$ are meaningful. However, this latter sum is evidently the coefficient of x^{n+m-i} in $g(x)f(x)$, which must therefore equal $f(x)g(x)$, as their corresponding coefficients agree, and commutativity has been established. The second, and final, item needing verification is (vi)′. We defer its proof in order to introduce a most useful concept, that of *degree*.

The above method of "sensibly rewriting" polynomials guarantees that any nonzero polynomial $f(x)$ can be associated with a uniquely determined $(n + 1)$-tuple $(a_0, a_1, a_2, \ldots, a_n)$ with entries in S such that $a_0 \neq 0$; put differently, $f(x)$ is *uniquely* expressible as $a_0x^n + a_1x^{n-1} + \ldots + a_n$ with $a_0 \neq 0$. Given such an expression (with nonzero *leading term*), we shall say that $f(x)$ has *degree* n, and write $\deg(f) = n$. For example, the typical first-degree polynomial (that is, degree = 1) is $ax + b$, with $0 \neq a \in S$ and $b \in S$; such polynomials are also called *linear* because of their interpretations as graphs in analytic geometry when $S = \mathbf{R}$. (Polynomials of degree 2, 3, 4, 5 are called *quadratic, cubic, quartic, quintic*, respectively.) The polynomials of degree 0 are simply the nonzero elements of S. As for the polynomial 0, its degree is the symbol $-\infty$ (various authors have other conventions for deg(0)) satisfying $(-\infty) + n = n + (-\infty) = (-\infty) + (-\infty) = -\infty$ for each nonnegative integer n. The *fundamental fact about degrees* is: for any polynomials $f(x)$ and $g(x)$, *the degree of the product* $f(x)g(x)$ *is the sum* $\deg(f) + \deg(g)$. If either $f(x)$ or $g(x)$ are 0, the rules for manipulating $-\infty$ make this clear. For the other case, $\deg(f) = n \geq 0$, $\deg(g) = m \geq 0$, $f(x) = \sum a_i x^{n-i}$ with $a_0 \neq 0$, $g(x) = \sum b_i x^{m-i}$ with $b_0 \neq 0$ and $f(x)g(x) = \sum c_i x^{n+m-i}$ with $c_0 = a_0 b_0$; since S is a domain, (vi)′ for S assures that $c_0 \neq 0$, whence $f(x)g(x)$ has degree $n + m$, as claimed.

Now, to finish the proof that $S[x]$ is a domain: the only stumbling block would be a counterexample to (vi)′. This would entail nonzero polynomials $f(x)$ and $g(x)$ such that $f(x)g(x) = 0$. However, the degree of such a product cannot be $-\infty$ since, by the above fundamental fact, that degree is the sum of the nonnegative integers $\deg(f)$ and $\deg(g)$. This completes the proof that $S[x]$ is a domain.

One should not wish for more, as $S[x]$ is not a field, even when S is itself a field. (Otherwise, $x^{-1} \in S[x]$, say $x^{-1} = \sum a_i x^{n-i}$ with each $a_i \in S$, so that multiplication by x throughout produces $1 = a_0 x^{n+1} + a_1 x^n + \ldots + a_n x$, violating the definition of equality for polynomials. Fear not, though, for x^{-1} survives where it belongs, in the quotient field of $S[x]$.) Consequently, division, or more precisely, divisibility, is not as routine in $S[x]$ as the other operations have been. Pursuing that insight leads to the rest of this section—and beyond!

Division for polynomials is an operation that most readers carried out —*very* often—in high-school algebra. For example, the command "divide $7x^5 - 2x^3 + 9$ by $2x^2 + 4x - 8$" was typically met by work arranged as follows:

$$\begin{array}{r|l}
 & (7/2)x^3 - 7x^2 + 27x - 82 \\
\hline
2x^2 + 4x - 8 & 7x^5 - 2x^3 + 9 \\
 & \underline{7x^5 + 14x^4 - 28x^3} \\
 & -14x^4 + 26x^3 + 9 \\
 & \underline{-14x^4 - 28x^3 + 56x^2} \\
 & 54x^3 - 56x^2 + 9 \\
 & \underline{54x^3 + 108x^2 - 216x} \\
 & -164x^2 + 216x + 9 \\
 & \underline{-164x^2 - 328x + 656} \\
 & 544x - 647
\end{array}$$

The result of using the *divisor* $2x^2 + 4x - 8$ and the *dividend* $7x^5 - 2x^3 + 9$ is thus the data consisting of the *quotient* $(7/2)x^3 - 7x^2 + 27x - 82$ and the *remainder* $544x - 647$. One may summarize, say in the quotient field of $\mathbf{Q}[x]$, by writing

$$\frac{7x^5 - 2x^3 + 9}{2x^2 + 4x - 8} = \frac{7}{2}x^3 - 7x^2 + 27x - 82 + \frac{544x - 647}{2x^2 + 4x - 8}.$$

As we next show, it is the equivalent "cross-multiplied" version of such an equation which can be most easily established in general.

To set the stage, S is (still) an integral domain; $f(x) = a_0x^n + \ldots + a_n \in S[x]$ will play the role of dividend; $g(x) = b_0x^m + \ldots + b_m \in S[x]$ is to be the divisor; and $g(x)$ is subjected to the extra condition that b_0 possesses a multiplicative inverse in S. (In particular, the last condition guarantees that $g(x) \neq 0$. If S is a field and $g(x) \neq 0$, the condition is then automatically satisfied.) The *division algorithm* asserts that, *under the above conditions,* there exist uniquely determined polynomials $q(x)$ and $r(x)$ in $S[x]$ (called the *quotient* and *remainder*, of course) such that both $f(x) = q(x)g(x) + r(x)$ and $\deg(r) < \deg(g)$. (In case $f(x) = q(x)g(x)$, so that $r(x) = 0$, the required inequality of degrees asserts that $-\infty < m$, an intuitively acceptable property which we (retroactively) assign to $-\infty$.)

We've already seen an *application* of the division algorithm, namely the above computation with $r(x) = 544x - 647$. Now, for its *proof!* Tackling the harder part first, we shall first produce satisfactory $q(x)$ and $r(x)$. This is a snap if $n < m$, for then choosing $q(x) = 0$ and $r(x) = f(x)$ suffices. Accordingly, we may further suppose that $n \geq m$. If we are confronted with a counterexample (for which no suitable pair $q(x)$, $r(x)$ may be produced), we may as well suppose the optimal case: n is the smallest possible value for the degree of the aspiring dividend in *any* counterexample in $S[x]$. (The point used here is the *well-ordering property*: Any nonempty set of positive integers possesses a smallest element.) Then, inspired by the first step in the computation which produced $\frac{7}{2}x^3$, we create $h(x) = f(x) - a_0(b_0)^{-1}x^{n-m}g(x)$, and observe that $\deg(h) < n$. (Why?) Consequently, $h(x)$ has too low a degree to be a counterexample, whence polynomials $q_1(x)$, $p(x) \in S[x]$ exist such that both $h(x) = q_1(x)g(x) + p(x)$ and $\deg(p) < m$. Then $q(x) = q_1(x) + a_0(b_0)^{-1}x^{n-m}$ and $r(x) = p(x)$ *are* satisfactory, contradicting the supposed presence of a counterexample.

The remaining part of the proof of the division algorithm concerns uniqueness of $q(x)$ and $r(x)$, and will be handled by a simple degree argument. If $f(x) = q(x)g(x) + r(x) = q_2(x)g(x) + r_2(x)$ with q, r, q_2, $r_2 \in S[x]$ and both $r(x)$ and $r_2(x)$ of degree(s) less than m, then consider the polynomial $F(x) = (q(x) - q_2(x)) \cdot g(x)$. If $q(x) - q_2(x)$ is nonzero, say of degree $d \geq 0$, the fundamental fact about degrees yields that $\deg(F) = d + m \geq m$. However, $F(x)$ is also expressible as $r_2(x) - r(x)$, and so $\deg(F) < m$. (What general principle about the degree of a sum has just been used? Prove it!) The contradiction $m \leq \deg(F) < m$ forces $q(x) - q_2(x) = 0$, whence $q(x) = q_2(x)$ and (by additive cancellation in the expressions for $f(x)$) also $r(x) = r_2(x)$, completing the proof.

The *algorithmic* aspect of the division algorithm is exemplified by the passage from $f(x)$ to $h(x)$ in the first part of the above proof. By iterating such passages, one subtracts from $f(x)$ a suitable sum of multiples of $g(x)$, the result eventually being $r(x)$, of degree less than $\deg(g)$. The reader is urged to use this constructive (or reductionist) perspective to reprove the existential (first) part of the division algorithm, and to convince him- or herself that the resulting algorithm is precisely what is built into the high-school method reviewed above. Once again, the "new math" seeks here to understand better the known valid computations, not to discard them.

Having treated division, let's consider divisibility of polynomials over (the domain) S. For once, it will be convenient to write $f, g, \ldots$ instead of $f(x), g(x), \ldots$ in the following discussion. If f and g belong to $S[x]$ and $g \neq 0$, we say that g *divides* f and write $g \mid f$ in case $f = qg$ for some $q \in S[x]$, that is, in case the remainder (in the sense of the division algorithm) is 0 when f is the dividend and g is the divisor. Taking $S = \mathbf{Q}$, we have the example $g \mid f$, where $f(x) = 2x^3 + 6x^2 - x - 3$ and $g(x) = x^2 - \frac{1}{2}$, the intervening quotient being $q(x) = 2x + 6$. An earlier computation gives an example of *nondivisibility*, denoted $g \nmid f$ with $f(x) = 7x^5 - 2x^3 + 9$ and $g(x) = 2x^2 + 4x - 8$. In general, if $f \in S[x]$ is any nonzero polynomial, then $f \mid 0$ and $f \mid f$. Moreover, transitivity holds: if $f \mid g$ and $g \mid h$, then $f \mid h$. (Why? Use associativity of multiplication in $S[x]$ to prove this.)

Two important properties of $\mid$ are recorded next. First is the *linear combination principle*. (Readers who have studied vectors or linear algebra will understand the choice of terminology.) This states that if $f \mid g$ and $f \mid h$, then $f \mid (pg + qh)$ for *any* polynomials $p, q \in S[x]$. (Indeed, if $g = Gf$ and $h = Hf$, then $pg + qh = (pG + qH)f$.) The second has to do with *associates*, that is, polynomials f and g such that $f \mid g$ and $g \mid f$. Certainly, if $0 \neq f \in S[x]$, then sf is associated to f whenever s is any element of S possessing a multiplicative inverse in S. (Check!) The useful fact is the converse: if f and g are associates, then $f = sg$, where $s \in S$ and $s^{-1} \in S$. Its proof is instructive. As $f \mid g$, there is some $G \in S[x]$ such that $g = Gf$ and, by the fundamental fact about degrees, $\deg(g) = \deg(G) + \deg(f) \geq \deg(f)$. By the same token, $g \mid f$ yields $H \in S[x]$ satisfying $f = Hg$ and $\deg(f) = \deg(H) + \deg(g) \geq \deg(g)$. Thus $\deg(f) = \deg(g)$, and $\deg(G) = \deg(H) = 0$: this fact means that G and H are *constants*, that is, nonzero elements of S. Moreover, $1f = f = Hg = H(Gf) = (HG)f$, and so condition (vi) for the domain $S[x]$ permits cancellation which yields $1 = HG$, showing that H is a suitable value for s, and completing the proof.

It now becomes convenient to specialize to the case in which S is a field, which, for sentimental reasons, we shall denote instead by F. Notice, by the preceding remarks, that any nonzero $f \in F[x]$ is associated to a unique *monic* polynomial $g \in F[x]$. (By definition, $g(x) = a_0x^m + \ldots + a_m$, of degree m, is said to be *monic* in case its leading coefficient, a_0, is 1.) We next consider a most important monic polynomial.

A *common divisor* of polynomials f and g in $F[x]$ will be understood to be any $h \in F[x]$ such that $h \mid f$ and $h \mid g$. Note that any nonzero element of F is a common divisor of any f and g. *A greatest common divisor* (g.c.d.) of f and g is, by definition, a common divisor d of f and g such that *any* common divisor of f and g must divide d. For example, one can show (in case $F = \mathbf{Q}$) that $d(x) = 2x - 4$ is a greatest common divisor of $f(x) = x^2 - 4$ and $g(x) = x^2 + x - 6$. In general, it is not a priori clear that each pair f, g has a g.c.d.—but it is nevertheless true, thanks to a spectacular algorithm which we shall get to shortly. First, a "normalizing" remark. The characterization of associates given above easily shows: if d is a g.c.d. of f and g, then a polynomial $e \in F[x]$ is also a g.c.d. of f and g if and only if d and e are associates. (Provide the details.) Thus, if f and g have any g.c.d. at all, they have one which is monic (by the comment in the preceding paragraph); that uniquely determined monic g.c.d. of f and g will be denoted (f, g) and called *the* g.c.d. of f and g. For example, $(x^2 - 4, x^2 + x - 6) = x - 2$.

There are good reasons why an algorithm to compute (f, g) is desirable. To name but one, the computations of invariant factors (for the rational canonical form) and elementary divisors (for the Jordan canonical form) of square matrices are often necessary in order to solve certain systems of ordinary differential equations which arise in applications. Such computations may be done via g.c.d.s, which is where an algorithm for g.c.d.s may be utilized, or by somehow factoring the matrices' characteristic polynomials. However, the latter can be achieved algorithmically only over *some* fields, such as $\mathbf{Q}$. (The factoring algorithm for $\mathbf{Q}[x]$ is given in Section 3 of Chapter 4. It is time-consuming by hand but, as in the case of any true algorithm, can be—and has been—programmed for computer. It is also rather deep, depending on Section 2 of Chapter 2 and Section 2 of Chapter 4.)

We shall now present an effective way to compute (and prove the existence of) the g.c.d. (f, g) over *any* field F. The case $f = g = 0$ is pointless. If $f \neq 0 = g$, then (f, g) is *the* monic polynomial associated to f; similarly, the case $f = 0 \neq g$ is trivial. To handle the case in which both f and g are nonzero, here's the (as promised) spectacular g.c.d. *algorithm* of Euclid.

It may be supposed that $\deg(f) \geq \deg(g)$, by relabelling the polynomials f and g, since finding (f, g) is clearly the same as finding (g, f). (If it's so clear, prove it!) Then, by the division algorithm (applicable since F is a field), we get:

$$f = q_1 g + r_1, \quad \text{where } \deg(r_1) < \deg(g);$$

if $r_1 \neq 0$, then

$$g = q_2 r_1 + r_2, \quad \text{where } \deg(r_2) < \deg(r_1);$$

$$\vdots$$

if $r_n \neq 0$, then

$$r_{n-1} = q_{n+1} r_n + r_{n+1}, \quad \text{where } \deg(r_{n+1}) < \deg(r_n).$$

Since the successive remainders have degrees which descend strictly, some $r_{n+1} = 0$. (Why? Use the well-ordering property.) The "spectacular" fact is that r_n is a g.c.d. of f and g (so its monic associate *is* (f, g)) *and* by solving the division-algorithm equations, one obtains r_n, and hence (f, g), expressed as a linear combination of f and g, that is, in the form $\alpha f + \beta g$ for *constructed* polynomials α and β in $F[x]$.

Let's prove the facts which were just asserted. (An illustrative example will follow.) First, $r_n \mid r_{n-1}$ since $r_{n+1} = 0$; as $r_{n-2} = q_n r_{n-1} + r_n$, the linear combination principle (6 paragraphs back) shows that $r_n \mid r_{n-2}$; and, by continuing to move *upward* through the algorithm's equations with similar applications of the linear combination principle, one ultimately obtains $r_n \mid g$ and $r_n \mid f$. In other words, r_n is a common divisor of f and g. Now if d were *any* common divisor of f and g, the linear combination principle may be used to move *downward* through the equations, with the successive results that $d \mid r_1$, $d \mid r_2, \ldots$ and, ultimately, $d \mid r_n$. Hence, r_n is indeed a g.c.d. of f and g. Its monic associate is then a monic g.c.d. of f and g, that is, (f, g). For the final assertion, calculate $r_n = r_{n-2} - q_n r_{n-1} = r_{n-2} - q_n(r_{n-3} - q_{n-1} r_{n-2})$, a linear combination of r_{n-2} and r_{n-3} which, by similar *upward* motion through the list of equations, becomes expressed as linear combinations of r_{n-3} and r_{n-4}, of r_{n-4} and $r_{n-5}, \ldots$ and, ultimately, of f and g, completing the proof.

For example, consider (f, g), where $f(x) = x^3 - 6x^2 + x + 4$ and $g(x) = x^2 - 4$. The algorithm's equations are:

$$f = (x - 6)(x^2 - 4) + (5x - 20)$$

$$g(x) = x^2 - 4 = \left(\frac{x}{5} + \frac{4}{5}\right)(5x - 20) + 12$$

$$5x - 20 = \left(\frac{5}{12} x - \frac{5}{3}\right)12 + 0.$$

Here $n = 2$, $r_n = 12$, and so (f, g) is the monic associate of 12, namely 1. (Cases in which the g.c.d. is 1 are of great theoretical impact and are discussed later in this section.) Moreover, $12 = g(x) - ((x/5) + (4/5)) \cdot$

$(f(x) - (x-6)g(x))$, whence $1 = (f, g) = \alpha f + \beta g$, where $\alpha(x) = (-x/60) - (1/15)$ and $\beta(x) = (1/60)(x^2 - 2x - 19)$.

A word of caution: The field F was not specified in the preceding example, but was assumed to be $\mathbf{Q}$. What if F had been the finite field $\mathbf{Z}_3$ instead? In that case, the polynomials are $f(x) = x^3 + x + 1$ and $g(x) = x^2 - 1$, but there is no harm in continuing to view f and g via their original expressions. The point to be emphasized here is that the algorithm for (f, g) terminates faster when $F = \mathbf{Z}_3$ than it did for $F = \mathbf{Q}$, since the remainder r_2 (= 12 as before) is 0 in $\mathbf{Z}_3$. Accordingly, for the given f and g, when $F = \mathbf{Z}_3$, we have $n = 1$, $r_1 = 5x - 20 = 2x - 2 = 2(x-1)$, and so $(f, g) = x - 1$ in this case. The moral is that (f, g) *sometimes* depends on the ambient field F.

And *sometimes* it doesn't! Specifically, one has the following result, which has great significance for the applications of similarity theory in linear algebra which were alluded to earlier. Let L be a subfield of the field F, and consider polynomials f and g in $L[x]$. Let d be the g.c.d. of f and g when the relevant field is perceived as L; then, by its very definition, $d \in L[x]$. Similarly, let D be the g.c.d. of f and g when the field containing the polynomial's coefficients is perceived as F; of course, $D \in F[x]$. The point is that $d = D$ and so, in particular, $D \in L[x]$. (This does not contradict the *caveat* of the preceding paragraph, since neither $\mathbf{Q}$ nor $\mathbf{Z}_3$ is a subfield of the other.) One application of this observation would be that the g.c.d. in $\mathbf{R}[x]$ (or in $\mathbf{C}[x]$ or in $\mathbf{Q}(\sqrt{2})[x]$ or . . .) of the polynomials $x^3 + x + 1$ and $x^2 - 1$ is (still) 1. By way of proof in the general case, notice that corresponding equations in the g.c.d. algorithms constructing d and D coincide, because of the uniqueness of quotients and remainders asserted in the earlier division algorithm. Thus, d and D are each monic associates of the same r_n, forcing $d = D$, as claimed.

The reader is urged to gain facility in applying the g.c.d. algorithm, as it will be used in Section 3 of Chapter 2 in order to detect multiple roots. Most importantly, it is put to work in Section 4 of Chapter 5 to "separate" real roots, a theoretical prerequisite for sensible applications of the numerical root-approximation schemes introduced in Chapter 6.

Next, an important concept is introduced. Polynomials f and g in $F[x]$ will be said to be *relatively prime* (to one another) in case $(f, g) = 1$, the g.c.d. being taken in $F[x]$. For example, $x^2 - 4$ and $x^2 + x - 6$ are not relatively prime; $x^3 - 6x^2 + x + 4$ and $x^2 - 4$ are relatively prime when $F = \mathbf{Q}$, but fail to be relatively prime when $F = \mathbf{Z}_3$. By the linear combination principle and the g.c.d. algorithm, one may show by a degree argument that a given pair of polynomials f and g are relatively prime if and only if some linear combination $\alpha f + \beta g$ is a nonzero element of F. (Prove this.) Thus, in general, if $d = (f, g)$ is the g.c.d. of f and g (with, as usual, not both f and g equal to 0), then f/d and g/d are relatively prime, since $\alpha f + \beta g = d$ leads to $\alpha(f/d) + \beta(g/d) = 1$.

We close this section with five important applications of the g.c.d. theory, some of which were promised earlier. First, by analogy with the situation for **Z**, let's agree to call a polynomial $f \in F[x]$ of positive degree *prime* in case it has only the trivial factorizations; that is, in case $f = gh$, with g and h in $F[x]$, forces either g or h to be a constant (so that f is associated to the other factor). By a degree argument, any linear polynomial is prime. Viewed in $\mathbf{R}[x]$, $x^2 + 1$ is prime (if that isn't clear, glance at the factor theorem in Section 1 of Chapter 2); however, $x^2 + 1$ is *not* prime in $\mathbf{C}[x]$, since $x^2 + 1 = (x - i)(x + i)$. It is clear that any associate of a prime is itself prime. (Isn't it?) It is now possible to state two other characterizations of prime polynomials f, namely as positive-degree $f \in F[x]$ such that (*) whenever $f \mid gh$ for polynomials $g, h \in F[x]$, either $f \mid g$ or $f \mid h$ (or both); and (**) for each $g \in F[x]$, either $f \mid g$ or $(f, g) = 1$ (but, necessarily, not both).

Let's establish the equivalence of the three characterizations of prime polynomials. First, assume that f satisfies (*). To prove that f satisfies the original definition of prime, let $f = gh$; in particular, $f \mid gh$ whence, without loss of generality, $f \mid g$ by (*). As $\deg(g) \leq \deg(g) + \deg(h) = \deg(f) = \deg(g) - \deg(g/f) \leq \deg(g)$, we have $\deg(h) = 0$; that is, h is a constant, as desired. For the second part of the proof of equivalence, assume next that f is prime in the original sense, and argue in favor of (**) as follows. For any g, let $d = (f, g)$. If $d \neq 1$, the factorization $f = d(f/d)$ forces f/d to be a constant since f is prime; then, f is associated to d and, since $d \mid g$, we infer that $f \mid g$ also, yielding (**). Finally, to prove (**) $\Rightarrow$(*), suppose that f satisfies (**) and that f divides a product gh. If $f \nmid g$, then (**) guarantees that f and g are relatively prime, so that suitable polynomials α and β satisfy $\alpha f + \beta g = 1$, by the g.c.d. algorithm. Therefore, multiplying by h throughout leads to $h = (h\alpha)f + \beta(gh)$, whence $f \mid h$ by the linear combination principle, establishing (*) and completing the proof.

Our second application of the g.c.d. theory was promised in Section 1, and is made possible by the preceding application and the next remark. The reader must surely have noticed that **Z** sustains a theory of division and divisibility much like that in $F[x]$. In particular, there is a division algorithm for **Z** (we've already used it—see the proof identifying n-th roots in Section 2) and, consequently, a g.c.d. algorithm in **Z**. More to the point, the above characterizations of prime polynomials apply, *mutatis mutandis*, to yield analogous characterizations of prime integers. (Technically, **Z** and $F[x]$ are the archetypical *Euclidean domains*, with "degree arguments" in **Z** proceeding via absolute value rather than via deg. The interplay between **Z** and $F[x]$ is strong enough to produce, at the end of this section, the analogue for $F[x]$ of the fundamental theorem of arithmetic.) We next exploit the integer analogue of (**) to prove that $\mathbf{Z}_p$ is a field whenever p is a (positive) prime integer. (Do not shed tears for the original polynomial version of (**): it will see yeoman service in Section 6 of Chapter 2.) To the proof, then. The only serious item to check is that each member n in the list $1, 2, 3, \ldots, p-1$ of nonzero elements of $\mathbf{Z}_p$ possesses a multiplicative inverse (also in the list). Now, since $1 \leq n < p$, a degree argument in $\mathbf{Z}$ shows that p does not divide n, and so the integer version of (**) implies that the integer g.c.d. of n and p is 1. Thus, $\alpha n + \beta p = 1$, for suitable integers α and β. Moreover, the division algorithm for $\mathbf{Z}$ gives $\alpha = qp + r$, where $q \in \mathbf{Z}$ and $r \in \{0, 1, 2, \ldots, p-1\}$. After some algebraic simplification, the result is an equation $rn + (\beta + qn)p = 1$ in $\mathbf{Z}$ which, when read in $\mathbf{Z}_p$, says $rn = 1$; that is, $r = n^{-1}$ in $\mathbf{Z}_p$, completing the proof. To illustrate, we find 3^{-1} in $\mathbf{Z}_7$ as follows: $p = 7$, $n = 3$, we get $\alpha = -2$ and $\beta = 1$ from the g.c.d. algorithm (check!), $q = -1$, $r = 5$ and, indeed, $5 = 3^{-1}$ in $\mathbf{Z}_7$.

Our third application of g.c.d. theory concerns *primitive n*-th *roots of* 1 in $\mathbf{C}$. Recall from Section 2 that, if n is a positive integer ≥ 2, then the n n-th roots of 1 are $1, \omega, \omega^2, \ldots,$ and ω^{n-1}, where $\omega = \operatorname{cis}(2\pi/n)$. A complex number τ is said to be a *primitive* n-th root of 1 in case $1, \tau, \tau^2, \ldots, \tau^{n-1}$ also lists the n-th roots of 1. Clearly, a primitive n-th root of 1 is just $\tau = \omega^b$ for some $1 \leq b \leq n-1$ such that ω is a power of τ. The key fact to be proved next is that if $1 \leq b \leq n-1$, then ω^b is a primitive n-th root of 1 if and only if the integer g.c.d. of b and n is 1. (The number of such b, that is, the number of primitive n-th roots of 1, is then computed as $\phi(n)$, the value of Euler's totient function, whose formula may be found in any text on elementary number theory.) For example (taking $n = 12$ and $\omega = \operatorname{cis}(\pi/6)$), the primitive 12-th roots of 1 are ω, ω^5, ω^7, and ω^{11}. For the general proof, note that: the integer g.c.d. of b and n is $1 \Leftrightarrow cb + mn = 1$ for some integers c and $m \Leftrightarrow (2\pi/n) - (2\pi cb/n) = 2\pi m$ for some integers c and $m \Leftrightarrow$ (by comparing args:) $\omega = \omega^{cb} = (\omega^b)^c$ for some $c \in \mathbf{Z}$. To complete the proof, observe that $c = qn + r$, with $q \in \mathbf{Z}$ and $0 \leq r \leq n-1$, yields $(\omega^b)^c = \omega^{b(qn+r)} = (\omega^n)^{bq} \cdot (\omega^b)^r = (\omega^b)^r$, since $\omega^n = 1$.

The fourth application redeems a pledge made in Section 1. Its method will be looked at more generally in Section 6 of Chapter 2, with a geometric payoff in Section 3 of Chapter 3. Specifically, if $\alpha = \sqrt[3]{2}$, we must show that $\mathbf{Q}(\alpha) = \{a + b\alpha + c\alpha^2 : a, b, c \text{ elements of } \mathbf{Q}\}$ is a field. As remarked earlier, (viii) is the only item at issue. Let $e = a + b\alpha + c\alpha^2$ be any nonzero element of $\mathbf{Q}(\alpha)$. In order to produce $e^{-1} \in \mathbf{Q}(\alpha)$, and hence verify (viii), it is helpful to observe that the polynomials $f(x) = x^3 - 2$ and $g(x) = cx^2 + bx + a$ are relatively prime. Now this would follow from (**) if $f(x)$ were prime in $\mathbf{Q}[x]$. (It is, but that may not be obvious. Peek at Chapter 4.) Summoning up the g.c.d. algorithm for an end-run, we produce the linear combination $H(x)f(x) + G(x)g(x) = 1$, where $H(x)$ and $G(x)$ are rather complicated, but explicit, polynomials which, we trust, the reader will obtain. (Now, after a pause to celebrate the explication of $H(x)$ and $G(x)$, that is, after the reader has carried out this application of the g.c.d. algorithm, he/she may wish to know why earlier remainders are nonzero or what to do for the degenerate cases in which b or c is 0. Instead of a tedious case analysis, we defer until Chapter 4's available.) Now, "plug in" α. (This familiar technique has some subtleties: see Section 1 of Chapter 2.) The result, with the usual notation, is $G(\alpha)e = 1$, since $f(\alpha) = \alpha^3 - 2 = 0$, whence the quantity denoted $G(\alpha)$ is a suitable e^{-1}. For example, $(\alpha + 3)^{-1} = \frac{1}{29}(\alpha^2 - 3\alpha + 9)$.

Our final application is enshrined in more abstract approaches as the assertion that "any Euclidean domain is a unique factorization domain (UFD)." We treat it as the analogue in $F[x]$ of the fundamental theorem of arithmetic. One of its practical consequences is the well-definedness of the multiplicity of a root (see Section 1 of Chapter 2.) The statement to be proved is: for any field F, any nonzero polynomial $f(x) \in F[x]$ may be written in an essentially unique way (that is, unique apart from the order of the factors) as $f(x) = cp_1(x)^{m_1}p_2(x)^{m_2} \ldots p_k(x)^{m_k}$, where c is a constant, the $p_i(x)$'s are distinct monic prime polynomials in $F[x]$, and each m_i is a positive integer. (If $\deg(f) = 0$, we permit the degenerate case $k = 0$.)

The proof is quite instructive. Existence of such a product decomposition for $f(x)$ follows from the well-ordering property: if $f(x)$ were of minimal degree amongst polynomials not so decomposable, then, since $f(x)$ is associated to some monic polynomial, $f(x)$ cannot be prime. Accordingly, $f(x) = g(x)h(x)$ for certain nonconstants $g(x)$ and $h(x)$. By the fundamental fact about degrees, $\deg(g)$ and $\deg(h)$ are each less than $\deg(f)$, whence $g(x)$ and $h(x)$ each have product decompositions of the stipulated kind. (Why?) Multiplying together such decompositions (and simplifying) produces a product decomposition expressing $f(x)$, contrary to hypothesis. This completes the proof of the existential assertion. (What if $f(x)$ is constant? Where did the above proof assume that $\deg(f) > 0$?) As for uniqueness, (*) of application 1 comes to the rescue. Indeed, the constant factor involved in the decomposition must be the coefficient of the leading term of $f(x)$, and so our task is evidently to show that if $p_1p_2 \ldots p_k = q_1q_2 \ldots q_n$ for certain monic primes $p_1, \ldots, q_n$ (possibly listed with repetition), then $k = n$ and, after relabelling, $p_i = q_i$ for each i. The way that (*) enters is this. Since $p_1 \mid p_1p_2 \ldots p_k$, we have $p_1 \mid q_1q_2 \ldots q_n$, and so (*) gives $p_1 \mid q_i$ for some i. However q_i, being prime, has only the trivial factorizations, so that p_1 and q_i are associates. As each is monic, p_1 and q_i are therefore equal. Cancel them from $p_1p_2 \ldots p_k = q_1q_2 \ldots q_n$ and iterate the entire process, stripping away the factors $p_2, p_3, \ldots, p_n$ in succession. The proof is complete.

Four remarks about the above result. First *given* the decompositions for polynomials f and g into products of primes, there is a simple way to compute (the corresponding product expression for) their g.c.d. (f, g). We leave the details to the reader—and the exercises—with the hint that another name for g.c.d. is "highest common factor."

Second, the following analogue of Euclid's proof that there are infinitely many positive prime integers shows, over any field F, that there are infinitely many monic prime polynomials in $F[x]$. Indeed, if $f_1(x), \ldots, f_n(x)$ were all the monic primes ($n \geq 1$ since the polynomial x is prime), the linear combination principle and a degree argument reveal that the polynomial $1 + f_1(x) \cdot f_2(x) \cdot \ldots \cdot f_n(x)$ is bereft of prime divisors, contrary to the above result.

Third, the above result permits us to view primes as the irreducible building blocks of $F[x]$. (Indeed, *irreducible* is an oft-used synonym for "prime.") Cataloguing the primes in $F[x]$, for arbitrary F, has not been done. Of course, for a finite field F, one could in principle write down the (finite) list $\mathfrak{L}$ of all nonconstant nonzero polynomials over F of degree less than of a given $f(x) \in F[x]$; then $f(x)$ would be prime if and only if it did not appear in the list of products of two (possibly equal) entries in $\mathfrak{L}$. For infinite fields: it is known that the linear polynomials are the only primes in $\mathbf{C}[x]$ (see Section 6 of Chapter 3); and the only additional primes in $\mathbf{R}[x]$ are the quadratics with negative discriminant (see Section 1 of Chapter 5).

Finally, a theoretical point. $S[x]$ could have been defined in case S was a nondomain such as $\mathbf{Z}_6$. In fact, the division algorithm carries over *verbatim* in this case. However, there is no suitable g.c.d. algorithm for $\mathbf{Z}_6[x]$ because the leading coefficients of remainders may fail to have multiplicative inverses. (At least, if S *were* a domain, the algorithm proceeds in the larger mathematical system $F[x]$, where $F = \text{qf}(S)$. But $\mathbf{Z}_6$ is not a subsystem of *any* field.) More damaging than unavailability of familiar tools in $\mathbf{Z}_6[x]$ is failure of uniqueness of factorization via monic primes: consider $(x-2)(x-3) = x(x-5)$ in $\mathbf{Z}_6[x]$! One moral is that the tools developed thus far may be expected to be of great use *because* they do not apply universally. Fortunately, they will apply where and when we need them.

EXERCISES 1.3

S continues to denote a domain, and F a field.

1. Simplify the following polynomials:
   (a) $2(-3x^4 - ix^3 + ½) - (9x^4 + (6 - i)x^3 - x)$, with $F = \mathbf{C}$;
   (b) $(x^2 - \pi x + 9)\cdot(2x - 7)$, with $F = \mathbf{R}$.
2. Find the quotient $q(x)$ and the remainder $r(x)$ in the following cases of divisor $g(x)$ and dividend $f(x)$:
   (a) $f(x) = 2x^3 - 3x + 8$, $g(x) = 2x - 9$, with $F = \mathbf{Q}(\sqrt[3]{2})$;
   (b) f and g as in (a), with $F = Z_3$;
   (c) $f(x) = 2x^2 - x + (1/3)$, $g(x) = 3x^2 + x + ½$, with $F = \mathbf{Q}$;
   (d) f as in (c), $g(x) = ix^3 - \sqrt{2}$, with $F = \mathbf{C}$.

3. (a) Let $f(x)$, $g(x)$, and $h(x)$ be polynomials in $S[x]$. If f and g are associates and if g and h are associates, prove that f and h are associates.
(b) If $f(x) \in S[x]$ possesses a multiplicative inverse in $S[x]$, prove that $f(x)$ is a constant.
(c) Find two linear polynomials in $\mathbf{Z}_6[x]$ whose product is 0.
4. (a) For f, g, h in $F[x]$ and not all 0, define the greatest common divisor (f, g, h) to be $((f, g), h)$. What properties does this notion of g.c.d. have? Show that the same object is defined by either of $(f, (g, h))$ and $((f, h), g)$.
(b) Define what ought to be meant by the greatest common divisor of any set of nonzero polynomials $f_1, f_2, f_3, \ldots$ in $F[x]$. (Which properties should such an object satisfy?) Prove that such an object always exists and is, moreover, a linear combination of finitely many of the f_i. (*Hint*: $\deg((f_1, \ldots, f_n))$ is a decreasing sequence.)
(c) For all f, g, and h in $F[x]$, prove that (fg, fh) and $f\cdot(g, h)$ are associates. Generalize.
5. What is the g.c.d. $(x^2 + 3x - 4, x^3 + 2x^2 - 7x - 14)$ if $F = \mathbf{Q}$? If $F = \mathbf{Z}_2$? If $F = \mathbf{Z}_3$?
6. (a) If f and g are nonzero polynomials in $F[x]$, prove that (f, g) may be characterized as the monic linear combination of f and g with minimal degree.
(b) If f, g, and h are in $F[x]$ and $(f, h) = 1 = (g, h)$, prove that $(fg, h) = 1$.
7. (a) Prove that $(x^3 - 6x^2 + x + 4, x^5 - 6x + 1) = 1$ in $\mathbf{Q}[x]$, using the g.c.d. algorithm.
(b) Prove that $(x^3 - 2x^2 - 2x - 2, x^4 - 3x^2 + x - 2) = 1$ in $\mathbf{R}[x]$.
(c) Is $(x^2 - 4, x^2 + x - 6) = x - 2$ for all fields F?
8. Let p be a positive prime number. Prove that $\operatorname{char}(S) = p$ if and only if $(x + 1)^p = x^p + 1$ in $S[x]$. (*Hint*: for the "only if" half, prove a lemma about the binomial coefficients in the expansion of $(x + 1)^p$.)
9. Find the expression for (f, g) as a product of primes in each of the following cases:
(a) $f(x) = (2x^2 - 7)^3(x^2 + 1)^4(3x - 7)$, $g(x) = (2x^2 - 7)^4(x^2 + 1)^2(7x - 3)$ with $F = \mathbf{Q}$;
(b) f, g as in (a), with $F = \mathbf{Z}_3$;
(c) f, g as in (a), with $F = \mathbf{Z}_7$;
(d) $f(x) = (x^2 - 4)^8$, $g(x) = (x^2 + 3x - 10)^3$, with $F = \mathbf{C}$;
(e) f, g as in (d), with $F = \mathbf{Z}_2$.
10. (a) Using the integer g.c.d. algorithm, find 4^{-1} in $\mathbf{Z}_7$.
(b) Find 6^{-1} in $\mathbf{Z}_{13}$.
(c) If $\alpha = \sqrt[3]{7}$, find rational numbers a, b, and c such that $(\alpha^2 + 1)^{-1} = a + b\alpha + c\alpha^2$.
11. How many primitive 8-th roots of 1 are there in $\mathbf{C}$? Identify them.
12. Consider the elements $x^3/(x + 1)$ and $(x + 1)/x^3$ in the quotient field of $S[x]$. Prove that neither of these elements lies in $S[x]$. (*Hint*: Review the proof that $x^{-1} \notin S[x]$.) Which two values of $\operatorname{char}(S)$ permit $(x^2 - 6x - 2)/(x - 2) \in S[x]$?

13. Factor $x^4 + x^3 + x + 2$ as a product of two monic quadratic polynomials in $\mathbf{Z}_3[x]$.
14. If f and g are nonzero polynomials in $F[x]$, prove that (f, g) may be characterized as the monic common divisor of f and g of maximal degree.

*4. Partial Fractions

One pleasant application of the divisibility theory developed in the preceding section is the justification of the partial-fraction procedure given in most texts on calculus as a very useful technique for performing the indefinite integration of rational functions. Our concern here is to prove that any rational function can, if fact, be expressed as a suitable sum of partial fractions. Specifically, we prove: if F is any field and if $f(x)$ and $g(x)$ are any polynomials in $F[x]$ with $g(x) \neq 0$, then $f(x)/g(x)$ is the sum of a polynomial over F and finitely many fractions of the form

$$\frac{\text{polynomial over } F \text{ of degree} < m}{\text{power of an } m\text{-th-degree prime polynomial in } F[x]} \cdot$$

Our proof is algorithmic, given the factorization of $g(x)$ as a product of primes in $F[x]$. There is a similar effectiveness to the ways that most calculus texts produce the coefficients in the numerators of the intervening fractions, but we stress that few, if any, of those texts explain why an expression of $f(x)/g(x)$ via such fractions is possible. (For completeness, we also note that those other texts' methods often appeal tacitly to the identity theorem, given below in Section 1 of Chapter 2, to compute the required coefficients.) Finally, for calculus, $F = \mathbf{R}$ and so, as explained in Section 3 (with the proof deferred to Chapter 5), $m \leq 2$ in this case.

First, the division algorithm gives $q(x)$ and $r(x)$ such that $f = qg + r$ and $\deg(r) < \deg(g)$. Then $f(x)/g(x) = q(x) + (r(x)/g(x))$, and our task is reduced to the explication of $r(x)/g(x)$. In other words, without loss of generality, $f(x) \neq 0$ and $\deg(f) < \deg(g)$. Next, consider $g(x) = p_1(x)^{m_1}p_2(x)^{m_2} \ldots p_k(x)^{m_k}$, where the p_i's are primes in $F[x]$, pairwise nonassociated, and each $m_i > 1$. Set

$$h(x) = p_2(x)^{m_2} \ldots p_k(x)^{m_k}.$$

Then $g = p_1^{m_1} \cdot h$ and, using the highest common factor idea, $(p_1^{m_1}, h) = 1$. The g.c.d. algorithm supplies polynomials G and H such that $Gp_1^{m_1} + Hh = 1$, whence multiplying by f gives $f = (fG)p_1^{m_1} + (fH)h$. Applications of the division algorithm yield $fG = G_1h + R_1$ and $fH = H_1 \cdot p_1^{m_1} + R_2$, for suitable polynomials G_1, R_1, H_1, and R_2 over F such that $\deg(R_1) < \deg(h)$ and $\deg(R_2) < \deg(p_1^{m_1}) = m_1 \cdot \deg(p_1)$. An easy computation reveals that

$$f - R_1 \cdot p_1^{m_1} - R_2 \cdot h = (G_1 + H_1)p_1^{m_1} \cdot h.$$

If this expression is nonzero, the fundamental fact about degrees implies that its degree d is at least $\deg(p_1^{m_1} \cdot h) = \deg(g)$ which (check this) exceeds $\max(\deg(f), \deg(R_1) + \deg(p_1^{m_1}), \deg(R_2) + \deg(h))$. But this max is greater than or equal to d, whence $d > d$, preposterous! The important upshot is $f - R_1 \cdot p_1^{m_1} - R_2 \cdot h = 0$, whence $f/g = (R_2/p_1^{m_1}) + (R_1/h)$. The first of the terms on the right-hand side is the type we'd like to reduce the consideration to. As h is factored in terms of $k - 1$ nonassociated primes, iteration expresses R_1/h appropriately, and so $f/g = \sum_{i=1}^{k}(T_i/p_i^{m_i})$ for suitable polynomials T_i over F (note that $T_1 = R_2$) such that $\deg(T_i) < \deg(p_i^{m_i})$.

Thus, consideration is reduced to the case $k = 1$, that is, $g(x) = p(x)^m$ such that p is irreducible, $m \geq 1$, and $\deg(f) < \deg(p^m) = mn$, where $n = \deg(p)$. So much for the denominators; now, for the numerators. The division algorithm (again) gives polynomials Q and R over F satisfying $f = Qp + R$ and $\deg(R) < n$; thus, $f/g = (R/p^m) + (Q/p^{m-1})$. The first term on the right-hand side is of the desired form, and we claim that iteration reduces Q/p^{m-1} to a sum of such terms. This is almost apparent: $m - 1$ *is* less than m; we need only check that $\deg(Q) < (m - 1)n$; and this holds since $\deg(Q) = \deg((f - R)/p) = \deg(f - R) - \deg(p) \leqslant \max(\deg(f), \deg(R)) - n \leqslant (mn - 1) - n = n(m - 1) - 1 < (m - 1)n$, completing the proof.

In all candor, it must be admitted that, although the above proof was needed in order to establish, once and for all, the existence of partial-fraction decompositions, our proof's algorithm is much slower in applications than those met in calculus courses.

EXERCISES 1.4

1. Go to any calculus text, open it to the section on partial fractions, and do comfortably many applications of this section's algorithm, using the text's integrands as suitable f/g. Or: *don't* go to your library, but use your imagination to rig up your own exercises.
2. Here's an analogue of base-n expansion in ordinary arithmetic. If $p(x)$ is a prime polynomial over the field F, prove that any $f(x) \in F[x]$ can be expressed as $f(x) = f_0(x) + f_1(x)p + f_2(x)p^2 + \ldots + f_m(x)p^m$, for suitable polynomials f_i over F, with $\deg(f_i) < \deg(p)$ for each i. (*Hint*: select m so that $\deg(f) < m \cdot \deg(p)$, and use this section's decomposition of f/p^m.) Use a degree argument to show that the f_i in such an expression are uniquely determined.

*5. More General Algebraic Structures and an Application to Trigonometric Identities

It is a sound maxim that the general is an outgrowth of the concrete. In that sense, our study of concrete structures such as **Z**, $F[x]$, and $\mathbf{Q}(\sqrt{2}, \sqrt{3})$ may be viewed as an introduction to the topics normally encountered in a modern undergraduate text on abstract algebra, namely groups, rings, and vector spaces. Indeed, **Z** and $F[x]$ (and the somewhat discredited $\mathbf{Z}_6[x]$) are examples of *rings*, algebraic structures satisfying conditions (i), (iii)–(v), and the strengthening of (vii) consisting of both distributive laws. Such a set of axioms has a reasonable claim to be studied since, as noted in Section 1, it permits proof of results such as $a0 = 0$. More to the point, structures besides domains receive uniform treatment as rings; these include division rings (such as the quaternions, **H**) exhibiting noncommutativity of multiplication, and certain rings of matrices. (Readers acquainted with matrix algebra will note that matrix rings may have noncommutative multiplication and may also fail to satisfy (vi)). Glancing at texts such as [1] and [10] (see the list of references at the back of the book) will convince the reader that our treatment of g.c.d.s in Section 3, together with the material in Section 2 of Chapter 4, is essentially the study of Euclidean domains and unique factorization domains (though not of their relative, principal ideal domains—about which, more shortly). As for the "modern" topics of factor ring and homomorphism: the former has already been encountered in computing with the structures $\mathbf{Z}_n$ and will have another concrete embodiment in constructing splitting fields in Section 6 of Chapter 2; instances of the latter include conjugation in **C** (see Section 2) and the evaluation and "overhead-barring" procedures in Section 1 of Chapter 2.

As for *groups*, structures with one binary operation satisfying variants of (i) and (iii)–(v), examples abound: take the additive structure of any ring! Such groups are also *abelian*, that is, commutative; other examples of abelian groups are afforded by the multiplicative structure on the set of n-th roots of 1 in **C**, for any fixed $n \geq 1$. The discussion of primitive n-th roots in Section 3, not to mention some of the axiomatic gymnastics in Section 1, may be viewed as a fragment of abstract group theory. Finally, it should be mentioned that instances of nonabelian groups arise naturally; for example, a nonabelian group of permutations underlies the topic of Section 5 of Chapter 2. With respect to *vector spaces*, we remark only that the treatment of $\mathbf{Q}(\sqrt{2}, \sqrt{3})$ in Section 1 showed it to be of *dimension* 4 by exhibiting a **Q**-*basis* of it. For an elaboration of these notions (and significant applications), see Section 6 of Chapter 2 and Section 3 of Chapter 3.

How does the foregoing overview of contemporary algebra relate to the second part of this section's title? Just so: with the aid of the theory of principal ideal domains, a proof has recently been given, in [14], for the following rule of thumb, familiar to all students of trigonometry. *Any trigonometric identity may be established by first expressing all its entries in terms of* $\sin\theta$ *and* $\cos\theta$ *and then simplifying the result with the help of the Pythagorean identity,* $\sin^2\theta + \cos^2\theta = 1$. It is our purpose here to present a more elementary, algorithmic proof of this result, using a bit of calculus and the division algorithm developed in Section 3.

A precise statement of the result must await the introduction of the notion of a polynomial in two variables. In general, if S is any domain, a *polynomial over S in the* (commuting, algebraically independent) *variables x* and y is formally a finite sum of *monomials* $sx^m y^n$, where $s \in S$, and the exponents m and n are nonnegative integers. Equality is defined in the obvious way (polynomials that *look* different, *are* different). As in the case with one variable, addition of polynomials is defined componentwise; and multiplication's definition is forced by the requirements of commutativity, associativity, and distributivity, together with the laws of exponents. Using a suitable notion of degree (provide details), one may show that $S[x, y]$, the set of all polynomials over S in the variables x and y, is a domain, containing S as a subdomain. Indeed $S[x,y]$ may be viewed either as $(S[y])[x]$ or as $(S[x])[y]$ where, for example, $(S[y])[x]$ is the domain of polynomials over $S[y]$ in the variable x. For instance, the element $f(x,y) = 3x^2y - 7xy^2 + 2x - 3y + \pi \in \mathbf{R}[x,y]$ may be viewed as the polynomial $(3y)x^2 + (-7y^2 + 2)x + (-3y + \pi)$ in $(\mathbf{R}[y])[x]$ or as $(-7x)y^2 + (3x^2 - 3)y + (2x + \pi)$ in $(\mathbf{R}[x])[y]$. Notice that $(S[x,y])[z]$ may reasonably be called the set (domain) of *polynomials over S in the* three (commuting, algebraically independent) *variables x,* y, and z. We leave to the reader the task of defining the notion of a polynomial over S in arbitrarily many variables.

Now we may give a precise statement of the result to be proved. Let $f(x, y) \in \mathbf{R}[x, y]$ be such that $f(\sin\theta, \cos\theta)$, the real-valued function obtained from $f(x, y)$ by replacing each occurrence of x (respectively y) by $\sin\theta$ (respectively $\cos\theta$), is the zero function; that is, $f(\sin\theta, \cos\theta) = 0$ for each real number θ. Then there exists a polynomial $g(x, y) \in \mathbf{R}[x, y]$ satisfying $f(x, y) = (x^2 + y^2 - 1)g(x, y)$; in particular, the trigonometric identity $f(\sin\theta, \cos\theta)$ factors as the product $(\sin^2\theta + \cos^2\theta - 1)g(\sin\theta, \cos\theta)$.

An example will clarify matters. Consider the trigonometric identity: $3\sin^4\theta\tan\theta + 6\sin^2\theta + 3\sin^2\theta\cos^2\theta\tan\theta = 6 + 3\sin^2\theta\tan\theta - 6\cos^2\theta$. As $\tan\theta = \sin\theta/\cos\theta$, it suffices to prove the following identity: $3\sin^5\theta + 6\sin^2\theta\cos\theta + 3\sin^3\theta\cos^2\theta = 6\cos\theta + 3\sin^3\theta - 6\cos^3\theta$. The latter identity "is" just $H(\sin\theta, \cos\theta) = 0$, where $H(x, y) = 3x^5 + 6x^2y + 3x^3y^2 - 6y - 3x^3 + 6y^3$. The reader should verify that the algorithm given below actually yields $H(\sin\theta, \cos\theta) = (\sin^2\theta + \cos^2\theta - 1)(3\sin^3\theta + 6\cos\theta)$; that is, in the above notation, $f = H$ and $g = 3x^3 + 6y$. Contrast this with the usual high-school solution of the given trigonometric identity!

Now, to our proof. Let $f(x, y) \in \mathbf{R}[x, y]$ be assumed, as above, to produce the trigonometric identity $f(\sin\theta, \cos\theta) = 0$. View $f \in (\mathbf{R}[y])[x]$. Although $\mathbf{R}[y]$ is not a field (and $(\mathbf{R}[y])[x]$ is not a principal ideal domain), the division algorithm established in Section 3 permits us to divide $f(x, y)$ by $x^2 + y^2 - 1$. (Technically, the method applies because, when $x^2 + y^2 - 1$ is viewed in $(\mathbf{R}[y])[x]$, its leading coefficient is 1, which certainly has a multiplicative inverse in $\mathbf{R}[y]$.) The upshot is

$$f(x, y) = (x^2 + y^2 - 1)\, g(x, y) + h(x, y),$$

where $h(x, y)$, viewed in $(\mathbf{R}[y])[x]$, is either 0 (which is what we wish to prove!) or has degree at most 1. Substitute $x = \sin\theta$ and $y = \cos\theta$ to get

$$0 = f(\sin\theta, \cos\theta) = (\sin^2\theta + \cos^2\theta - 1)g(\sin\theta, \cos\theta) + h(\sin\theta, \cos\theta)$$
$$= 0 + h(\sin\theta, \cos\theta) = h(\sin\theta, \cos\theta).$$

(Once again, we caution the reader that the substitution process involves a subtlety having to do with commutativity of multiplication (which, fortunately, is satisfied here): see Section 1 of Chapter 2 for details.) No further algebra is needed in the proof; the rest of the proof will be "analytic."

Our task is reduced to showing that $h(x, y)$ is the zero polynomial. By the restriction on degree given above by the division algorithm, we may write $h(x, y) = F_1(y)x + G_1(y)$, for suitable members $F_1(y)$ and $G_1(y)$ of $\mathbf{R}[y]$. As $h(\sin\theta, \cos\theta) = 0$, we have

$$F_1(\cos\theta)(\sin\theta) + G_1(\cos\theta) = 0.$$

Setting $\theta = \pi/2$ in this equation gives $F_1(0) + G_1(0) = 0$, while setting $\theta = -\pi/2$ gives $-F_1(0) + G_1(0) = 0$. Thus $F_1(0) = 0 = G_1(0)$; that is, the constant terms of $F_1(y)$ and $F_2(y)$ are each 0. Accordingly, there are polynomials $F_2(y)$ and $G_2(y)$ in $\mathbf{R}[y]$ such that $F_1(y) = yF_2(y)$ and $G_1(y) = yG_2(y)$. Substitution into the previously displayed equation leads to $(\cos\theta)T(\theta) = 0$, where

$$T(\theta) = F_2(\cos\theta)(\sin\theta) + G_2(\cos\theta).$$

Thus, $T(\theta) = 0$ for all real numbers θ for which $\cos\theta \neq 0$. Recall that $\cos\theta = 0$ only at certain *isolated* values of θ, namely, $\pm\pi/2$, $\pm 3\pi/2$, $\pm 5\pi/2, \ldots$. Moreover, $T(\theta)$ is a continuous function of θ, since the class of continuous functions is closed under composition, products, and sums and contains $\sin\theta$, $\cos\theta$ and all polynomial functions. Consequently, if $\cos(a) = 0$, then $T(a) = \lim_{\theta\to a} T(\theta) = \lim_{\theta\to a} 0 = 0$. If we let $h_1(x, y) = F_2(y)x + G_2(y)$, then the preceding argument gives $h_1(\sin\theta, \cos\theta) = T(\theta) = 0$ for *each* real θ.

Treating $h_1(x, y)$ just as we considered $h(x, y)$, we find polynomials $F_3(y)$ and $G_3(y)$ such that $F_2(y) = yF_3(y)$, $G_2(y) = yG_3(y)$; and $h_2(x, y) = F_3(y)x + G_3(y)$ satisfies $h_2(\sin\theta, \cos\theta) = 0$. Iteration of this procedure leads to polynomials satisfying, in particular, $F_n(y) = yF_{n+1}(y)$ and $G_n(y) = yG_{n+1}(y)$ for each $n \geq 1$. Thus

$$F_1(y) = yF_2(y) = y^2F_3(y) = \ldots y^nF_{n+1}(y).$$

If $F_1(y)$ is not the zero polynomial, let m be its degree, select an integer $n > m$, and observe from the last displayed equation that $m = n + c$, where c is the (nonnegative) degree of $F_{n+1}(y)$. This contradiction shows that $F_1(y) = 0$. Similarly $G_1(y) = 0$, and so $h(x, y) = 0$, to complete the proof.

Finally, we remark that $\mathbf{C}[x, y]$ may be used (it *was* in [14]), rather than $\mathbf{R}[x, y]$, in our proofs since $\cos(z)$ is nonzero for each nonreal $z \in \mathbf{C}$.

EXERCISES 1.5

1. Go to any text on trigonometry, open it to the section on identities, and do comfortably many applications of this section's algorithm, expressing each identity as a multiple of $\sin^2\theta + \cos^2\theta - 1$. (For multiple-angle items, such as $\sin(4\theta)$, argue as in #5, Exercises 1.2.) Or: *don't* go to your library, but use your imagination to rig up your own exercises.
2. (a) Express $f(x,y) = 2x^2y^3 - x^3y^2 + 7x^2 - 9y + \pi$ first as an element of $(\mathbf{R}[x])[y]$, then as an element of $(\mathbf{R}[y])[x]$.
   (b) Express $f(x,y,z) = 2x^2z^2y - 3xz^2y - 17yz + 14x + i$ as an element of $(\mathbf{C}[x,y])[z]$.

2

The Roots of the Matter or the Matter of Roots

We have now reached the heart of our subject's theory. This chapter will develop the mathematical facts of life for *arbitrary* polynomials over *arbitrary* fields. (Later chapters treat *specific* low-degree polynomials and *specific* coefficient fields.) Section 1 introduces the important process of *evaluation* (or "substitution") and the concept of *root,* both on the basis of the division algorithm developed in Chapter 1, leading to the qualitative blockbuster that a nonzero n-th-degree polynomial over a field has at most n roots. (Changing "at most" to "in some sense, exactly" is the subject of the optional sixth section on *splitting fields*.) Section 1 also contains the theory underlying the numerical root-approximation methods of Horner (see Section 5 of Chapter 6) and Graeffe (Sections 7 and 8 of Chapter 6). The calculus' view of polynomial as function leads to Section 2's interpolation result indicating the extent to which polynomials are characterized by their values; and the formalism of differential calculus combines with Chapter 1's g.c.d. algorithm to produce a catalogue of roots' *multiplicities,* in Section 3. The observation that a monic polynomial is characterized by its roots (and their multiplicities) leads to the study of the so-called *elementary symmetric* polynomials in Section 4, with consequences (Newton's formulas) that constitute the theory underlying the numerical root-approximation method of Bernoulli (see Section 6 of Chapter 6). Related combinatorial topics concerning "not-so-elementary" symmetric polynomials, *homogeneous* polynomials, and *formal power series* are treated in the optional fifth and seventh sections.

1. The Factor Theorem and the Remainder Theorem

Throughout this section, we deal with a *domain* S (which, in later applications, will typically be a field). Consider a polynomial $f(x)$ in $S[x]$. The process of *evaluating* $f(x)$ at a number r, resulting in a number $f(r)$, is a familiar one, and was in fact explicitly used already, in Sections 3 and 5 of Chapter 1. A couple of examples, to reinforce the notion: if $f(x) = x^3 - \pi x^2 + \frac{1}{2}x - 7$ with $S = \mathbf{R}$, then $f(0) = 0^3 - \pi 0^2 + \frac{0}{2} - 7 = -7$ and $f(-2) = (-2)^3 - \pi(-2)^2 + \frac{1}{2}(-2) - 7 = -4\pi - 16$. It will also be helpful to allow "numbers beyond S" to be substituted for x in certain evaluations. For the above f, this would permit $f(i) = i^3 - \pi i^2 + \frac{i}{2} - 7 = \frac{-i}{2} + \pi - 7$, as well as computations like $f(\frac{4}{x}) = (\frac{4}{x})^3 - \pi(\frac{4}{x})^2 + \frac{1}{2}(\frac{4}{x}) - 7 = (64/x^3) - (16\pi/x^2) + (2/x) - 7$ and $f(2x-1) = (2x-1)^3 - \pi(2x-1)^2 + \frac{2x-1}{2} - 7 = 8x^3 - (12+4\pi)x^2 + (7+4\pi)x - (\pi + \frac{17}{2})$. (The reader should check the preceding calculation, for example, by the binomial theorem. Proficiency at such computations will be essential for speedy applications of Horner's method in Chapter 6; for the fastest way to perform many such substitutions, see the method of *synthetic division* in Section 1 of Chapter 6.) In general $f(r)$, the evaluation of $f(x)$ at r, will be permitted whenever $r \in T$ for a domain T of which S is a subdomain. The above examples involving $S = \mathbf{R}$ used $\mathbf{C}$ and the quotient field of $\mathbf{R}[x]$ as suitable values of T.

Now we come to a crucial point which is often, regrettably, sloughed over in more elementary treatments. Specifically, for each element r belonging to a suitable T (a domain containing S as a subdomain), the following two properties hold for $f, g \in S[x]$: if $h(x) = f(x) + g(x)$, then $h(r) = f(r) + g(r)$; and if $p(x) = f(x)g(x)$, then $p(r) = f(r)g(r)$. The first of these is rather obvious, and we leave its verification to the reader. The second is much less obvious: in fact, it would fail if multiplication in T

were not commutative! (To illustrate the last comment, recall—or believe! —that the division ring $\mathbf{H}$, whose multiplication is not commutative, contains elements i and j such that $i^2 = j^2 = -1$ and $ij \neq ji$. Inside $\mathbf{H}[x]$, the polynomial $p(x) = x^2 + 1$ is the product of $f(x) = x - i$ and $g(x) = x + i$; but if evaluation is permitted in the context of the nondomain $\mathbf{H}$, then $p(j) = j^2 + 1 = 0$ although $f(j)g(j) = (j - i)(j + i) = ji - ij \neq 0$.) Sobered by this example, the reader will certainly wish to prove the assertion about $p(r)$—use the definition of multiplication in $S[x]$ and recall that $p(r)$ is defined by substituting r into the "expanded" version of $p(x)$—and may possibly wish to rethink his/her reason for accepting the statement about $h(x)$! It will be convenient to refer to the assertions about $h(x)$ and $p(x)$ as the *fundamental* (*homomorphism*) *properties of evaluation*.

To fix notation, let $f(x) \in S[x]$ and let $r \in S$. (If one has $r \in T$ in the above sense, that's covered by this set-up too: replace S by T, viewing $f \in T[x]$.) The *remainder theorem* asserts: there exists a (unique) polynomial $q(x) \in S[x]$ such that $f(x) = q(x)(x - r) + f(r)$. Put differently: the remainder upon "dividing $f(x)$ by $x - r$" is just $f(r)$. This observation speeds up certain procedures enormously. For example, you no longer *have* to divide $x - 5$ into $3x^3 - \pi x + i$ in order to find the remainder: it's just $3(5)^3 - \pi 5 + i = 375 - 5\pi + i$. To prove the remainder theorem, first get $f(x) = q(x)(x - r) + R(x)$, with q and R in $S[x]$ and $\deg(R) < \deg(x - r) = 1$, by the division algorithm. (Checking back to Section 3 of Chapter 1, you'll find that the division algorithm is applicable because the leading coefficient of $x - r$ is 1, which has a multiplicative inverse in S.) Now, $\deg(R)$ is either $-\infty$ or 0; that is, $R(x) = R \in S$. Applying the fundamental properties of evaluation to $f(x) = q(x)(x - r) + R$ produces $f(r) = q(r) \cdot (r - r) + R = q(r) \cdot 0 + R = 0 + R = R$, as claimed.

Which brings us, finally, to the subject of roots. For S, $f(x)$, and r as above, we say that r is a *root* of $f(x)$ in case $f(r) = 0$. For example, i is a root of $x^2 + 1$; $\sqrt{2} + \sqrt{3}$ is a root of $x^4 - 10x^2 + 1$; and π is not a root of $x^3 - 6$. An immediate corollary of the remainder theorem is the *factor theorem* which asserts: r is a root of $f(x)$ if and only if $(x - r) \mid f(x)$, where the vertical bar denotes "divides in $S[x]$," as in Section 3 of Chapter 1. (Prove the factor theorem, using the uniqueness property of quotients and remainders.)

Which brings us to the fundamental decomposition result for polynomials over a domain S. To wit: any nonzero polynomial $f(x) \in S[x]$ can be written in an essentially unique way (that is, unique apart from the order of factors) as $f(x) = (x - r_1)^{m_1} \cdot (x - r_2)^{m_2} \cdot \ldots \cdot (x - r_k)^{m_k} \cdot g(x)$, where the r_i are distinct elements of S, each m_i is a positive integer, and $g(x)$ is a polynomial over S which has no roots *in* S. (Degenerate cases are permitted. For example, if $f(x) = x^2 + 1$ and $S = \mathbf{R}$, then $k = 0$ and $g(x) = x^2 + 1$; in case $f(x) = x^2 - 1$ and $S = \mathbf{Q}$, then $g(x) = 1 \in S$.) The proof that such a decomposition exists is easily given. Indeed, if n

$= \deg(f)$: the degenerate case $n = 0$ is accommodated with $k = 0$; any minimal-degree counterexample $f(x)$ would necessarily have a root $r \in S$, lest $k = 0$ intervene with $g = f$; then the factor theorem would produce $q \in S[x]$ such that $f(x) = q(x)(x - r)$; and, by the fundamental fact about degrees, $\deg(q) = n - 1$, whence $q(x)$ admits a suitable decomposition, whose product with $x - r$ explodes the assumption that $f(x)$ fails to be suitably decomposed. As for the proof that such a decomposition is essentially unique: consider a $g(x)$ appearing in such a decomposition of $f(x)$, let $F = \mathrm{qf}(S)$, and study *the* expression of $g(x)$ as a product of monic prime polynomials and a constant in $F[x]$; no such monic coincides with any $x - r_i$ which appeared in the description of $f(x)$ involving $g(x)$, by the factor theorem, since r_i is not a root of $g(x)$. Accordingly, uniqueness of the stipulated decomposition $f(x) = (x - r_1)^{m_1} \cdot \ldots \cdot (x - r_k)^{m_k} \cdot g(x)$ follows from the uniqueness assertion in Section 3 of Chapter 1 regarding factorization of $f(x)$ in $F[x]$.

More can be said about the r_i and m_i appearing above. Indeed, by the second fundamental property of evaluation, each $r \in S$ leads to $f(r) = (r - r_1)^{m_1}(r - r_2)^{m_2} \ldots (r - r_k)^{m_k} g(r)$. As $g(x)$ is supposed not to have any roots in S, the fact that S is a domain (specifically, condition (vi)′) guarantees that $f(r) = 0$ if and only if some $r - r_i = 0$; that is, $r_1, r_2, \ldots, r_k$ *are precisely the roots of* $f(x)$ *in the domain* S. The positive integer m_i is called the *multiplicity* of r_i as a root of $f(x)$. (Note that, although $g(x)$ may have a root in some domain T containing S, the multiplicity of r_i as a root of $f(x)$ is still m_i when T is the coefficient domain of reference, since $g(r_i) \neq 0$.) If $m_i = 1$, then r_i is called a *simple root* of $f(x)$; if $m_i > 1$, call r_i a *multiple root* of $f(x)$. For example, if $f(x) = x^2 - 1 = (x - 1)(x + 1)$ and $S = \mathbf{Q}$, then $k = 2$, and both $r_1 = 1$ and $r_2 = -1$ are simple roots; however, if $S = \mathbf{Z}_2$, then $k = 1$, and $r_1 = 1$ is a multiple root with multiplicity 2, since $x^2 - 1 = (x - 1)^2 \in \mathbf{Z}_2[x]$. Because of such pathological examples, we expect effective handling of multiplicities only for characteristic 0; a combined algebraic-analytic approach to such will be given in Section 3.

There is an accepted way of counting the roots of a polynomial $f(x) \in S[x]$ in the domain S: each root r_i, of multiplicity m_i, is counted m_i times; that is, one time for each factor $x - r_i$ appearing in the above decomposition of $f(x)$. (The roots are said to be "counted with their multiplicities.") Even with such a generous overcount, we now have the promised blockbuster: *if* $f(x)$ *is a nonzero polynomial, of degree* n, *over the domain* S, *then* $f(x)$ *has at most* n *roots* (counted with their multiplicities, in any domain T which contains S as a subdomain). The proof, fittingly enough, uses the fundamental fact about degrees: if $f(x) = (x - r_1)^{m_1} \ldots (x - r_k)^{m_k} g(x)$ as above, then $f(x)$ has $\sum m_i$ roots according to our counting convention, but $n - \sum m_i = \deg(f) - \deg((x - r_1)^{m_1} \ldots (x - r_k)^{m_k}) = \deg(g) \geq 0$, whence $\sum m_i \leq n$, as claimed.

The preceding result is simply invalid when S is replaced by a non-domain, with the notion of root similarly extended. For instance, the quadratic polynomial $x^2 + 1 \in \mathbf{H}[x]$ has infinitely many roots in the division ring $\mathbf{H}$ of quaternions, including i, j, and ij. A commutative system (ring) such as $\mathbf{Z}_8$ which is not a domain admits the pathology that $x^2 - 1 \in \mathbf{Z}_8[x]$ has the *four* roots 1, 3, 5, and 7 in $\mathbf{Z}_8$, although $4 > \deg(x^2 - 1)$. (Question: where does the proof of the blockbuster break down for $\mathbf{Z}_8$? *Hint*: see the remark concerning $\mathbf{Z}_6[x]$ at the close of Section 3 of Chapter 1.)

An important consequence of the blockbuster is the *identity theorem*: if $f(x)$ and $g(x)$ are polynomials over a domain S, each of degree at most n, such that $f(r_i) = g(r_i)$ for some collection of $n + 1$ *distinct* elements r_i (in some domain containing S as a subdomain), then $f(x) = g(x)$. The proof is exceedingly easy, but interesting: by the first fundamental property of evaluation, $h(x) = f(x) - g(x)$ satisfies $h(r_i) = f(r_i) - g(r_i) = 0$; as $h(x)$ has degree at most n and possesses more than n roots, blockbuster shows that $h(x) = 0$, that is, $f(x) = g(x)$. In particular (taking $g = 0$), if a polynomial $f(x) \in S[x]$ of degree at most n has more than n roots in the domain S, then $f(x) = 0$.

The last assertion suggests that a polynomial may be, in some sense, determined by its roots. Certainly, this would hold for a monic n-th-degree polynomial $f(x)$ with n roots (counting multiplicities) in some domain T since, in the above notation, monic $f(x) = (x - r_1)^{m_1} \ldots (x - r_k)^{m_k} g(x)$ and $\sum m_i = n$ forces $g(x) = 1$. The fact—which we will call *son of blockbuster*—is: any nonzero polynomial of degree n over a domain S has *exactly* n roots (counting multiplicities) in some field containing S as a subdomain. (Section 6 proves this fact.) Consequently, *any nonzero polynomial $f(x)$ over a domain S is characterized by its leading coefficient, c, and its roots*, the latter being viewed in a sufficiently "large" field F containing S; one then writes $f(x) = c(x - r_1) \ldots (x - r_n)$ and refers to $r_1, \ldots, r_n$ as "(all) the (n) roots of $f(x)$ (in F)." For example, the quintic polynomial $f(x) = 2x^5 - 4x^4 + 4x^3 - 8x^2 + 2x - 4 \in \mathbf{Z}[x]$ factors as $f(x) = (x - 2)\cdot(x - i)^2(x + i)^2 \in \mathbf{C}[x]$ and so the roots of $f(x)$ in this case are 2, i, i, $-i$, and $-i$.

One upshot of the identity theorem deserves special attention, namely the relation between our view of polynomials (as abstract expressions of a certain form) and the view (prevalent in high-school algebra and calculus) of polynomials as functions. Of course, any polynomial $f(x) \in S[x]$ induces a function $S \to S$ whose value at any $r \in S$ is merely $f(r)$. In seeking a reconciliation between the two views of polynomials, one may sensibly ask: what harm or confusion (if any) is there in "identifying" a polynomial with the function that it induces? The answer, surprisingly enough, depends on S. Indeed, if S is a domain, then: the requirement that no pair of distinct polynomials in $S[x]$ can induce the same function $S \to S$ is equivalent to the requirement that S be infinite. Before proving this, note that one consequence is the response "no harm at all" whenever char(S) $= 0$; in particular, students of calculus (where S is **R**, and sometimes **C**) have witnessed a satisfying reconciliation. Enough sentiment: to the proof. One direction is easy: if S is infinite and polynomials $f(x)$ and $g(x)$ in $S[x]$ induce the same function, let $n \geq \max(\deg(f), \deg(g))$, and select $n+1$ distinct elements $r_1, \ldots, r_{n+1}$ in S; as $f(r_i) = g(r_i)$ for each i by hypothesis, the identity theorem yields $f(x) = g(x)$, as desired. For the reverse direction, $S = \{r_1, \ldots, r_n\}$ is finite. In this case, note that $f(x) = (x - r_1) \ldots (x - r_n)$ and $g(x) = 0$ are distinct (compare degrees) polynomials which, by the second fundamental property of evaluation, each induce the function $S \to S$ which is identically zero.

We next turn to the matter of building new polynomials from a given one, where the roots of the new polynomial bear a simple algebraic relation to the old roots. Applications of these constructions will be given, in Chapter 6, to the methods of Horner and Graeffe. There are four methods to be discussed, each beginning with a nonzero polynomial $f(x) = c(x - r_1) \ldots (x - r_n) \in S[x]$, where, as usual, $r_1, \ldots, r_n$ are all the roots of $f(x)$ in some field containing S, and $c \in S$. It is convenient to treat the first three methods together: these concern finding an n-th-degree polynomial over S whose roots are

(1) $r_1 + k, r_2 + k, \ldots, r_n + k$, where k is a specified constant in S;
(2) $kr_1, kr_2, \ldots, kr_n$, where k is a nonzero constant in S;
(3) $1/r_1, 1/r_2, \ldots, 1/r_n$, in case each r_i is nonzero.

The methods in question prescribe the following polynomials: for (1), $f(x - k)$; for (2), $k^n f(x/k)$; and for (3), $x^n f(1/x)$. For example, if $f(x)$ is the polynomial $x^2 - 4x + 13$ over **Z** (with roots given by the quadratic formula as $r_1 = 2 + 3i$ and $r_2 = 2 - 3i$ in **C**), then one may verify via the quadratic formula that

(1) a polynomial over **Z** with the roots $r_1 + 3$, $r_2 + 3$ is $f(x-3) = (x-3)^2 - 4(x-3) + 13 = x^2 - 10x + 34$;

(2) a polynomial over **Z** with the roots $3r_1$, $3r_2$ is $3^2 f(x/3) = 9[(x/3)^2 - 4(x/3) + 13] = x^2 - 12x + 117$; and

(3) a polynomial over **Z** with the roots $1/r_1$, $1/r_2$ is $x^2 f(1/x) = x^2[(1/x)^2 - 4(1/x) + 13] = 13x^2 - 4x + 1$.

Verification of the methods in general is easy. Indeed, if $f(x) = c(x - r_1) \ldots (x - r_n)$ as above, then (1) $f(x-k) = c(x - [r_1 + k]) \ldots (x - [r_n + k])$; (2) $k^n f(x/k) = c[k((x/k) - r_1)] \ldots [k((x/k) - r_n)] = c(x - kr_1) \ldots (x - kr_n)$; and (3) $x^n f(1/x) = c[x((1/x) - r_1)] \ldots [x((1/x) - r_n)] = c[-r_1(x - (1/r_1))] \ldots [-r_n(x - (1/r_n))] = \pm c(r_1 r_2 \ldots r_n)(x - (1/r_1)) \cdot \ldots (x - (1/r_n))$.

The fourth, and final, method to be discussed here concerns producing a polynomial over S with the roots $r_1^2, r_2^2, \ldots, r_n^2$. It will be necessary to work inside $S[x, y]$, the domain of polynomials over S in the (commuting, algebraically independent) variables x and y. (The reader in need of a review of this concept should read the fourth paragraph of Section 5 in Chapter 1.) Observe that $f(y)f(-y) = c^2(y - r_1) \ldots (y - r_n)(-y - r_1) \ldots (-y - r_n) = (-1)^n c^2(y^2 - r_1^2) \ldots (y^2 - r_n^2)$. Thus, some $g(y) \in S[y]$ has the property that $f(y)f(-y) = g(y^2)$, and the required polynomial in $S[x]$ with roots $r_1^2, \ldots, r_n^2$ is evidently $g(x)$. For example, consider once again $f(x) = x^2 - 4x + 13$. To find a polynomial $g(x) \in S[x]$ with roots r_1^2, r_2^2 (that is, $-5 + 12i$, $-5 - 12i$), compute $f(y)f(-y) = (y^2 - 4y + 13)(y^2 + 4y + 13) = (y^2)^2 + 10(y^2) + 169$, and so setting $g(x) = x^2 + 10x + 169$ suffices. (Check, using the quadratic formula.)

Two remarks about the above methods: First, they can be combined so as to produce polynomials with roots whose relation to the given polynomial's roots is, algebraically, *quite* involved. By way of example, let $f(x)$ again be $x^2 - 4x + 13$, with roots r_1 and r_2. To find $g(x) \in \mathbf{Z}[x]$ with roots $(r_1 - 1)/(r_1 - 2)$ and $(r_2 - 1)/(r_2 - 2)$, apply method (1) to $f(x)$ with $k = -2$; then method (3) to the resulting polynomial; and then (1) again to *that* resulting polynomial, with $k = 1$, yielding the desired $g(x) = 9x^2 - 18x + 10$. (Work out the intermediate steps that were sketched. Could a *different* combination of methods have been used?) For another illustration, to find a polynomial with roots $1/r_1^2, \ldots, 1/r_n^2$, combine (3) with the fourth method, in *either* order. (Check that, for the above $f(x)$, one thereby obtains, as desired, $169x^2 + 10x + 1$.) The second remark is that alternate methods of building polynomials (equivalent to the above four methods) may be developed using elementary symmetric polynomials: see the examples in Section 4 below.

To close the section, we select two topics from the themes in later chapters, involving "special" roots. The first concerns nonreal complex roots of real polynomials: the fact is that such occur in conjugate pairs. More precisely, if $f(x) \in \mathbf{R}[x]$ and z is a complex root of $f(x)$, then its conjugate z^* is also a root of $f(x)$. (Indeed, their multiplicities coincide, but a proof of this additional fact is deferred to Section 3.) A proof has *already* been given: see the concluding observation in Section 2 of Chapter 1; another proof is sketched in Exercise 7 below.

Finally, we come to the topic of *integral roots* (that is, roots in $\mathbf{Z}$). The following observations will yield a more streamlined approach to the subject, especially to Eisenstein's theorem and Gauss's lemma (in Sections 1 and 2 of Chapter 4), than is customary in books at this level. They center around the process called reduction modulo p, where p is a positive prime integer. One instance of such reduction is already familiar: although the set of elements of the field $\mathbf{Z}_p$ has been denoted $\{0, 1, \ldots, p-1\}$, we have unabashedly sanctioned statements like "$5 = 2$ and $-20 = -2$ in $\mathbf{Z}_3$." In order to make some important points, it is now time, not for penance, but for greater care. Specifically, we relabel so that, at those moments when precision is most necessary, we write $\mathbf{Z}_p = \{\bar{0}, \bar{1}, \bar{2}, \ldots, \overline{p-1}\}$; and, in general, any $r \in \mathbf{Z}$ has $\bar{r}$, its *reduction modulo p*, satisfying sanctioned statements now written to resemble "$\bar{5} = \bar{2}$ and $-\overline{20} = -\bar{2}$ in $\mathbf{Z}_3$." (Readers who have studied elementary number theory will recognize the equivalence relation, $\equiv \pmod{p}$, and its resulting equivalence classes lurking in our cavalier treatment of $\mathbf{Z}_p$.) Indeed, *reduction modulo p will be defined for* (not necessarily constant) *integral polynomials as well*: if $f(x) = a_0x^n + \ldots + a_n$ is any polynomial in $\mathbf{Z}[x]$, then its reduction modulo p is the polynomial

$$\bar{f}(x) = \overline{a_0}\,x^n + \overline{a_1}\,x^{n-1} + \ldots + \overline{a_n} \in \mathbf{Z}_p[x].$$

(Certain purists would use a new variable, other than x, over $\mathbf{Z}_p$.) For example, if $f(x) = 14x^2 - 8x + 6$ and $p = 5$, then $\bar{f}(x) = \bar{4}x^2 + \bar{2}x + \bar{1}$ (or $\bar{4}x^2 - \bar{3}x + \bar{1}$ or, feeling cavalier again, $4x^2 + 2x + 1$ or, as before, $14x^2 - 8x + 6$) in $\mathbf{Z}_5[x]$. One potent consequence of the new notation is apparent: if $f(x) \in \mathbf{Z}[x]$ and $r \in \mathbf{Z}$, then $\overline{f(r)} = \bar{f}(\bar{r}) \in \mathbf{Z}_p$. (Proof?!) In particular, if r is an integral root of an integral polynomial f, then $\bar{r}$ is a root of $\bar{f}$.

Underlying the preceding calculation are some results which, by analogy with the earlier study of evaluation, will be called the *fundamental (homomorphism) properties of reduction modulo p*. These assert, for any (possibly constant) f, g in $\mathbf{Z}[x]$: if $h(x) = f(x) + g(x)$, then $\bar{h}(x) = \bar{f}(x) + \bar{g}(x)$; and if $H(x) = f(x)g(x)$, then $\bar{H}(x) = \bar{f}(x)\bar{g}(x)$. (Proof?!) Observe that analogous homomorphism properties of the conjugation process were at work in establishing the result concerning conjugate pairs which was recalled two paragraphs ago. The homomorphism properties of reduction modulo p will greatly ease our later treatment of the aforementioned results of Eisenstein and Gauss, but now, let's put the preceding theory to work, by proving: if $f(x) = a_0x^n + \ldots + a_n \in \mathbf{Z}[x]$ is such that both $f(0)$ and $f(1)$ are odd, then $f(x)$ does not have an integral root. (Application: $x^3 + x - 1$ has no integral root. If you'd like to know what *sort* of roots it *does* have, see Sections 1 and 2 of Chapter 3.) For the proof, suppose, on the contrary, that $f(r) = 0$ for some $r \in \mathbf{Z}$. As $a_n = f(0)$ is odd and $r(a_0r^{n-1} + \ldots + a_{n-1}) + a_n = f(r) = 0$, it follows that a_n is an integral multiple of r, and so r is also odd. Now, reduce modulo $p = 2$. Then $\bar{r} = \bar{1}$, whence

$$\begin{aligned}\bar{f}(\bar{r}) &= \overline{a_0}\,\bar{r}^n + \overline{a_1}\,\bar{r}^{n-1} + \ldots + \overline{a_n}\\ &= \overline{a_0} + \overline{a_1} + \ldots + \overline{a_n} = \overline{a_0 + a_1 + \ldots a_n}\\ &= \overline{f(1)} = \bar{1} \neq \bar{0},\end{aligned}$$

contradicting $\bar{f}(\bar{r}) = \overline{f(r)} = \bar{0}$, to complete the proof.

The remainder of this section (exclusive of the exercises!) may be omitted on a first reading. Much more can be said about the relation between r being a root of $f(x)$ and $\bar{r}$ being a root of $\bar{f}(x)$: see Exercise 9 below. For now, we observe a triviality, observe a nontriviality, and then balance the latter with a pearl. The triviality: if $r \in \mathbf{Z}$ and $f(x) \in \mathbf{Z}[x]$ are such that $\bar{r}$ is a root of $\bar{f}(x)$ when reduction modulo p is considered for *each* p, then r is a root of $f(x)$. (Proof: $\bar{f}(\bar{r}) = \bar{0}$ means $f(r)$ is an integral multiple of each of the infinitely many p's, contradicting the fundamental theorem of arithmetic.) The nontriviality: there exist polynomials $f(x) \in \mathbf{Z}[x]$ such that (1) for infinitely many primes p, the reduction $\bar{f}(x)$, of $f(x)$ modulo p, has a root in $\mathbf{Z}_p$, and (2) for infinitely many p, $\bar{f}(x)$ does not have a root in $\mathbf{Z}_p$. One such polynomial is $f(x) = x^2 - 2$. Students of number theory will recognize, as a consequence of the Legendre–Gauss law of quadratic reciprocity, that in this case the primes p in (1) are those of the form $8k \pm 1$ (with $1 \leq k \in \mathbf{Z}$) and $p = 2$; the p's in (2) are those of the form $8k \pm 3$. The infinitude of such sets of primes is a consequence of a celebrated result of Dirichlet concerning the dispersion of primes in arithmetic progressions.

The pearl is the following charming result of Schur, which may be regarded as a far-reaching extension of Euclid's result that the set of prime integers is infinite. If $f(x)$ is a nonconstant polynomial in $\mathbf{Z}[x]$, then there exist infinitely many positive prime integers p such that $\bar{f}(x)$, the reduction of $f(x)$ modulo p, has a root in $\mathbf{Z}_p$. For the proof, note by the blockbuster that the set of values $f(r)$, as r traverses $\mathbf{Z}$, contains something not in the set $\{0, -1, 1\}$, since the polynomials $f(x)$, $f(x) + 1$, and $f(x) - 1$ each have only finitely many roots (and $\mathbf{Z}$ is infinite). By the fundamental theorem of arithmetic, some prime p divides some value $f(r)$, whence $\bar{r}$ is a root of the reduction of $f(x)$ modulo *that* p. Now, if the result fails, let $p_1, p_2, \ldots, p_n$ be *all* the distinct primes corresponding to which $\bar{f}(x)$ has a root. Observe that if the constant term d ($= f(0)$) of $f(x)$ were equal to 0, then $\bar{f}(\bar{0}) = \bar{0}$ for *each* prime p, contradicting the supposed denial of the result; hence $d \neq 0$. Thus, the polynomial $h(x) = f(p_1 p_2 \ldots p_n dx) \in \mathbf{Z}[x]$ may be expressed as $h(x) = dg(x)$, where $g(x) \in \mathbf{Z}[x]$ is a nonconstant polynomial with constant term 1; moreover, each *other* coefficient of $g(x)$ is an integral multiple of $p_1 p_2 \ldots p_n$. Arguing as above via blockbuster applied to the values of $g(x)$, we infer that $\bar{g}(x)$ has a root in some $\mathbf{Z}_q$; that is, $g(k)$ is an integral multiple of q, for some $k \in \mathbf{Z}$ and some prime q. As q divides $dg(k) = f(p_1 p_2 \ldots p_n dk)$, it follows that the reduction, $\bar{f}(x)$, modulo q has a root in $\mathbf{Z}_q$; that is, q is one of the p_i. Consequently, since q divides each coefficient of $g(x)$ except the constant term 1, we see that q divides $g(k) - 1$. (Notice use of the integer version of what was termed the linear combination principle in Section 3 of Chapter 1.) However, the choice of k and q guarantees that q also divides $g(k)$ and so, again by the linear combination principle, $1 = g(k) - [g(k) - 1]$ is an integral multiple of the prime q, the desired contradiction.

EXERCISES 2.1

1. (a) Let $f(x) = x^3 + (i - 1)x^2 - ix + 2 \in \mathbf{C}[x]$. Evaluate $f(0), f(1), f(-1)$, $f(i), f(2i - 1)$, and $f(x - 2)$.
(b) Let $f(x) = x^3 + 2x + 1 \in \mathbf{Z}_3[x]$. Evaluate $f(0)$, $f(1)$, $f(-1)$, and $f(x + 1)$.
2. (a) Without carrying out the evaluation process, find $f(r)$ in case $f(x) = x^4 + ix - 7 \in \mathbf{C}[x]$ and $r = -1$. (*Hint*: remainder theorem.)
(b) Without carrying out the division process, find the remainder in $\mathbf{Q}[x]$ when $x^3 - 6$ is divided by $x + 7$.
(c) Repeat (b) with the divisor $2x - 3$. (*Hint*: relate to the case with divisor $x - (3/2)$.)
(d) Repeat (b) working inside $\mathbf{Z}_{11}[x]$.
3. (a) For an arbitrary domain S with $f(x) \in S[x]$ and $r \in S$, prove that $x - r \mid (f(x) - f(r))$.
(b) For an arbitrary domain S and positive integer n, prove that $x - 1$ divides $x^n - 1$.
(c) For any domain S with $\text{char}(S) \neq 2$ and any positive integer n, prove that $x + 1 \mid (x^n + 1)$ if and only if n is odd. What happens if $\text{char}(S) = 2$?
(d) If F is any field and $f(x)$ is an irreducible polynomial in $F[x]$, with $\deg(f) \geqslant 2$, prove that $f(x)$ has no root in F.
4. Compute the multiplicity of r as a root of $f(x)$ in $S[x]$ for the following cases:
(a) $r = 1, f(x) = x^4 - 1, S = \mathbf{Z}$;
(b) $r = 1, f(x) = x^3 - 2x^2 + x, S = \mathbf{R}$;
(c) r and f as in (a), $S = \mathbf{Z}_2$;
(d) r and f as in (b), $S = \mathbf{Z}_2$.
5. (a) Define what ought to be meant by a *root* of a polynomial in n variables.
(b) Let S be an infinite domain. Prove that each nonzero n-variable polynomial $f(x_1, \ldots, x_n) \in S[x_1, \ldots, x_n]$ induces a function $S \times S \times \ldots \times S \to S$ which is not identically 0 (where n is a positive integer). (*Hint*: if f were a counterexample with minimal n, then $f = \Sigma p_i(x_n)^{m-i}$ for certain $p_i \in S[x_1, \ldots, x_{n-1}]$ and, by minimality of n, there exist $r_1, r_2, \ldots, r_{n-1}$ in S so that $(r_1, r_2, \ldots, r_{n-1})$ is not a root of p_0. Consider $f(r_1, r_2, \ldots, r_{n-1}, x_n) \in S[x_n]$.)
(c) Let $F = \{r_1, r_2, \ldots, r_n\}$ be a finite field with n elements. If $f(x)$ and $g(x)$ are polynomials in $F[x]$ which induce the same function $F \to F$, prove that $f(x) - g(x)$ is divisible by $(x - r_1)(x - r_2) \ldots (x - r_n)$.
6. Consider $f(x) = x^3 - x + 2 \in \mathbf{Z}[x]$, with the roots r_1, r_2, r_3 (say, in $\mathbf{C}$). Find the (if possible, monic) polynomial in $\mathbf{Z}[x]$ with the following roots:

(a) $r_1 + 8, r_2 + 8, r_3 + 8$;
(b) $3r_1, 3r_2, 3r_3$;
(c) $r_1/10, r_2/10, r_3/10$ (*Hint*: $r/10 = kr$, with $k = 1/10$);
(d) r_1^2, r_2^2, r_3^2;
(e) r_1^4, r_2^4, r_3^4;
(f) $1/r_1^2, 1/r_2^2, 1/r_3^2$;
(g) $(4r_1 + 3)/(2r_1 + 1), (4r_2 + 3)/(2r_2 + 1), (4r_3 + 3)/(2r_3 + 1)$ (*Hint*: $(4r + 3)/(2r + 1) = 2 + (1/(2r + 1))$).

7. Use polar form and De Moivre's theorem to find another proof that whenever $z \in \mathbf{C}$ is a root of $f(x) \in \mathbf{R}[x]$, then z^* is also a root of $f(x)$. (*Hint*: if $z = r\,\text{cis}\,\theta$ with $0 \leqslant r \in \mathbf{R}$ and $0 \leqslant \theta < 2\pi$, then $z^* = r\,\text{cis}(-\theta)$. Deduce the existence of real numbers u and v such that $f(z) = u + vi$ and $f(z^*) = u - vi$.)
8. (a) Prove that $x^4 + x^3 - 6x + 7 \in \mathbf{Z}[x]$ has no integral roots.
(b) Repeat (a) for $x^n + nx + 1$, where n is any odd integer greater than 1.
(c) Give an example of a polynomial $f(x) \in \mathbf{Z}[x]$ such that $f(x)$ has no integral root, $f(1)$ is odd, and $f(0)$ is even. (*Hint*: $\sqrt{2}$ is irrational.)
(d) Prove that $f(x) \in \mathbf{Z}[x]$ has no integral root in case none of $f(0), f(1)$, and $f(-1)$ is an integral multiple of 3. (*Hint*: use reduction modulo 3 to show, via the three given conditions, that any putative integral root r would need to satisfy $\overline{r} \neq \overline{0}, \overline{r} \neq \overline{1}$, and $\overline{r} \neq -\overline{1}$, respectively.)
(e) Prove that $f(x) = x^4 + 3x - 8$ has no integral root.
9. (a) If $f(x) = 2x + 1 \in \mathbf{Z}[x]$, prove that $\overline{f}(x)$, the reduction of $f(x)$ modulo p, has a root in $\mathbf{Z}_p$ for each $p \neq 2$. Generalize. Comment also on the behavior of $g(x) = 4x + 2$.
(b) For arbitrary positive integer n and prime p, find $f(x) \in \mathbf{Z}[x]$ and $r \in \mathbf{Z}$ such that $f(r) \neq 0 \in \mathbf{Z}$, $\overline{f}(\overline{r}) = 0 \in \mathbf{Z}_p$, and $\deg(f) = n$. (*Hint*: $\overline{p} = 0 \in \mathbf{Z}_p$.)
10. Let $f(x) = x^2 - 3 \in \mathbf{Z}[x]$.
(a) Prove that infinitely many primes p exist such that $\overline{f}(x)$ has a root in $\mathbf{Z}_p$.
(b) Find at least two primes p such that $\overline{f}(x)$ does not have a root in $\mathbf{Z}_p$. (Find more if you know the law of quadratic reciprocity!)
11. If $r_1, \ldots, r_d$ are all the roots of $f(x) \in \mathbf{C}[x]$ then, for each n, construct a complex polynomial with the roots $r_1^n, \ldots, r_d^n$. (*Hint*: if $\omega = \text{cis}(2\pi/n)$, Consider $f(y)f(\omega y)f(\omega^2 y) \ldots f(\omega^{n-1} y)$.)
12. Let m and n be positive integers, with integer g.c.d. d. Prove, over any field, that $(x^m - 1, x^n - 1) = x^d - 1$. (*Hint*: what possible roots might $((x^m - 1)/(x^d - 1), (x^n - 1)/(x^d - 1))$ have?)

*13. (a) Prove, for any positive integer m, that $10^m - 1$ is an integral multiple of 9. (*Hint*: use #3(b).) Deduce the rule of *casting out nines:* if $a_0, a_1, \ldots, a_m$ are arbitrary integers, then $(\Sigma a_i 10^i) - (\Sigma a_i)$ is an integral multiple of 9.
(b) Generalize (a) to the context of base-n expansions (n possibly unequal to 10).

2. Lagrange Interpolation

The view of polynomial as function, which was mentioned in the preceding section, leads irresistibly to the question whether a polynomial may, in some sense, be characterized by its values (that is, by the values assumed by the function which is induced by the given polynomial). In the strict sense, the answer is "of course, no," since the zero function on a finite field with n elements may be induced by (both the polynomial 0 and) a suitable n-th-degree polynomial. However, last section's identity theorem (not to mention #5(c) of Exercises 2.1) suggests that a positive result may be available if a satisfactory relation exists between the *number* of values considered and the *degree* of the inducing polynomial. In fact, there is such a (best possible) result, it is known as *Lagrange's interpolation theorem* (*formula*), and it may be stated as follows. If $a_1, a_2, \ldots, a_{n+1}$ are $n+1$ distinct elements of a field F and if $b_1, b_2, \ldots, b_{n+1}$ are any elements of F (possibly with repetition), then there is exactly one polynomial $f(x) \in F[x]$ such that (1) $\deg(f) \leq n$ and (2) $f(a_i) = b_i$ for each $i = 1, 2, \ldots, n+1$. This result plays a key role in the algorithm for factoring in $\mathbf{Q}[x]$: see Section 3 of Chapter 4.

From an appropriately abstract point of view, the theorem of Lagrange is a corollary of a generalization of elementary number theory's so-called *Chinese remainder theorem.* However, we prefer to give the following direct, constructive (algorithmic, programmable) proof, which will also serve to explain the term "formula" often used in describing Lagrange's result. The idea behind the proof is to write

$$f(x) = b_1 f_1(x) + b_2 f_2(x) + \ldots + b_{n+1} f_{n+1}(x)$$

where the f_i's are polynomials over F satisfying: $\deg(f_i) \leq n$ and $f_i(a_i) = 1$ for each i, and $f_i(a_j) = 0$ whenever $i \neq j$. It is clear (isn't it?) that, given such f_i, the stated construction for $f(x)$ *does* satisfy (1) and (2), as desired. To produce such intervening f_i, consider the product

$$f_i(x) = \prod_{j \neq i} \frac{x - a_j}{a_i - a_j} .$$

(To make sure that the notation is not impeding communication, we write out two samples: if $n = 3$, then $f_1(x) = ((x - a_2)/(a_1 - a_2))((x - a_3)/(a_1 - a_3))((x - a_4)/(a_1 - a_4))$ and $f_3(x) = ((x - a_1)/(a_3 - a_1))((x - a_2)/(a_3 - a_2))((x - a_4)/(a_3 - a_4))$.) It is clear (isn't it?) that the above product construction of f_i has the desired properties, and so a suitable $f(x)$ *has* been constructed. Moreover, no other polynomial can satisfy (1) and (2), thanks *precisely* to the identity theorem, which completes the proof.

Here are two illustrations of Lagrange's formula. First, the unique polynomial $f(x) \in \mathbf{Q}[x]$, of degree less than or equal to 3, such that

$f(0) = 0, f(1) = 1, f(-1) = 1$, and $f(3) = 9$ is

$$f(x) = 0\,\frac{x-1}{0-1}\frac{x+1}{0+1}\frac{x-3}{0-3} + 1\,\frac{x-0}{1-0}\frac{x+1}{1+1}\frac{x-3}{1-3}$$

$$+\ 1\,\frac{x-0}{-1-0}\,\frac{x-1}{-1-1}\,\frac{x-3}{-1-3} + 9\,\frac{x-0}{3-0}\frac{x-1}{3-1}\frac{x+1}{3+1}$$

which is (isn't it?), after simplification, just $f(x) = x^2$, as the reader might well have guessed! Similarly, *the* polynomial $g(x) \in \mathbf{Z}_3[x]$, of degree less than 2, such that $g(\bar{1}) = \bar{0}$ and $g(\bar{0}) = \bar{1}$ is

$$g(x) = \bar{0}\left(\frac{x-\bar{0}}{\bar{1}-\bar{0}}\right) + \bar{1}\left(\frac{x-\bar{1}}{\bar{0}-\bar{1}}\right) = -x + \bar{1}$$

(or, if you wish to write it thus, $\bar{2}x + \bar{1}$).

The reader will note that, for the case $F = \mathbf{R}$ and $n \leq 2$, the existence of a suitable Lagrangian *interpolant* $f(x)$ amounts to the geometric insight that any set of at most three points in the Euclidean plane is contained in either a straight line or a parabola. Since a line with positive slope is the graph of an increasing function and no parabola can be the graph of an increasing function, it becomes of interest, even in case $n = 2$, to ask whether "increasing" data may be interpolated by a globally increasing polynomial (function). Specifically, we may ask: given real numbers $a_1 < a_2 < \ldots < a_{n+1}$ and $b_1 < b_2 < \ldots < b_{n+1}$, is there a polynomial $h(x) \in \mathbf{R}[x]$ such that $h(a_i) = b_i$ for each i and the induced function is increasing at each real number? The answer is "yes": such an interpolant (typically non-Lagrangian), inducing a function which is increasing on all of $\mathbf{R}$, must exist; a proof is given in [20]. (Unfortunately, it is known that such $h(x)$ is not uniquely determined by the above requirements; moreover, deg(h) may not be prescribed or bounded in terms of n.) The weaker result in which the interpolant's induced function is only supposed increasing on the closed bounded interval $[a_1, a_{n+1}]$ was first proved in [28]; a variant of this weaker result is actually equivalent to the fundamental approximation result of Weierstrass mentioned in the introduction.

EXERCISES 2.2

1. Produce $f(x) \in \mathbf{Q}[x]$, of degree at most the specified n, with the given interpolation properties:
   (a) $n = 2; f(0) = 4, f(2) = 6$, and $f(4) = 12$.
   (b) $n = 3; f(0) = 1, f(1) = 4, f(-1) = -2$, and $f(3) = 10$.
2. Produce $f(x) \in \mathbf{C}[x]$, of degree at most the specified n, with the given interpolation properties:
   (a) $n = 2; f(0) = 4, f(2) = 6$, and $f(4) = 12$.
   (b) $n = 1; f(\pi - i) = ½$, and $f(i - 2) = 3\pi$.

3. Produce $f(x) \in \mathbf{Z}_3[x]$, of degree at most the specified n, with the given interpolation properties:
(a) $n = 2; f(\overline{0}) = \overline{0}, f(\overline{1}) = \overline{2}$, and $f(\overline{2}) = \overline{2}$.
(b) $n = 1; f(\overline{1}) = \overline{2}$, and $f(\overline{2}) = \overline{2}$.
4. (a) If $a_1, \ldots, a_{n+1}$ are $n + 1$ distinct elements of a field F, if $b_1, \ldots, b_{n+1}$ are any elements of F, and if $f(x) \in F[x]$ is the polynomial over F, of degree at most n, such that $f(a_i) = b_i$ for each i, then for any $g(x) \in F[x]$, prove: $g(a_i) = b_i$ for each i if and only if $(x - a_1) \ldots (x - a_{n+1})$ divides $(f(x) - g(x))$. (*Hint for the only if half*: if $r(x)$ is the remainder obtained by dividing $g(x)$ by $(x - a_1) \ldots (x - a_{n+1})$, compute $r(a_i)$, using the fundamental properties of evaluation.)
(b) Let F, a_i, b_i be as in (a). Select any nonzero element $c \in F$ and any positive integer $m > n$. Prove that there exists $g(x) \in F[x]$ such that $\deg(g) = m$, the leading coefficient of $g(x)$ is c, and $g(a_i) = b_i$ for each i. (*Hint*: consider $f(x) + cx^{m-n-1}(x - a_1) \ldots (x - a_{n+1})$.) Prove that $g(x)$ is uniquely determined if $m = n + 1$. What if $m > n + 1$?
5. If F is a finite field, prove that each function $F \to F$ is induced by a polynomial in $F[x]$.
6. (An exercise to show that this section's topic was properly couched amidst fields, in that domains do not admit so nice an interpolation theorem:) Prove that there is no polynomial $f(x) \in \mathbf{Z}[x]$ satisfying the conditions $f(0) = 4$, $f(2) = 6$, and $f(4) = 12$. (*Hint*: If there were such an $f(x)$, identify its constant term. Is $f(x)$'s coefficient of x even or odd?)

3. The Derivative and Multiple Roots

The procedure to be given in Section 4 of Chapter 5 for separating the roots of a given polynomial $f(x) \in \mathbf{R}[x]$ is best treated in case those roots are simple (that is, of multiplicity 1). This section presents an algorithm which reduces to that case. Its algorithmic features are inherited in part from Chapter 1's g.c.d. algorithm and in part from the differentiation formulas of elementary calculus. We begin by recasting those formulas in a broader setting.

Let S be a domain. If $f(x) = a_0x^n + a_1x^{n-1} + \ldots + a_n$ is a polynomial over S, then the *derivative* of $f(x)$ is *defined* to be the polynomial $f'(x) \in S[x]$ given by $f'(x) = na_0x^{n-1} + (n-1)a_1x^{n-2} + \ldots + a_{n-1}$. Applications of this definition are familiar from the calculus; for instance, if $f(x) = -2x^3 + \pi x - 4$, $g(x) = x$, and $h(x) = 9$ are given, then $f'(x) = -6x^2 + \pi$, $g'(x) = 1$, and $h'(x) = 0$. Care must be taken when interpreting the derivative's formula when char(S) is not 0; for example, if $S = \mathbf{Z}_2$ and

$f(x) = x^2 + 1$ is considered as a polynomial over S, then $f'(x) = 2x = 0$, since $2 = 0$, (that is, $\bar{2} = \bar{0}$) in $\mathbf{Z}_2$. Accordingly, the best that one may say in general is that if $f(x) \neq 0$, then $\deg(f') \leq \deg(f) - 1$. However, our intended serious uses of differentiation which occur in Chapter 5 will be made in the context of $S = \mathbf{R}$, which is covered by the following general observation. If S is any domain of characteristic 0 and $f(x) \in S[x]$ has degree $n \geq 1$, then $\deg(f') = n - 1$. (Proof: With the above notation, $a_0 \neq 0$ and, as the hypothesis on characteristic assures $n \neq 0$ in S, we thus have $n \cdot a_0 \neq 0$ by domain property (vi)′.)

DETOUR: readers who will have need to differentiate polynomials only in case S is **R** or **C** (and those cases suffice for the later work in Chapters 5 and 6) may omit the heavy theory of the next two paragraphs, at least on a first reading. The point is that the differentiation rules given below are presumed known from the calculus's view of polynomial as function, and we've seen, via the identity theorem, that such a view coincides with ours over domains of characteristic 0, such as **R** and **C**.

It is interesting to note that the above formula for $f'(x)$ may be reinterpreted, in the context of an arbitrary domain S, in terms similar to the difference quotient approach to differentiation in the calculus. The key observation is this regard concerns the domain $S[x, h]$ of polynomials in two variables over S. (Students of calculus will appreciate that the second variable has been denoted h, rather than y, only for reasons of tradition.) To wit: the binomial theorem easily shows that, for each positive integer k there exists $g_k \in S[x, h]$ such that (1) $(x + h)^k = x^k + h \cdot g_k(x, h)$, and (2) the evaluation of $g_k(x, h)$, viewed in $(S[x])[h]$, obtained by formally setting h equal to 0, is the polynomial $g_k(x, 0) = kx^{k-1} \in S[x]$. (Prove this! Why does the binomial theorem apply?) In other words, the difference quotient $((x + h)^k - x^k)/h$ in the quotient field of $S[x, h]$ is a polynomial in $S[x, h]$ whose "value at $h = 0$" is just kx^{k-1}. Hence, if $f(x) = a_0x^n + \dots + a_n \in S[x]$, then the polynomial $g(x, h) \in S[x, h]$ given by $g(x, h) = a_{n-1}g_1(x, h) + a_{n-2}g_2(x, h) + \dots + a_0g_n(x, h)$ satisfies $(f(x + h) - f(x))/h = g(x, h)$ and $g(x, 0) = f'(x)$.

One now has a choice as to how to establish five familiar rules of differential calculus in the context of an arbitrary domain S. A direct approach using the definition of $f'(x)$ is possible (see Exercise 1 below), but we shall employ the difference quotient approach, so as to make our proofs follow along lines of the corresponding arguments in the calculus. All five rules pertain to $f(x)$ and $F(x)$ in $S[x]$, satisfying $f(x + h) = f(x) + hg(x, h)$ and $F(x + h) = F(x) + hG(x, h)$ as above, for suitable g and G in $S[x, h]$.

(a) *Product rule*: If $H(x) = f(x)F(x)$, then $H'(x)$ is given by $H'(x) = f(x)F'(x) + F(x)f'(x)$.

Proof: $H(x + h) - H(x) = f(x + h)F(x + h) - f(x)F(x)$ [by the second fundamental property of evaluation] $= f(x+h)(F(x+h) - F(x)) + F(x)\cdot(f(x+h) - f(x)) = h(f(x+h)G(x,h) + F(x)g(x,h))$, whence $H'(x) = f(x)G(x,0) + F(x)g(x,0) = f(x)F'(x) + F(x)f'(x)$, as asserted.

(b) *Constant factor rule*: If $s \in S$ and $H(x) = sF(x)$, then $H'(x) = sF'(x)$.

Proof: Apply (a) with $f(x)$ replaced by s, noting that the derivative of s is 0.

(c) *Sum rule*: If $H(x) = f(x) + F(x)$, then $H'(x) = f'(x) + F'(x)$.

Proof: Easy exercise for the reader.

(d) *Chain rule*: If $H(x) = f(F(x))$, then $H'(x) = f'(F(x)) \cdot F'(x)$.

Proof: $H(x+h) - H(x) = f(F(x+h)) - f(F(x)) = f(F(x) + hG(x,h)) - f(F(x)) = hG(x,h) \cdot g(F(x), hG(x,h))$ [explain!] whence $H'(x) = G(x,0) \cdot g(F(x), 0 \cdot G(x,0)) = F'(x) \cdot g(F(x),0) = F'(x) \cdot f'(F(x))$, as desired.

(e) *Power rule*: If n is a positive integer and $H(x) = F(x)^n$, then $H'(x) = nF(x)^{n-1}F'(x)$.

Proof: Apply (d), with $f(x) = x^n$.

The reader is assumed to have gained facility in handling (a)–(e) during courses on calculus.

There is a fortunate connection between differentiation and multiplicity, which we proceed to develop. Recall from Section 1 that r (in S) is a root of $f(x) \in S[x]$ with multiplicity m (≥ 1) in case $f(x) = (x-r)^m g(x)$ for some $g(x) \in S[x]$ such that $g(r) \neq 0$. (It is actually an easy exercise to show that this coincides with the earlier description of multiplicity: use the second fundamental property of evaluation and the factor theorem.) The promised fortunate connection is just this: if r is a root of $f(x)$ with multiplicity m (≥ 1) and if the domain S has characteristic 0, then *the multiplicity of r as a root of the derivative $f'(x)$ is precisely $m-1$.* More generally, if F is a field and $p(x)$ is an irreducible factor of some $f(x) \in F[x]$, we say that the multiplicity of $p(x)$ as a factor of $f(x)$ is m in case $p(x)^m \mid f(x)$ and $p(x)^{m+1} \nmid f(x)$. The fortunate connection is then a consequence of the following lemma: *If the field F has characteristic* 0 *and if $p(x)$ is an irreducible factor of some $f(x) \in F[x]$ of multiplicity m, then the multiplicity of $p(x)$ as a factor of $f'(x)$ is precisely $m-1$.* For its proof, note that the hypothesis gives $f(x) = p(x)^m g(x)$ for some $g(x) \in F[x]$ such that $p \nmid g$. By applying the product and power rules for differentiation and then simplifying, we obtain $f'(x) = p(x)^{m-1}(mg(x)p'(x) + p(x)g'(x))$. According to the linear combination principle, the proof will be complete once it is known that $p(x) \nmid mg(x)p'(x)$. As $p(x) \nmid g(x)$, property (*) of irreducibles, given in Section 3 of Chapter 1, reduces our task to establishing $p(x) \nmid mp'(x)$. However, this last detail follows from the fundamental fact about degrees, since $\deg(m \cdot p'(x)) = \deg(m) + \deg(p'(x)) = 0 + (\deg(p) - 1) < \deg(p)$.

We insert three questions for the reader concerning the preceding lemma and its proof: how should one interpret the lemma's assertion in case $m = 1$; why is the fortunate connection in fact a consequence of the lemma; and what does the argument yield in cases of nonzero characteristic (for which the last step of the given proof fails)???

Next, to indicate a connection between g.c.d.s and multiplicity, we isolate a fragment of the later algorithm; apart from motivational purposes, this is beneficial since the fragment is valid for arbitrary characteristics. Specifically, if $f(x)$ is a polynomial of positive degree over a field F, then: $f(x)$ *has multiple roots* (that is, has at least one multiple root) if and only if $(f(x), f'(x)) \neq 1$. For the proof, suppose first that $d(x) = (f, f')$ has positive degree. The son of blockbuster in Section 1 then provides a root r of $d(x)$ in some field L containing F. As $d \mid f$, r is also a root of $f(x)$, whence $f(x) = (x - r)g(x)$ for some $g(x) \in L[x]$, by the factor theorem. Differentiation leads to $f'(x) = (x - r)g'(x) + g(x)$, whence $f'(r) = g(r)$ [why?], $x - r \mid g(x)$ [why?], and $(x - r)^2 \mid f(x)$, showing that r is the desired multiple root of $f(x)$. Conversely, if some element r in a field L containing F is a multiple root of $f(x)$, write $f(x) = (x - r)^2 h(x)$ for some $h(x) \in L[x]$. Differentiate as in the proof of the lemma, obtaining $f'(x) = (x - r)[2h(x) + (x - r)h'(x)]$. Since one consequence of the g.c.d. algorithm in Section 3 of Chapter 1 was that $d(x)$ may be computed over *any* field containing F as a subfield, working over L gives $x - r \mid d(x)$, whence $\deg(d) > 0$ by the fundamental fact about degrees, completing the proof.

Now, we are ready to state this section's main algorithmic result on multiplicities. Let $f(x)$ be a polynomial of positive degree over a field F of characteristic 0. Define a sequence $f_i \in F[x]$ by $f_0 = f$, $f_1 = (f_0, f_0')$, $f_2 = (f_1, f_1')$, . . . ; let $g_i = f_{i-1}(x) \cdot f_{i+1}(x)/f_i(x)^2$, for $i = 1, 2, 3, \ldots$. Then for each i, $g_i(x) \in F[x]$, each root of $g_i(x)$ which happens to lie in F is a simple root of $g_i(x)$, and the roots of $f(x)$ in F with multiplicity i are precisely the (simple) roots of $g_i(x)$ in F.

Before embarking upon the proof, let's illustrate the algorithm by showing that 1 is the only root of $f(x) = x^5 - 4x^4 + 6x^3 - 6x^2 + 5x - 2 \in \mathbf{Q}[x]$ which has multiplicity 2 (that is, in the usual jargon, is a *double root*). It will be necessary to find g_2, and so we begin by computing f_1, f_2 and f_3. Now $f_1(x) = (f(x), f'(x)) = (x^5 - 4x^4 + 6x^3 - 6x^2 + 5x - 2,\ 5x^4 - 16x^3 + 18x^2 - 12x + 5)$ which, by an application of the g.c.d. algorithm in Section 3 of Chapter 1, is found to be simply $x - 1$. (Check this.) Then $f_2(x) = (f_1(x), f_1'(x)) = (x - 1, 1) = 1$; and, of course, $f_3(x) = (f_2(x), f_2'(x)) = (1, 0) = 1$. Accordingly, $g_2(x) = f_1(x)f_3(x)/f_2(x)^2 = x - 1$, whose only root is 1, whence by the algorithm's assertion, 1 is the only double root of $f(x)$. Notice for this example that $f_k(x) = 1$ for each $k \geq 2$, so that $g_k(x) = 1$ for each $k \geq 3$, whence each root of $f(x)$ other than 1 must be a simple root.

For the proof, introduce the following notation: for each $i = 1, 2, 3, \ldots$, let $r_1^{(i)}, r_2^{(i)}, \ldots, r_{t_i}^{(i)}$ be all the distinct roots of $f(x)$ in F which have multiplicity i. As $f(x)$ has only finitely many roots, we may consider $i = 1, 2, 3, \ldots, q$, where no root of $f(x)$ in F has multiplicity exceeding q. Define, for each i, the polynomial $w_i(x) = (x - r_1^{(i)})(x - r_2^{(i)}) \cdot \ldots \cdot (x - r_{t_i}^{(i)}) \in F[x]$. Apparently, $\deg(w_i) = t_i$ for those i such that $t_i \neq 0$; in case $t_i = 0$, interpret $w_i(x)$ to be the constant polynomial 1. Then $f(x) = w_1(x)^1 \cdot w_2(x)^2 \cdot \ldots \cdot w_q(x)^q \cdot h(x)$, where $h(x) \in F[x]$ has no roots in F. Set ($f_0 = f$ and) $h_0 = h$. By the above lemma, $r_\lambda^{(\mu)}$ is a root of $f_0'(x)$ with multiplicity $\mu - 1$; thus, $f_0'(x) = w_2(x)^1 \cdot w_3(x)^2 \cdot \ldots \cdot w_q(x)^{q-1} \cdot k(x)$, where $k(x) \in F[x]$ is such that no $r_\lambda^{(\mu)}$ is a root of $k(x)$. Hence, as we set $f_1 = (f_0, f_0')$, the result is $f_1 = w_2(x)^1 \cdot w_3(x)^2 \cdot \ldots \cdot w_q(x)^{q-1} \cdot h_1$, where $h_1 \in F[x]$ is "rootless" in F. (This is evident when you recall the idea behind #9 in Exercises 1.3, that is, the computation of g.c.d.s via the polynomial analogue of the fundamental theorem of arithmetic.) The situation is now ripe for iteration, the general output ($j = 1, 2, \ldots, q$) being $f_j(x) = (f_{j-1}, f_{j-1}') = w_{j+1}(x)^1 \cdot w_{j+2}(x)^2 \cdot \ldots \cdot w_q(x)^{q-j} \cdot h_j$, where $h_j \in F[x]$ is rootless in F. (So $f_{q-1} = w_q h_{q-1}$ and $f_q = h_q$, in particular.) For each $j \leq q - 1$, $g_j(x) = f_{j-1} f_{j+1} / f_j^2$ simplifies to $w_j(h_{j-1} h_{j+1} / h_j^2)$. By the construction of the w's and the rootless nature of the h's, the assertion about g_j will therefore follow once it is known that $h_j^2 \mid h_{j-1} h_{j+1}$. *This* will be established by showing: if $p(x) \in F[x]$ is an irreducible factor of h_j with multiplicity m, then $p^{2m} \mid h_{j-1} h_{j+1}$. Indeed, since $p^m \mid h_j \mid f_j \mid f_{j-1}'$, the above lemma yields that $p^{m+1} \mid f_{j-1}$; as p is irreducible and rootless in F, condition (*) in Section 3 of Chapter 1 then leads to $p^{m+1} \mid h_{j+1}$. In the same way, $p^{m-1} \mid (f_j, f_j') = f_{j+1}$ leads to $p^{m-1} \mid h_{j+1}$, whence $p^{2m} = p^{m+1} p^{m-1} \mid h_{j-1} h_{j+1}$, as desired. Finally, the assertions about g_j for $j \geq q$ follow from a sympathetic reading of the preceding argument, and we leave the attendant bookkeeping as an easy exercise for the reader.

We close this section with an admonition and an application. First, be advised that the preceding algorithm is simply inapplicable in case F has nonzero characteristic. For instance, it was seen in Section 1 that $f(x) = x^2 - \bar{1} \in \mathbf{Z}_2[x]$ has the double root $\bar{1} \in \mathbf{Z}_2$, although the formalities of the above algorithm would produce $g_1(x) = \bar{1}$ and $g_2(x) = \bar{1}$ in this case. Next, the application: it was observed in Section 1 that if z is a complex root of some polynomial $f(x) \in \mathbf{R}[x]$, then the conjugate z^* is also a root of $f(x)$. We can now prove even more, namely that z and z^* have the same multiplicities as roots of $f(x)$: indeed, one merely notes that z^* will necessarily be a root of the *same* g_j (in $\mathbf{R}[x]$) that z is a root of.

EXERCISES 2.3

1. (a) Go to any calculus text and do *its* exercises on differentiating polynomials. Or: follow the spirit of #1 in Exercises 1.4 (respectively 1.5).
(b) If $f(x) = (x^2 - 3)^{49} \in \mathbf{R}[x]$, use the chain rule to compute $f'(2)$.
2. Working directly from the text's definition of the derivative, obtain new proofs of the differentiation formulas (a)–(e).
3. Let S be a domain of characteristic 0, and consider a polynomial $f(x)$ over S of degree $n \geqslant 1$.
(a) By analogy with the calculus, define the "higher derivatives" $f''(x)$, $f^{(3)}(x), \ldots, f^{(n)}(x)$.
(b) Prove *Maclaurin's formula*: $f(x) = f(0) + f'(0)x + (f''(0)x^2/2!) + \ldots + (f^{(n)}(0)x^n/n!)$.
(c) Prove *Taylor's theorem*: for each $h \in S$, $f(x+h) = f(h) + f'(h)(x-h) + (f''(h)/2!)(x-h)^2 + \ldots + (f^{(n)}(h)/n!)(x-h)^n$. (*Hint*: Apply the result in (b) to the polynomial $g(x) = f(x+h)$. Observe that $g^{(k)}(x) = f^{(k)}(x+h)$ for each k, by virtue of the chain rule; hence find $g^{(k)}(0)$. Finally, substitute $x - h$ in place of x.)
4. (a) Find the multiplicity of $x^2 + 1$ as an irreducible factor in $\mathbf{R}[x]$ of $x^6 + x^4 - x^2 - 1$.
(b) Repeat (a), this time working in $\mathbf{Z}_2[x]$.
(c) Find the multiplicity of 2 (in $\mathbf{Q}$) as a root of $x^5 - 4x^4 - 16x^3 + 8x^2 + 16x - 32$ in two different ways (that is, by appropriate factoring and by this section's algorithm.)
5. Does $x^3 - 2x^2 - x + 2$ have repeated roots when viewed as a polynomial in $\mathbf{Q}[x]$? What if the coefficients are considered in $\mathbf{Z}_2$?
6. Working in $\mathbf{C}[x]$, find the algorithm's sequence $g_1, g_2, g_3, \ldots$ for the following choices of $f(x)$: $x^4 - 2x^3 - 3x^2 + 10x - 4$; $x^3 - 7x^2 + 15x - 9$; $x^4 - 4ix^3 - 5x^2 + 4ix + 4$; and $8x^4 - 12x^3 + 6x^2 - x$.
7. Find the multiplicity of $2 - i$ (in $\mathbf{C}$) as a root of $x^3 + 3x^2 - 23x + 35$. Generalize.

4. Elementary Symmetric Polynomials and Newton's Formulas

This section introduces the so-called *elementary symmetric polynomials.* These are certain combinatorially defined polynomials in several variables which permit new utilization of Section 1's result that a nonzero polynomial (in one variable) over a domain is determined by its roots and leading term. Their study paves the way for Chapter 6's treatment of the Bernoulli iteration method for computing numerical approximations of roots; a foretaste of that topic is provided in this section through the analysis of a quadratic example. Another consequence will be the development of alternatives to the four "polynomial-building" methods in Section 1. Although this section is "self-contained," the reader is alerted that its themes are pursued more deeply in the optional Sections 5 and 7 below. Throughout, we shall draw polynomial coefficients from a domain denoted R (our usual symbol "S" being reserved, lower case, to denote the symmetry of the protagonists).

Consider $R[x_1, x_2, \ldots, x_n, x]$, the domain of polynomials over R in the $n + 1$ (commuting, algebraically independent) variables $x_1, x_2, \ldots, x_n$, and x. One such polynomial of paramount importance is $(x - x_1)(x - x_2) \ldots (x - x_n)$ which, after expansion, may be written as $x^n - s_1x^{n-1} + s_2x^{n-2} + \ldots + (-1)^n s_n$, for certain polynomials $s_i = s_i(x_1, \ldots, x_n) \in R[x_1, \ldots, x_n]$. We call s_i the i-th *elementary symmetric polynomial* (*in* $x_1, \ldots, x_n$); by equating corresponding coefficients, we obtain the formulas

$$s_1(x_1, \ldots, x_n) = x_1 + x_2 + \ldots + x_n = \sum x_i,$$

$$s_2(x_1, \ldots, x_n) = x_1x_2 + \ldots + x_1x_n + x_2x_3 + \ldots + x_{n-1}x_n = \sum x_ix_j,$$

$$s_3(x_1, \ldots, x_n) = x_1x_2x_3 + \ldots + x_{n-2}x_{n-1}x_n = \sum x_ix_jx_k,$$

$$\vdots$$

$$s_n(x_1, \ldots, x_n) = x_1x_2 \ldots x_n.$$

Observe that the number of monomials occurring (with coefficient 1) in the expression for s_i is the same as the number of ways of selecting a subset with i elements from a set with n elements, that is, the combinatorial symbol $\binom{n}{i} = n!/(n-i)!i!$. For example, $s_2(x_1, x_2, x_3) = x_1x_2 + x_1x_3 + x_2x_3$ has $\binom{3}{2} = 3!/1!2! = 1 \cdot 2 \cdot 3/1 \cdot 1 \cdot 2 = 3$ constituent monomials.

Two comments about notation: in order to provide a uniform statement for Newton's formulas below, it will be convenient to interpret the symbol s_0 as the constant polynomial 1; and, lest too many monomials intervene, note that the *indices* $i, j, k, \ldots$ involved in the summations of the elementary symmetric polynomials are assumed *distinct* and *in increasing order.*

Next, two comments about terminology. First, the s_i are *symmetric* polynomials in the variables $x_1, x_2, \ldots, x_n$ in the sense that any permutation (that is, rearrangement or relabelling) of $x_1x_2, \ldots, x_n$ leaves s_i unchanged. Indeed, such a rearrangement merely brings about a permutation of the factors $x - x_1, \ldots, x - x_n$, thus leaving the polynomial

$$(x - x_1)(x - x_2) \ldots (x - x_n),$$

and its coefficients, unchanged. Secondly, the s_i are elementary both in the sense of being readily computable and in the deeper sense of yielding descriptions of arbitrary symmetric polynomials. This deeper sense will be stated precisely and proved in Section 5; it will also be illustrated by the computations in the examples two paragraphs hence.

Before giving the promised illustration, we shall first apply the preceding theory to polynomials in *one* variable. Let $f(x) = a_0x^n + \ldots + a_n \in R[x]$ be any polynomial over R of positive degree. If $r_1, \ldots, r_n$ are "all the roots of $f(x)$," recall that $f(x) = a_0(x - r_1)(x - r_2) \ldots (x - r_n)$. However, the definition of the elementary symmetric polynomials leads to $a_0(x^n - s_1x^{n-1} + \ldots + (-1)^n s_n) = a_0(x - x_1)(x - x_2) \ldots (x - x_n)$, whence the (homomorphism property of) the substitution of the value r_i for the corresponding x_i shows that $a_0(x^n - s_1(r_1, \ldots, r_n)x^{n-1} + \ldots + (-1)^n \cdot s_n(r_1, \ldots, r_n)) = f(x)$. Accordingly, the equating of corresponding coefficients then produces the important set of equations

$$-a_0s_1(r_1, \ldots, r_n) = a_1,$$

$$a_0s_2(r_1, \ldots, r_n) = a_2,$$

$$\vdots$$

$$(-1)^n a_0 s_n(r_1, \ldots, r_n) = a_n.$$

An equally useful formulation of these equations is: $s_i(r_1, r_2, \ldots, r_n) = (-1)^i \cdot (a_i/a_0)$. As the reader may have foreseen, $s_i(r_1, r_2, \ldots, r_n)$ is called the i-th *elementary symmetric function* of (the roots) $r_1, r_2, \ldots, r_n$; with traditional laxity, we shall often denote it, too, by s_i, suppressing reference to the roots entirely. Use of the above equations often speeds calculations, as in the problem of detecting the "missing" root of $f(x) = 8x^3 - 36x^2 + 46x - 15$, given the roots $r_1 = \frac{3}{2}$ and $r_2 = \frac{5}{2}$. One adequate but lengthy approach would be to find r_3 by obtaining $x - r_3$ as the quotient when $f(x)$ is divided by $8(x - \frac{3}{2})(x - \frac{5}{2})$. A far simpler tack is: $r_3 = (r_1 + r_2 + r_3) - r_1 - r_2 = s_1 - r_1 - r_2 = (-a_1/a_0) - r_1 - r_2 = \frac{36}{8} - \frac{3}{2} - \frac{5}{2} = \frac{4}{2} = \frac{1}{2}$. Before proceeding, the reader may wish to try similar problems, such as #1 and #2 of Exercises 2.4.

To illustrate "polynomial building" as a consequence of the above theory, consider the polynomial $f(x) = x^3 - x + 2 = (x - r_1)(x - r_2)(x - r_3) \in \mathbf{Z}[x]$, a familiar friend studied in #6 of Exercises 2.1. We shall produce the monic polynomials in $\mathbf{Q}[x]$ with the following sets of roots:

(a) $r_1 + 8, r_2 + 8, r_3 + 8$;
(b) $3r_1, 3r_2, 3r_3$;
(c) r_1^2, r_2^2, r_3^2;
(d) $1/r_1, 1/r_2, 1/r_3$.

Denote the sought polynomial(s) by $g(x) = x^3 - t_1x^2 + t_2x - t_3$; the t_i are, of course, the elementary symmetric functions of the roots of $g(x)$. These will be computed in terms of the elementary symmetric functions of the r_i, that is, in terms of $s_1 = 0$, $s_2 = -1$, and $s_3 = -2$. If the following computations seem unmotivated (or ingenious, though gruelling), note that the (optional) next section develops an appropriate algorithm.

For (a):

$$t_1 = (r_1 + 8) + (r_2 + 8) + (r_3 + 8) = (r_1 + r_2 + r_3) + 24$$
$$= s_1 + 24 = 0 + 24$$
$$= 24;$$
$$t_2 = (r_1 + 8)(r_2 + 8) + (r_1 + 8)(r_3 + 8) + (r_2 + 8)(r_3 + 8)$$
$$= (r_1r_2 + r_1r_3 + r_2r_3) + 16(r_1 + r_2 + r_3) + 192$$
$$= s_2 + 16s_1 + 192 = -1 + 0 + 192$$
$$= 191;$$
$$t_3 = (r_1 + 8)(r_2 + 8)(r_3 + 8) = r_1r_2r_3 + 8(r_1r_2 + r_1r_3 + r_2r_3) + 64(r_1 + r_2 + r_3)$$
$$+ 512 \quad = s_3 + 8s_2 + 64s_1 + 512 = -2 - 8 + 0 + 512 = 502;$$

and so $g(x) = x^3 - 24x^2 + 191x - 502$ in this case.

For (b):

$$t_1 = 3r_1 + 3r_2 + 3r_3 = 3(r_1 + r_2 + r_3) = 3s_1$$
$$= 0;$$
$$t_2 = (3r_1)(3r_2) + (3r_1)(3r_3) + (3r_2)(3r_3)$$
$$= 9(r_1r_2 + r_1r_3 + r_2r_3) = 9s_2$$
$$= -9;$$
$$t_3 = (3r_1)(3r_2)(3r_3) = 27r_1r_2r_3 = 27s_3$$
$$= -54;$$

and so $g(x) = x^3 - 9x + 54$ in this case.

For (c):

$$t_1 = r_1^2 + r_2^2 + r_3^2 = (r_1 + r_2 + r_3)^2 - 2(r_1r_2 + r_1r_3 + r_2r_3)$$
$$= s_1^2 - 2s_2 = 0 + 2$$
$$= 2;$$
$$t_2 = r_1^2r_2^2 + r_1^2r_3^2 + r_2^2r_3^2 = (r_1r_2 + r_1r_3 + r_2r_3)^2 - 2(r_1^2r_2r_3 + r_1r_2^2r_3 + r_1r_2r_3^2)$$
$$= s_2^2 - 2r_1r_2r_3(r_1 + r_2 + r_3) = s_2^2 - 2s_3s_1 = 1 - 0$$
$$= 1;$$

$$t_3 = r_1^2 r_2^2 r_3^2 = (r_1 r_2 r_3)^2 = s_3^2$$
$$= 4;$$

and so $g(x) = x^3 - 2x^2 + x - 4$ in this case.

For (d):

$$t_1 = (1/r_1) + (1/r_2) + (1/r_3) = (r_2 r_3 + r_1 r_3 + r_1 r_2)/r_1 r_2 r_3 = s_2/s_3 = 1/2;$$
$$t_2 = (1/r_1)(1/r_2) + (1/r_1)(1/r_3) + (1/r_2)(1/r_3) = (r_3 + r_2 + r_1)/r_1 r_2 r_3$$
$$= s_1/s_3 = 0;$$
$$t_3 = (1/r_1)(1/r_2)(1/r_3) = 1/r_1 r_2 r_3 = 1/s_3 = -1/2;$$

and so $g(x) = x^3 - \frac{1}{2}x^2 + \dfrac{1}{2}$ in this case.

The reader may now wish to try similar problems, such as #3, #4, and #5 of Exercises 2.4.

Because of the anticipated utility of the root-squaring approximation technique in Chapter 6, the above solution of (c) suggests the introduction of the k-th *power polynomial*

$$p_k = p_k(x_1, x_2, \ldots, x_n) = x_1^k + x_2^k + \ldots + x_n^k$$

for each $k = 1, 2, 3, \ldots$. It will also be convenient to interpret p_0 to be n. (We shall further call $p_k(r_1, \ldots, r_n) = r_1^k + \ldots + r_n^k$ the k-th *power function of* (typically, the roots) $r_1, r_2, \ldots, r_n$, and often denote it merely by p_k: more traditional laxity!) The above solution of (c) indicates how to express p_2 in terms of s_i. A similar ad-hoc calculation yields

$$\begin{aligned}
p_3(x_1,x_2,x_3,x_4) &= x_1^3 + x_2^3 + x_3^3 + x_4^3 \\
&= (x_1 + x_2 + x_3 + x_4)^3 \\
&\quad - 3(x_1^2x_2 + x_1^2x_3 + x_1^2x_4 + x_2^2x_1 + x_2^2x_3 + x_2^2x_4 \\
&\quad + x_3^2x_1 + x_3^2x_2 + x_3^2x_4 + x_4^2x_1 + x_4^2x_2 + x_4^2x_3) \\
&\quad - 6(x_1x_2x_3 + x_1x_2x_4 + x_1x_3x_4 + x_2x_3x_4) \\
&= s_1^3 - 3(x_1x_2 + x_1x_3 + x_1x_4 + x_2x_3 + x_2x_4 + x_3x_4)(x_1 + x_2 + x_3 + x_4) \\
&\quad + 9(x_1x_2x_3 + x_1x_2x_4 + x_1x_3x_4 + x_2x_3x_4) \\
&\quad - 6(x_1x_2x_3 + x_1x_2x_4 + x_1x_3x_4 + x_2x_3x_4) \\
&= s_1^3 - 3s_2s_1 + 3(x_1x_2x_3 + x_1x_2x_4 + x_1x_3x_4 + x_2x_3x_4) \\
&= s_1^3 - 3s_2s_1 + 3s_3.
\end{aligned}$$

Fortunately, there is a general way to solve for the p_k in terms of the s_i (and conversely), as a result of the following useful results known as *Newton's formulas*:

(1) If $k \geqslant n$, then $p_k s_0 - p_{k-1}s_1 + \ldots + (-1)^{n-1}p_{k-n+1}s_{n-1} + (-1)^n \cdot p_{k-n}s_n = 0$.

(2) If $1 \leqslant k \leqslant n$, then $p_k s_0 - p_{k-1}s_1 + p_{k-2}s_2 + \ldots + (-1)^{k-1}p_1 s_{k-1} + (-1)^k k s_k = 0$.

Of course, the p_k and s_i in (1) and (2) may be interpreted either as polynomials in $x_1, x_2, \ldots, x_n$ or as the corresponding functions (of some $r_1, \ldots, r_n$).

The proof of Newton's formulas will be given later in this section. Now, we digress briefly to introduce the Bernoulli iteration method of root approximation and indicate how it is readily implemented using Newton's formulas. As data, consider $f(x) \in \mathbf{C}[x]$, with roots $r_1, \ldots, r_n$ such that $|r_1| > |r_2| \geq \ldots \geq |r_n|$; our task is to approximate r_1. (Such problems arise frequently in applications of linear algebra where one needs to find the eigenvalue of maximum modulus of a given matrix A; in such cases, $f(x)$ is taken to be the characteristic polynomial of A.) Employing the standard limit theorems which most readers will have studied in calculus, we compute

$$\lim_{k\to\infty} \frac{p_{k+1}}{p_k} = \lim_{k\to\infty} \frac{r_1^{k+1} + \ldots + r_n^{k+1}}{r_1^k + \ldots + r_n^k}$$

$$= \lim_{k\to\infty} \frac{r_1^{k+1}\left(1 + \left(\frac{r_2}{r_1}\right)^{k+1} + \ldots + \left(\frac{r_n}{r_1}\right)^{k+1}\right)}{r_1^k\left(1 + \left(\frac{r_2}{r_1}\right)^{k} + \ldots + \left(\frac{r_n}{r_1}\right)^{k}\right)}$$

$$= r_1\left(\frac{1 + 0 + \ldots + 0}{1 + 0 + \ldots + 0}\right) = r_1,$$

since $\lim_{k\to\infty}(r_i/r_1)^k = 0$ for each $i \geq 2$. (To allay the fears of readers who may have encountered limit computations over $\mathbf{R}$ but not over $\mathbf{C}$, the later applications will only involve $f(x) \in \mathbf{R}[x]$ and real r_i.)

The preceding observation that p_{k+1}/p_k "approaches" r_1 as k becomes "large" may remind some readers of the computations necessary to apply calculus's ratio test for convergence of infinite series. Bernoulli iteration also admits an analogue of the root test for the special case where all the roots r_i are (real and) nonnegative, that is, in case $r_1 > r_2 \geq \ldots \geq r_n \geq 0$. Indeed, one then has

$$\lim_{k\to\infty} \sqrt[k]{p_k} = \lim_{k\to\infty} \sqrt[k]{r_1^k}\ \sqrt[k]{1 + \left(\frac{r_2}{r_1}\right)^k + \ldots + \left(\frac{r_n}{r_1}\right)^k} = r_1 \cdot 1 = r_1.$$

As the following example shows, the root-squaring technique of Section 1 also leads to direct calculations of $\sqrt[2^m]{p_{2^m}}$ without recourse to Newton's formulas; thus, taking values of $\sqrt[2^m]{p_{2^m}}$ for large m can be expected to yield better approximations of r_1.

To illustrate, consider the quadratic $f(x) = x^2 - 10x + 9 \in \mathbf{R}[x]$ with roots r_1 $(= 9)$, r_2 $(= 1)$ and elementary symmetric functions $s_0 = 1$, $s_1 = 10$, and $s_2 = 9$. According to Newton's formulas, ($p_0 = n = 2$ and)

$$p_1 = p_1 s_0 = (-1)^{1+1} \cdot 1 \cdot s_1 = 10,$$

$$p_2 = p_2 s_0 = p_1 s_1 + (-1)^{2+1} \cdot 2 \cdot s_2 = 82,$$

$$p_3 = p_3 s_0 = p_2 s_1 + (-1)^{2+1} p_1 s_2 = 730,$$

$$p_4 = p_4 s_0 = p_3 s_1 + (-1)^{2+1} p_2 s_2 = 6562.$$

The approximations to the largest root r_1, as generated by the p_{k+1}/p_k method, are then $p_2/p_1 = 8.2$, $p_3/p_2 \doteq 8.9024$, and $p_4/p_3 \doteq 8.9890$. On the other hand, use of the $\sqrt[k]{p_k}$ method generates the corresponding approximations 10, $\sqrt{p_2} \doteq 9.0554$, $\sqrt[3]{p_3} \doteq 9.0041$, and $\sqrt[4]{p_4} \doteq 9.0003$. Since the above derivation of the Bernoulli technique leads to the inequalities

$$\frac{p_{k+1}}{p_k} < r_1 < \sqrt[k]{p_k} \quad \text{for all sufficiently large } k$$

in the case $r_1 > r_2 \geq \ldots \geq r_n \geq 0$ (provide the details), it would not require much courage to surmise that r_1 is between 8.99 and 9.00, to the accuracy of two decimal places.

Calculus students will recall that the root test is more powerful than the ratio test, and hence will not be surprised that the $\sqrt[k]{p_k}$ method seemed to converge faster to r_1 than the p_{k+1}/p_k method in the preceding example. Still faster convergence arises via root squaring in the following way. Denoting the roots of $x^2 - 10x + 9$ by r_1 and r_2 as before, we find as in Section 1 that the monic polynomial over $\mathbf{Z}$ with roots r_1^2 and r_2^2 is $x^2 - 82x + 81$ (check using the $f(y)f(-y)$ trick); now, $p_2 = r_1^2 + r_2^2$ is the negative of the first elementary symmetric function of this new polynomial, whence $p_2 = 82$ (incidentally checking the earlier application of Newton's formulas). Starting with this new polynomial and root squaring, we find that the monic polynomial over $\mathbf{Z}$ with roots r_1^4 and r_2^4 is $x^2 - 6562x + 6561$, whence $p_4 = r_1^4 + r_2^4 = 6562$ (also agreeing with the earlier use of Newton's formulas). Finally, applying root squaring to *this* new polynomial shows that r_1^8 and r_2^8 are the roots of $x^2 - 43046722x + 43046721$, whence $p_8 = 43046722$ (which is *new* information). Observe the rapid convergence to r_1 through only three steps of the "$\sqrt[2^m]{p_{2^m}}$, with root squaring, but without Newton" method: $\sqrt{p_2} \doteq 9.0554$, $\sqrt[4]{p_4} \doteq 9.0003$, and

$\sqrt[8]{p_8} \doteq 9.0000$ to an accuracy of four decimal places. (Indeed we have that $\sqrt[8]{p_8} \doteq 9.000000026$.) Further examples and technical niceties about rapidity of convergence will be deferred to Chapter 6. As for techniques whereby one might check that the conditions $r_1 > r_2 \geq \ldots \geq r_n \geq 0$ obtain, see (Section 3 above for the $r_1 > r_2$ part and) the root separation material in Sections 1–4 of Chapter 5.

It is now time to establish Newton's formulas. We turn first to (1), which admits a rather direct proof. Indeed, we begin with the identity

$$x^n - s_1 x^{n-1} + s_2 x^{n-2} + \ldots + (-1)^n s_n = (x - x_1)(x - x_2) \ldots (x - x_n);$$

next, for each $i = 1, 2, \ldots, n$, we substitute x_i for x and multiply by x_i^{k-n} (recalling that $k \geq n$ for formula (1)), resulting in the n equations

$$x_i^k - s_1 x_i^{k-1} + s_2 x_i^{k-2} + \ldots + (-1)^n s_n x_i^{k-n} = 0$$

arising from $i = 1, 2, \ldots, n$. After adding these n equations together and simplifying, we readily obtain (1).

Proofs of (2) are much harder. Several known proofs are *analytic*; that is, they use results from the calculus over **R**. One such proof appears in [2]; another, using power series, is given in Section 7 below. For the sake of completeness, we shall next provide an *algebraic* proof of (2), but advise that readers familiar with power series will probably find the proof in Section 7 to be much more appealing and natural.

We shall first require a definition and a lemma. The definition: a polynomial $f(x_1, \ldots, x_n) \in R[x_1, \ldots, x_n]$ is said to be *homogeneous of degree* q in case each of its constituent monomials $rx_{i_1}^{m_1}x_{i_2}^{m_2} \ldots x_{i_k}^{m_k}$ (with, of course, $0 \neq r \in R$, $1 \leq i_1 < \ldots < i_k = m$, and each $m_j \geq 0$) satisfies $m_1 + m_2 + \ldots + m_k = q$. For example, any constant is homogeneous of degree 0; $2x_1^3x_2^5 - \pi x_1^7 x_2 + ix_2^8 \in \mathbf{C}[x_1, x_2]$ is homogeneous of degree 8; and $x_1^8 - x_2^7$ is *not* homogeneous (regardless of the value of q).

The lemma: *if* $f \in R[x_1, \ldots, x_n]$ *is homogeneous of degree* $q < n$ *and if each evaluation obtained from* f *by replacing* $n - q$ *of the variables* $x_1, \ldots, x_n$ *by* 0 *becomes the zero polynomial* (*in the remaining* q *variables*), *then* $f(x_1, \ldots, x_n) = 0$. To begin its proof, note that $q > 0$, since constants are unaffected by substitutions. Now, if f has a nonzero constituent monomial $t = rx_{i_1}^{m_1} \ldots x_{i_k}^{m_k}$ (as above, with each $m_j > 0$), the evaluation of f obtained by replacing all the variables *except* $x_{i_1}, \ldots, x_{i_k}$ by 0 is a polynomial g which also has t as a constituent monomial. (Explain how the homogeneity of f has just been used.) However, g is a polynomial in k variables, and $n - k \geq n - q$ (since $q = m_1 + \ldots + m_k \geq 1 + \ldots + 1 = k$), whence g may be obtained by substitution into a polynomial which has been assumed to be 0. Inexorably, $g = 0$, contradicting the presence of t, and proving the lemma.

With an eye toward proving Newton's formula (2) using the lemma, we concoct the polynomial

$$f(x_1, \ldots, x_n) = p_k - p_{k-1}s_1 + p_{k-2}s_2 + \ldots + (-1)^{k-1}p_1s_{k-1} + (-1)^k ks_k$$

in $R[x_1, \ldots, x_n]$, where k is given, as in the statement of (2), such that $1 \leq k < n$. (Note that the case $k = n$ in (2) actually is a consequence of (1), which has already been established.) Since the s_i are symmetric and the p_i are (easy exercise) symmetric, it follows that f is also a symmetric polynomial; moreover, f is homogeneous of degree k. Let $t_j = s_j(x_1, \ldots, x_k, 0, \ldots, 0)$ and $q_j = p_j(x_1, \ldots, x_k, 0, \ldots, 0)$, considered as elements of $R[x_1, \ldots, x_k]$. Evidently, t_j and q_j are just the j-th elementary symmetric polynomial and the j-th power polynomial, respectively, in the variables $x_1, \ldots, x_k$. Thus, by (1), $q_k - q_{k-1}t_1 + q_{k-2}t_2 + \ldots + (-1)^{k-1}q_1t_{k-1} + (-1)^k kt_k = 0$; that is, $f(x_1, x_2, \ldots, x_k, 0, \ldots, 0) = 0$. Symmetry of f then guarantees that setting *any* $n - k$ variables to be 0 reduces f to the zero polynomial, whence the above lemma applies, showing $f = 0$, completing the proof of (2).

To close this section, we offer yet another view of the fact that a nonzero polynomial over a domain is determined by its roots and its leading term. Specifically: if $\alpha_1, \alpha_2, \ldots, \alpha_n$ and $\beta_1, \beta_2, \ldots, \beta_n$ are lists, possibly with repetition, drawn from a domain R of characteristic 0, such that $\alpha_1^k + \ldots + \alpha_n^k = \beta_1^k + \ldots + \beta_n^k$ for each $k = 1, 2, \ldots, n$, then the α's and the β's describe the same set of elements (including multiplicities, apart from their ordering). By the uniqueness of decompositions of the type studied in Section 1 (or, for that matter, in Section 3 of Chapter 1), it is enough to show that the monic polynomials $(x - \alpha_1) \ldots (x - \alpha_n)$ and $(x - \beta_1) \ldots (x - \beta_n)$ coincide, that is, that their corresponding elementary symmetric functions coincide. By Newton's formula (2), this in turn will follow once it has been shown that the corresponding power functions $p_1, p_2, \ldots, p_n$ coincide (how have we just used char$(R) = 0$?), but this latter condition is precisely the import of the hypothesis about the α's and β's. This completes the proof.

EXERCISES 2.4

1. Find the quartic polynomials $f(x) = a_0x^4 + a_1x^3 + a_2x^2 + a_3x + a_4 \in \mathbf{Q}[x]$, with roots r_1, r_2, r_3, and r_4, satisfying the following conditions:
   (a) $a_0 = -2, r_1 = ½ + i, r_2 = ½ - i, r_3 = 6, r_4 = 6$;
   (b) $a_0 = 2, a_1 = -1, r_1 = 9, r_2 = -½, r_3 = -4$;
   (c) $a_0 = -5$, with the elementary symmetric functions of the roots given by $s_1 = 3/2, s_2 = 16, s_3 = -10$, and $s_4 = -1/12$;
   (d) $a_0 = 2, a_1 = -4, a_2 = 3, r_1 = 6, r_2 = -7$;
   (e) $a_4 = 6$, with the s_i as in (c).
2. (a) If $f(x) = x^3 - x^2 + a_2x + a_3 \in \mathbf{R}[x]$ has $2 + i$ as a root, find all the roots of $f(x)$ and compute the coefficients a_2 and a_3.
   (b) Compute the roots of $4x^3 + 12x^2 + 11x + 3 \in \mathbf{Z}[x]$, given that those

roots are in arithmetic progression (that is, given that the roots are of the form $r_1 = r_1$, $r_2 = r_1 + k$, and $r_3 = r_2 + k$, for some $k \in \mathbf{C}$).

(c) What are you able to conclude from the information that $x^3 + 7x^2 + 14x + a_4 \in \mathbf{C}[x]$ has roots r_1, r_2, r_3 such that $r_1 = 4r_2$?

3. Let r_1, r_2, r_3 be all the roots of $2x^3 - 3x^2 - x + 2 \in \mathbf{Z}[x]$. Compute the following, by expressing them in terms of the elementary symmetric functions of the r_i:
   (a) $(1/r_1) + (1/r_2) + (1/r_3) = \Sigma(1/r_i)$;
   (b) $(1/r_1^2) + (1/r_2^2) + (1/r_3^2) = \Sigma(1/r_i^2)$;
   (c) $r_1^2r_2 + r_1^2r_3 + r_2^2r_3 + r_2^2r_1 + r_3^2r_1 + r_3^2r_2 = \Sigma r_i^2 r_j$;
   (d) $(1/(r_1 + 2)) + (1/(r_2 + 2)) + (1/(r_3 + 2)) = \Sigma(1/(r_i + 2))$;
   (e) $(1/(r_1^2 + 1)) + (1/(r_2^2 + 1)) + (1/(r_3^2 + 1)) = \Sigma(1/(r_i^2 + 1))$.
4. Let r_1, r_2, r_3 be the roots of $7x^3 - ix + \pi \in \mathbf{C}[x]$. Using the methods of this section, find the complex polynomials with the following roots:
   (a) $2/r_1, 2/r_2, 2/r_3$;
   (b) r_1r_2, r_1r_3, r_2r_3;
   (c) r_1^3, r_2^3, r_3^3.
5. Let r_1 and r_2 be the roots of $2x^2 - 3x + \pi \in \mathbf{R}[x]$. *Without* using the quadratic formula, find the real polynomials with the following roots:
   (a) $r_1 + 8, r_2 + 8$;
   (b) $(4r_1 + 3)/(2r_1 + 1), (4r_2 + 3)/(2r_2 + 1)$.
6. Consider the polynomial $x^2 - 8x + 13 \in \mathbf{R}[x]$, with roots r_1, r_2. (You may accept the fact that $0 < r_1 < r_2$.)
   (a) Use Newton's formulas to compute the power functions $p_1, p_2, \ldots, p_8$ of the roots r_i. Verify the values p_2, p_4, and p_8 by root squaring.
   (b) Evaluate p_{k+1}/p_k for $k = 1, 2, 3, \ldots, 7$.
   (c) Using a calculator with a square-root key (or using logarithm tables), estimate $\sqrt[n]{p_n}$ for $n = 2, 4, 8$.
   (d) Using the quadratic formula (and calculator/tables), estimate r_1 and r_2.
7. Repeat the first three parts of #6 for the real polynomial $3x^3 - 18x^2 + 30x + 14$. (For the analogue of the fourth part, see Sections 1 and 2 of Chapter 3.)
8. Repeat the first three parts of #6 for the real polynomial $x^4 - 14x^3 + 65x^2 - 110x + 52$. (For the analogue of the fourth part, see Section 4 of Chapter 3.)
9. If $n \geqslant 3$, use Newton's formulas to prove that $p_3 = s_1^3 - 3s_1s_2 + 3s_3$. Obtain a similar expression for p_4.

*5. The Fundamental Theorem for Symmetric Polynomials

In this section, we make precise the assertion in Section 4 that the elementary symmetric polynomials s_i "yield descriptions of arbitrary symmetric polynomials." Specifically, working in $R[x_1, \ldots, x_n]$, we shall give an algorithmic proof of the *fundamental theorem*: each symmetric polynomial (over the domain R, in the variables $x_1, \ldots, x_n$) is expressible as a (finite) sum of terms of the form $rs_1^{m_1}s_2^{m_2} \ldots s_n^{m_n}$, for suitable $r \in R$ and nonnegative integers m_i. After reading this section, the reader may well wish to apply its algorithm systematically to earlier problems such as those met in #3, #4, and #5 of Exercises 2.4.

Two new notions are needed for the proof. The first is the definition of *dictionary order*. If $f = r_1x_1^{m_1} \ldots x_n^{m_n}$ and $g = r_2x_1^{k_1} \ldots x_n^{k_n}$ are nonzero monomials, the required ordering is given by the following condition: $f < g$ in case $m_1 = k_1$, $m_2 = k_2, \ldots,$ $m_{i-1} = k_{i-1}$ and $m_i < k_i$ for some $i = 1, 2, \ldots, n$; that is, in case the "name" $(m_1, m_2, \ldots, m_n)$ precedes the name $(k_1, \ldots, k_n)$ in a dictionary based on the ordered alphabet $0, 1, 2, 3, \ldots$. The well-ordering principle, which has been used to advantage earlier (see the proof of the division algorithm in Section 3 of Chapter 1), readily shows that an infinite strictly descending chain of monomials, $\ldots < f_{n+1} < f_n < \ldots < f_2 < f_1$, cannot exist.

Any nonzero polynomial $f \in R[x_1, \ldots, x_n]$ may be written, in a unique way, as a sum of monomials, $f = f_1 + f_2 + \ldots + f_k$, such that $f_1 < f_2 < f_3 < \ldots < f_k$. (Why? Compare with Exercise 1 below. Which f_i is the "constant term" of f?) With the above notation f_k is called the *leading term* of f, and one writes $f_k = \mathfrak{L}(f)$. For example, $\mathfrak{L}(s_1) = x_1$, $\mathfrak{L}(s_2) = x_1x_2, \ldots,$ and $\mathfrak{L}(s_n) = x_1x_2 \ldots x_n$. A crucial fact to be noted is that $\mathfrak{L}(fg) = \mathfrak{L}(f) \cdot \mathfrak{L}(g)$, whenever f and g are nonzero polynomials: this is a descendant of the fundamental fact about degrees in Section 3 of Chapter 1, and the reader is asked to supply his/her own proof of it. In particular, note that $\mathfrak{L}(s_1^{m_1} \ldots s_n^{m_n}) = x_1^{m_1 + \ldots + m_n}x_2^{m_2 + \ldots + m_n} \ldots x_n^{m_n}$.

Before embarking on the proof, we isolate a part of it as a lemma: *If f is a symmetric polynomial (in $R[x_1, \ldots, x_n]$) and $\mathfrak{L}(f) = rx_1^{d_1} \ldots x_n^{d_n}$, then* $d_1 \geq d_2 \geq \ldots \geq d_n$. Indeed, if $d_i < d_j$ for some $1 \leq i < j \leq n$, interchanging x_i and x_j would show, by the symmetry of f, that f had a constituent monomial

$$t = rx_1^{d_1} \ldots x_{i-1}^{d_{i-1}}x_j^{d_i}x_{i+1}^{d_{i+1}} \ldots x_{j-1}^{d_{j-1}}x_i^{d_j}x_{j+1}^{d_{j+1}} \ldots x_n^{d_n}.$$

By the definition of dictionary order, we get $\mathfrak{L}(f) < t$, contradicting the "supremacy" of $\mathfrak{L}(f)$, thus proving the lemma.

Now, to prove the fundamental theorem, consider any nonzero symmetric polynomial f, let $\mathfrak{L}(f) = rx_1^{d_1} \ldots x_n^{d_n}$, and define integers $m_1, \ldots, m_n$ according to the rules $m_1 = d_1 - d_2$, $m_2 = d_2 - d_3, \ldots, m_{n-1}$

$= d_{n-1} - d_n$, and $m_n = d_n$. By the preceding lemma, each $m_i \geq 0$; moreover $\mathfrak{L}(rs_1^{m_1} \ldots s_n^{m_n}) = rx_1^{m_1+\cdots+m_n} \ldots x_n^{m_n} = rx_1^{d_1}x_2^{d_2} \ldots x_n^{d_n} = \mathfrak{L}(f)$. Thus, $f - rs_1^{m_1} \ldots s_n^{m_n}$ is a symmetric polynomial which, if nonzero, has leading term less (in the dictionary order) than $\mathfrak{L}(f)$. By the above application of the well-ordering property, finitely many iterations yield 0 as the difference $f - g$, where g is a sum of terms of the desired type. This completes the proof.

To illustrate the algorithm, consider $f(x_1, \ldots, x_n) = x_1^3 + x_2^3 + \ldots + x_n^3$, a symmetric polynomial (in fact, the 3rd-power polynomial, p_3) treated in the special case $n = 4$ by ad-hoc means in Section 4 (and presumably also studied in #9 of Exercises 2.4 via Newton's formulas). Now $\mathfrak{L}(f) = x_1^3 = x_1^3x_2^0 \ldots x_n^0$ and so, in the notation of the above proof, $d_1 = 3$ and $d_2 = \ldots = d_n = 0$. The algorithm's recipe calls for $m_1 = d_1 - d_2 = 3$, $m_2 = d_2 - d_3 = 0, \ldots, m_{n-1} = d_{n-1} - d_n = 0$, and $m_n = d_n = 0$, as well as the formation of $h = f - rs_1^{m_1} \ldots s_n^{m_n} = f - s_1^3 = f - (\sum x_i)^3 = f - [\sum x_i^3 + 3\sum x_i^2x_j + 6\sum x_ix_jx_k]$, whose leading term is $-3x_1^2x_2$. (What if char(R) = 3?) Applying the algorithm again, this time with $r = -3$, $d_1 = 2$, $d_2 = 1$, and $d_3 = \ldots = d_n = 0$, produces new exponents $m_1 = 1$, $m_2 = 1$, and $m_3 = \ldots = m_n = 0$, as well as the formation of $h - (-3)s_1s_2 = h + 3s_1s_2 = (-3\sum x_i^2x_j - 6\sum x_ix_jx_k) + 3(\sum x_i)(\sum x_ix_j) = 3\sum x_ix_jx_k$. This last espression is evidently $3s_3$ (there's no point in using the algorithm for something so obvious!), whence $f = h + s_1^3 = (-3s_1s_2 + 3s_3) + s_1^3$, the desired formula.

To facilitate the proof of the fundamental theorem of algebra in Chapter 3, Section 6, we record the following corollary: *If $f(x)$ is a polynomial of positive degree n over a field F such that "all the roots of f" are $r_1, r_2, \ldots, r_n$ (in some field of which F is a subfield), then for each symmetric polynomial $h(x_1, \ldots, x_n) \in F[x_1, \ldots, x_n]$, the evaluation $h(r_1, \ldots, r_n) \in F$.* For its proof, note that $h(x_1, \ldots, x_n)$ is a sum of terms of the form $rs_1(x_1, \ldots, x_n)^{m_1} \ldots s_n(x_1, \ldots, x_n)^{m_n}$, where $0 \neq r \in F$ and each $m_i \geq 0$, according to this section's main result; by the (homomorphism) properties of evaluation, $h(r_1r_2, \ldots, r_n)$ is then a sum of terms of the form $rs_1(r_1, \ldots, r_n)^{m_n} \ldots s_n(r_1, \ldots, r_n)^{m_n}$. However, all these terms (and hence their sum) lie in F since each elementary symmetric function $s_i(r_1, \ldots, r_n)$ is, apart from sign, a ratio of a pair of coefficients of $f(x)$. Apart from the announced use of the corollary, one should note that it was part of the foundation for certain early treatments of Galois theory (about which more will be said in Section 5 of Chapter 3).

Finally, we note that this section's main result extends to rational functions (that is, elements of the quotient field of $R[x_1, \ldots, x_n]$) which are symmetric in the obvious sense. A discussion of this extension would take us too far afield (no pun intended).

EXERCISES 2.5

1. Prove that dictionary order on nonzero monomials is transitive ($[f<g$ and $g<h] \Rightarrow f<h$), antisymmetric ($f<g \Rightarrow g \nless f$), and hence irreflexive ($f \nless f$).
2. Assuming that $n \geq 5$, express the symmetric polynomials $\Sigma x_i^3 x_j x_k$ and $\Sigma x_i^2 x_j^2 x_k$ in terms of the elementary symmetric polynomials $s_1, \ldots, s_n$, using this section's algorithm. (Take $R = \mathbf{Z}$.)
3. Prove that each nonzero symmetric polynomial can be expressed in *only one way* as a sum of nonzero monomials $rs_1{}^{m_1} \ldots s_n{}^{m_n}$. (*Hint*: Subtract contestents, obtaining, if possible, a nontrivial description of 0. Use multiplicativity of $\mathcal{L}$ to derive a contradiction.)
4. Prove that $f \in R[x_1, \ldots, x_n]$ is symmetric in case f is unchanged after each interchange of a pair of variables. (This exercise appears in group theory as the assertion that any permutation of a finite set is a product of suitably many transpositions.)

*6. Splitting Fields: n Roots Exist!

The main purpose of this section is to prove the result which was called son of blockbuster in Section 1, namely that any polynomial of positive degree n over a domain has n roots (counting multiplicities) in some suitably larger field. One upshot of the construction will be a more rigorous treatment of the finite algebraic structures $\mathbf{Z}_m$ introduced in Section 1 of Chapter 1 (and accorded somewhat less cavalier handling in Section 1 of this chapter). Before describing how to construct a "suitably larger field," we begin with an example showing that selecting any old bigger field just won't do.

Consider an integral monic cubic polynomial $f(x) = x^3 + px + q \in \mathbf{Z}[x]$ which, when viewed over $\mathbf{Q}$, is prime in $\mathbf{Q}[x]$. Examples of such abound since, by virtue of the factor theorem and the fundamental fact about degrees, primeness of $f(x)$ in $\mathbf{Q}[x]$ is equivalent to the condition that f has no rational roots. Explicit examples of such $f(x)$ are the polynomials $x^3 + 3x + 1$ and $x^3 + x - 1$ which were met in Section 1. (In truth, it was shown earlier only that these examples had no roots in $\mathbf{Z}$. The stronger assertion about their being rootless in $\mathbf{Q}$ will be *assumed* here and will be *proved* as a consequence of the rational root test in Section 1 of Chapter 4.) For "any old bigger field," we take $F = \mathbf{Q}(\sqrt{n})$ in the sense of Section 1 of Chapter 1, where n is any squarefree integer exceeding 1 (such as 2, 3, 5, 6, 7, 10, . . .).

The proof that $f(x)$ has no root in the field F is indirect: suppose, on the contrary, that $f(x)$ has the root $a + b\sqrt{n} \in F$ (where a and b are certain rational numbers). Expanding the equation $(a + b\sqrt{n})^3 + p(a + b\sqrt{n}) + q = 0$ and equating corresponding coefficients (why is this permissible? Recall the "pesky fact" in Section 1 of Chapter 1.), we derive $a^3 + 3ab^2n + pa + q = 0$ and $3a^2b + b^3n + pb = 0$. As $f(a) \neq 0$, the first of these equations guarantees that $b \neq 0$ whence, by property (vi)′ of domains, the second equation leads to $3a^2 + b^2n + p = 0$. Substituting for b^2n from *this* equation into the first derived equation produces $a^3 + 3a(-3a^2 - p) + pa + q = 0$, that is, $-8a^3 - 2pa + q = 0$, whence $-2a$ is a rational root of $f(x)$, the desired contradiction. (Readers acquainted with the rudiments of linear algebra will realize, after reading Section 3 of Chapter 3, that the preceding argument may, from a certain theoretical point of view, be replaced by the observation that 2 is not an integral multiple of 3.)

Now, to the proof of son of blockbuster (sometimes known as the theorem of Cauchy–Kronecker–Steinitz!). Our task is to show that if $f(x) \in S[x]$ is a polynomial of positive degree n over the domain S, then $f(x)$ has n roots in some field of which S is a subdomain. First, it does no harm (and will help) to replace S by its quotient field F. Secondly, we assert that it is enough to produce *one* root in this general situation. Indeed, if F is a subfield of a field L and $f(\alpha) = 0$ for some element $\alpha \in L$, the factor theorem gives $f(x) = (x - \alpha)g(x)$ for some polynomial $g(x)$ over L of degree $n - 1$ (another application of the fundamental fact about degrees); production of all n roots, that is, of the "other" $n - 1$ roots, then follows by virtue of the well-ordering property, that is, by starting with $g(x) \in L[x]$ and iterating the *one*-root analysis. Thirdly, our final piece of strategy is to use the unique factorization strategy of Section 3 of Chapter 1. Indeed, if we select any irreducible factor $p(x)$ of $f(x)$ in $F[x]$ and are able to produce a root α of $p(x)$ in a suitable field, then the second fundamental (homomorphism) property of evaluation shows that α will also be a root of $f(x)$. Accordingly, the three considerations of this paragraph have *reduced the task to finding one root of $f(x)$ in case S is a field F and $f(x)$ is irreducible in $F[x]$.*

Construction of the suitable field L will proceed via an explicit *factor ring* construction. As a set, L is just $\{\bar{g} : g \in F[x]\}$ where, for each polynomial $g \in F[x]$, we define $\bar{g} = \{g + qf \in F[x] : q \in F[x]\}$. Note that $\bar{g}$ is a certain subset of $F[x]$ which happens to contain g. Here's the key *observation* about this overhead-barring operation. For polynomials g_1 and g_2 in $F[x]$, either $\overline{g_1} = \overline{g_2}$ or $\overline{g_1}$ is disjoint from $\overline{g_2}$ (in the usual sense, that $\overline{g_1} \cap \overline{g_2}$ is empty); moreover, the condition $\overline{g_1} = \overline{g_2}$ is equivalent to the requirement that $f \mid (g_1 - g_2)$. The key observation's proof follows readily from the domain axioms satisfied by $F[x]$, and is left to the reader. Its *use* is to assure the well-definedness of the following binary operations in L:

$$\overline{g_1} + \overline{g_2} = \overline{g_1 + g_2} \quad \text{and} \quad \overline{g_1} \cdot \overline{g_2} = \overline{g_1 \cdot g_2}\,.$$

Taking $0 = \bar{0}$, $-\bar{g} = \overline{-g}$, and $1 = \bar{1}$, the reader will have no trouble in verifying that L inherits from $F[x]$ all the domain axioms, with the possible exception of (vi).

In fact, (vi) is *not* an exception since we next proceed to establish the stronger axiom (viii), namely the existence of a multiplicative inverse for each nonzero $\bar{g} \in L$. Now the above key observation shows that $f \nmid g$ since $\bar{g} \neq \bar{0} = 0$, and so (here is where the irreducibility of $f(x)$ comes in!) condition (**) in Section 3 of Chapter 1 forces $(f, g) = 1$. Then the g.c.d. algorithm produces polynomials q_1 and q_2 in $F[x]$ such that $q_1(x)f(x) + q_2(x)g(x) = 1$, whence $\overline{q_2 g} = \bar{1}$; that is, $\overline{q_2}\bar{g} = \bar{1}$ and $\overline{q_2}$ is a suitable value of $\bar{g}^{-1}$. Thus (viii) has been verified, and L is a field.

How are we to regard F as a subfield of L? The natural temptation is to identify each element $r \in F$ with the corresponding $\bar{r} \in L$. Fortunately no harm accrues from this view of F, since distinct r's produce distinct $\bar{r}$'s, the point being that the fundamental fact about degrees shows that the (positive-degree) polynomial $f(x)$ cannot divide a (nonzero, difference of r's) constant. (Purists might argue at this point that the subfield of L that has been concocted is not F but merely its algebraic copy, $\bar{F} = \{\bar{r} \in L : r \in F\}$. All objections cease, however, if one creates an altered L as the union of F with the set-theoretic complement of $\bar{F}$ in (the "old") L.)

Can you guess which root α of $f(x)$ has been created in L? It is just $\alpha = \bar{x}$! Indeed, if $f(x) = a_0x^n + \ldots + a_{n-1}x + a_n \in F[x]$, then the above explanation of how to view F as a subfield of L reinterprets $f(x)$ as $\overline{a_0}x^n + \ldots + \overline{a_{n-1}}x + \overline{a_n} \in L[x]$, whence $\bar{f}(\alpha) = \overline{a_0}\bar{x}^n + \ldots + \overline{a_{n-1}}\bar{x} + \overline{a_n} \in L$. However, the definition of the binary operations on L then gives $\bar{f}(\alpha) = \overline{a_0x^n + \ldots + a_{n-1}x + a_n} = \bar{f}$ which, by the key observation, is just $\bar{0} = 0 \in L$. This completes the proof of son of blockbuster.

More algebraic hay can be made. If $f(x)$ is an n-th-degree polynomial over a field F and if "all the roots of $f(x)$," namely $r_1, \ldots, r_n$, lie in some field L of which F is a subfield, then the *splitting field* (*of* $f(x)$ *over* F) is defined to be the field $F(r_1, \ldots, r_n)$. Recall from #9 of Exercises 1.1 that this subfield of L is the smallest subfield of L which contains both F and the roots r_i; the terminology is well-chosen inasmuch as $f(x)$ "splits" in $F(r_1, \ldots, r_n)[x]$ as the product of its (constant) leading coefficient and the linear factors $x - r_i$. The splitting fields over the field $\mathbf{Q}$ of the polynomials $x^2 - 10$, $x^3 - 1$, and $x^2 - 9$ respectively are $\mathbf{Q}(\sqrt{10}, -\sqrt{10}) = \mathbf{Q}(\sqrt{10})$, $\mathbf{Q}(1, \omega, \omega^2) = \mathbf{Q}(\omega) = \mathbf{Q}(\sqrt{3}\, i)$, and $\mathbf{Q}$, respectively. One defines the splitting field of a set of polynomials over a given field in the obvious way; for instance the splitting *field* over $\mathbf{Q}$ of the set $\{x^2 - 10, x^3 - 1, x^2 - 9\}$ is the field $\mathbf{Q}(\sqrt{10}, \sqrt{3}\, i)$. By further exploration of the Cauchy–Kronecker–Steinitz construction (along the lines to be begun in Section 3 of Chapter 3), one proves that splitting fields exist; and that any two splitting fields of a given set of polynomials over a given field have essentially the same structure (or, to employ the usual jargon, are *isomorphic*). This applies in particular to the *algebraic closure* $\mathcal{F}$ of a field F, which is by definition the splitting field over F of the set of *all* the polynomials in $F[x]$. It may be shown that $\mathcal{F}$ is *algebraically closed* in the sense that $\mathcal{F}$ is its own algebraic closure. One way to state the fundamental theorem of algebra (given a more traditional statement in Section 6 of Chapter 3) is that $\mathbf{C}$ is the algebraic closure of $\mathbf{R}$ (and, hence, $\mathbf{C}$ is algebraically closed).

Although the above construction of a root is somewhat deep and requires careful study, the reader may possibly have already noticed that two of its features have been met earlier. First, the verification of axiom (viii) four paragraphs ago proceeded by the "same" method which treated the case $F = \mathbf{Q}$, $f(x) = x^3 - 2$ in Section 3 of Chapter 1. Secondly, the overhead-barring construction of the field L five paragraphs ago leads to a new understanding of the overhead-barring approach which was employed in Section 1 in order to detect integral roots. Indeed, a new view of $\mathbf{Z}_m$ results, even if m is not a prime, in the following way. Fix an integer $m > 1$ (to be viewed as the analogue of $f(x)$ in the Cauchy–Kronecker–Steinitz construction). For each integer n, let $\bar{n} = \{n + qm \in \mathbf{Z} : q \in \mathbf{Z}\}$ and set $\mathbf{Z}_m = \{\overline{m} : m \in \mathbf{Z}\}$. One has the precise analogue of the earlier key observation: either $\overline{n_1} = \overline{n_2}$ or $\overline{n_1} \cap \overline{n_2}$ is empty; moreover, $\overline{n_1} = \overline{n_2}$ if and only if $n_1 - n_2$ is an integral multiple of m. Consequently, $\mathbf{Z}_m$ is the set $\{\bar{0}, \bar{1}, \bar{2}, \ldots, \overline{m-1}\}$ with m members. The definitions of addition and multiplication on $\mathbf{Z}_m$ are, by continuing the analogy with the Cauchy–Kronecker–Steinitz construction, just $\overline{n_1} + \overline{n_2} = \overline{n_1 + n_2}$ and $\overline{n_1} \cdot \overline{n_2} = \overline{n_1 \cdot n_2}$: compare with Section 1's fundamental (homomorphism) properties of reduction modulo m. Most importantly, the domain axioms other than (vi) are now evidently inherited by $\mathbf{Z}_m$ from $\mathbf{Z}$, a point that may have been troubling the careful reader. Looking back to Section 1 of Chapter 1, we confess that axioms (iii) and (vii) are difficult to verify—but definitely doable—by the earlier view of $\mathbf{Z}_m$; the present factor ring, overhead-barring, homomorphism approach has, properly, rendered such verification trivial.

EXERCISES 2.6

1. Prove the key observation about the overhead-barring process which was asserted in the fifth paragraph of this section.
2. Let F be a field, and let $f(x) \in F[x]$ be an irreducible polynomial of degree $n \geq 1$. Let L be the field built therefrom by the Cauchy–Kronecker–Steinitz construction. Prove that each member of L can be expressed, in a unique way, in the form $\overline{r_0} + \overline{r_1}\,\bar{x} + \overline{r_2}\,\bar{x}^2 + \ldots + \overline{r_{n-1}}\,\bar{x}^{n-1}$ for appropriate $\overline{r_i}$ $(= r_i)$ in F. (*Hint*: use the division algorithm and the fundamental fact about degrees.)
3. Identify the splitting field of $x^4 - 1$ over $\mathbf{Q}$; of $x^4 - 1$ over $\mathbf{R}$; of $x^2 + 1$ over $\mathbf{R}$; of $x^2 + 1$ over $\mathbf{C}$; and of $x^2 - 1$ over $\mathbf{Q}$.

4. Let **F** be the algebraic closure of the countable field **Q**. Assume the fundamental theorem of algebra, so that **F** may be viewed as a subfield of **C**. Then prove that:
(a) **Q**$[x]$ is denumerable (that is, is in one-to-one correspondence with the set of positive integers).
(b) Each element of **F** belongs to an appropriate subfield of **F** of the form **Q**$(r_1, r_2, \ldots r_n)$ which is a splitting field over **Q** of some finite set of polynomials in **Q**$[x]$.
(c) **F** is denumerable.
(d) **F** $\neq$ **C**.
5. Working inside $\mathbf{Z}_8$, as viewed in this section:
(a) list six different integers which are members of the set $\overline{3}$ (which is itself a member of $\mathbf{Z}_8$); and
(b) compute $\overline{2} + \overline{7}$, $\overline{2} \cdot \overline{7}$, and $\overline{2} - \overline{7}$.

*7. Formal Power Series, Newton Revisited, and the Number of Homogeneous Monomials

While reading Section 3 of Chapter 1, the reader may well have wondered if some decent mathematical sense may be attributed to an expression such as $a_0 + a_1x + a_2x^2 + \ldots$. In fact, musings about such "polynomials of possibly infinite degree" will be rigorized in this section by studying formal power series, leading in particular to a *natural* proof of Newton's formulas (which *were* proved, in a rather *difficult* way, in Section 4). Another upshot of studying power series will be the introduction of certain homogeneous polynomials h_i, which relate nicely to the elementary symmetric polynomials s_i and power polynomials p_i, and which give a neat conceptual approach to an important combinatorial problem.

To rigorize the whimsy concerning formal power series, let's work over a field F (the careful reader will be able to make the appropriate generalization to a domain for *some* of what follows) and proceed by analogy with the earlier introduction of polynomials. A *formal power series over F in the variable x* will be an expression $f(x) = a_0 + a_1x + a_2x^2 + \ldots$, arising from a sequence $a_0, a_1, a_2, \ldots$ of elements of F; such data is often summarized (pun intended) as $\sum_{n=0}^{\infty} a_nx^n$, but the $\sum$ should *not* be thought of as signifying addition in some infinite sense. The datum $\sum a_nx^n$ may, of course, be viewed as the list $(a_0, a_1, a_2, \ldots)$, so that the notions of equality and addition for formal power series are componentwise affairs. Continuing the analogy with polynomials, we define the product of $\sum a_nx^n$ and $\sum b_nx^n$ to be $\sum c_nx^n$, where $c_n = a_0b_n + a_1b_{n-1} + \ldots + a_{n-1}b_1 + a_nb_0$. In this way, $F[x]$ becomes a subsystem of $F[[x]]$, the latter being the notation for the set of formal power series over F in x, endowed with the above binary operations of addition and multiplication.

The analogy between $F[x]$ and $F[[x]]$ extends beyond the mere definitions to the following basic result: $F[[x]]$ is a domain. As one would expect from the proof in Section 3 of Chapter 1 that $F[x]$ is a domain, the assertion concerning $F[[x]]$ is an easy consequence of an analogue of the fundamental fact about degrees. Specifically: each nonzero $f(x) \in F[[x]]$ may be expressed in a unique way as $f(x) = x^n g(x)$ where n is a nonnegative integer, $g(x) \in F[[x]]$, and $g(x)$ possesses a multiplicative inverse in $F[[x]]$. (Why does such a representation for nonzero formal power series lead to domainhood for $F[[x]]$?) This, in turn, follows readily (how?) from the observation that *a given formal power series* $f(x) = \sum a_n x^n$ *has a multiplicative inverse in* $F[[x]]$ *if and only if* $a_0 \neq 0$. (To prove the "only if" part, equate corresponding constant terms in the equation $f(x)f(x)^{-1} = 1$. More importantly, to prove the "if" part, construct the coefficients of a suitable $f(x)^{-1} = \sum b_n x^n$ by solving for b_0 in $a_0 b_0 = 1$, for b_1 in $a_0 b_1 + a_1 b_0 = 0$, for b_2 in $a_0 b_2 + a_1 b_1 + a_2 b_0 = 0$, etc.) In particular, $(1 - rx)^{-1}$ exists in $F[[x]]$ regardless of the choice of $r \in F$; indeed, the above parenthetical argument leads to $(1 - rx)^{-1} = 1 + rx + r^2x^2 + r^3x^3 + \dots$. This example will not only gladden *aficionados* of geometric series, but will facilitate the utilization of this section's key identity.

We pause to record that the quotient field of the domain $F[[x]]$ is denoted $F((x))$ and is called *the field of formal Laurent series.* Of course, $F[[x]]$ is a proper subset of $F((x))$. (Why? Is x^{-1} a formal power series?) The above representation of nonzero formal power series extends naturally to a description of nonzero formal Laurent series (see Exercise 1 below). Another upshot is a sufficiently strong notion of "degree" which yields a proof that (like $\mathbf{Z}$ and $F[x]$) $F[[x]]$ is a Euclidean domain and hence a unique factorization domain, which (*un*like $\mathbf{Z}$ or $F[x]$) has, apart from associates, only one prime element. The interested reader should develop a theory of divisibility, g.c.d.s, associates, and primes for $F[[x]]$ which will shed light on the last assertion.

Next, we shall develop the promised *key identity.* Specialize to the case in which F is the quotient field of the domain $\mathbf{Q}[x_1, \dots, x_n]$ of polynomials over $\mathbf{Q}$ in the variables $x_1, \dots, x_n$. (Any field of coefficients other than $\mathbf{Q}$ would do as well.) Then the key identity asserts:

$$(1 - x_1x)(1 - x_2x) \dots (1 - x_nx) = 1 - s_1(x_1, \dots, x_n)x + \dots$$
$$+ (-1)^n s_n(x_1, \dots, x_n)x^n;$$

verification follows readily (doesn't it?) by expansion of the left-hand side.

To see the import in $F[[x]]$ of the key identity (whose statement only concerns $F[x]$), let $f(x)$ denote the polynomial in (either side of) the key identity, and (*Deus ex machina*!) consider $f'(x)/f(x)$. This is, according to the definition of derivative in Section 3, just

$$\frac{-s_1 + 2s_2x - 3s_3x^2 + \ldots + (-1)^n ns_n x^{n-1}}{1 - s_1x + s_2x^2 + \ldots + (-1)^n s_n x^n},$$

a thoroughly respectible denizen of the quotient field of $F[x]$. However, thanks to an iterated use of Section 3's product rule, $f'(x)/f(x)$ is also just

$$\frac{-x_1}{1-x_1x} - \frac{x_2}{1-x_2x} - \ldots - \frac{x_n}{1-x_nx},$$

which in $F((x))$ is "just"

$$-x_1(1 + x_1x + x_1^2x^2 + \ldots) - \ldots - x_n(1 + x_nx + x_n^2x^2 + \ldots) =$$
$$-(p_1(x_1, \ldots, x_n) + p_2(x_1, \ldots, x_n)x + p_3(x_1, \ldots, x_n)x^2 + \ldots).$$

Equating the two expressions for $f'(x)/f(x)$, crossmultiplying, simplifying, and equating corresponding coefficients of $1, x, x^2, \ldots, x^{n-1}$ in the resulting equal formal power series, we obtain the set of n equations

$$p_1 = s_1,$$
$$-p_1s_1 + p_2 = -2s_2,$$
$$p_1s_2 - p_2s_1 + p_3 = 3s_3,$$
$$\vdots$$
$$(-1)^{n-1}p_1s_{n-1} + (-1)^{n-2}p_2s_{n-2} + \ldots - p_{n-1}s_1 + p_n = (-1)^{n+1}ns_n,$$

which the reader will recognize as a rewriting of (2) of Newton's formulas. Unlike the earlier proof in Section 4, (2) has just been established without first deriving (1). But, what of (1)? Does the above Laurent-series approach snare it too? Yes, indeed: proceeding as above and equating corresponding coefficients of x^n, x^{n+1}, $x^{n+2}, \ldots$, we obtain a new proof of (1) of Newton's formulas. (Check.)

The reader who has studied power series functions in calculus should observe another way of interpreting the preceding argument. To wit: consider $x_1, \ldots x_n$ to be nonzero real numbers and $F = \mathbf{R}$ in the key identity; note that passing from $f(x)$ to $f'(x)/f(x)$ is (rather than a *Deus ex machina*) merely calculus's technique of logarithmic differentiation; check that the expansions of $(1 - x_ix)^{-1}$ are instances of the calculus's binomial theorem, valid for all real x such that $0 \leq |x| < \min\{|x_i|^{-1} : i = 1, \ldots, n\}$; recall that products of absolutely convergent infinite series—in particular, of power-series functions taken on the interior of their intervals of convergence—are computed formally as above; and finally, observe that the technique of equating corresponding coefficients works in the present analytic setting since each power-series function with nontrivial interval of convergence about the origin coincides with its Maclaurin series. Thus, it is a moot point whether the above unified proof of Newton's formulas is algebraic or analytic, for suitably interpreted, it is either!

Before indicating another use of power series, let us for a moment consider an affair of the heart. Imagine a traditional romantic who wishes to purchase a bouquet of seven roses from a florist who stocks (at least seven each of) red roses, pink roses, white roses, yellow roses, and peach roses. We ask: how many different bouquets, considered as unordered sets, may our romantic acquire?

It is apparent that a more general analysis is needed. Suppose that a selection of r objects is to be made from (at least r each of) n types of objects. The total number of such selections that are possible is denoted ${}_n\mathrm{H}_r$ (the letter H stands for "homogeneous": we'll see why that's appropriate shortly) and the *main result* is that ${}_n\mathrm{H}_r = \binom{n+r-1}{r}$, the usual combinatorial symbol that counts the number of ways in which a subset of r objects may be drawn from a universe consisting of $n + r - 1$ distinct members. In our romantic example, $n = 5$ and $r = 7$, whence, ${}_n\mathrm{H}_r = \binom{11}{7} = 11!/7!4!$ $= 11 \cdot 10 \cdot 9 \cdot 8/1 \cdot 2 \cdot 3 \cdot 4 = 330$ is the number of possible bouquets.

We shall sketch a proof of the formula for ${}_n\mathrm{H}_r$. Suppose that 1, 2, 3, . . . , n are the types of objects that may be selected. (In our example, "1" stands for "red rose," "2" stands for "pink rose," etc.) Describe each selection (in our example, bouquet) as a sequence of r digits in increasing order. (In our example, 1122245 describes a bouquet with two red roses, three pink roses, one yellow rose, and one peach rose.) Imagine the selections listed one beneath another thus:

$$\begin{array}{c} 1111\ldots11 \\ 1111\ldots12 \\ \vdots \\ nnnn,\ldots,nn \end{array}$$

This array has ${}_n\mathrm{H}_r$ rows and r columns. The set of ${}_n\mathrm{H}_r$ "strings" may be converted to another set of objects by adding 0 to the first digit in a string, 1 to the second digit, 2 to the third, . . . , and $r - 1$ to the last. Under such a transformation one obtains the listing

$$\begin{array}{cccccccc} 1 & 2 & 3 & 4 & \ldots & r-1 & r & \\ 1 & 2 & 3 & 4 & \ldots & r-1 & r+1 & \\ & & & & \vdots & & & \\ n & n+1 & n+2 & n+3 & \ldots & n+r-2 & n+r-1 & \end{array}$$

Each row in the new list may be viewed as a subset of r objects, listed in increasing order, drawn from the universe $\{1, 2, 3, \ldots, n + r - 1\}$. It is easy to see (do it!) that each such subset appears exactly once in the transformed list. As there $\binom{n+r-1}{r}$ such subsets and the set of them is, by the above transformation, in one-to-one correspondence with the initial list of ${}_nH_r$ selections, the proof is complete.

So much for romance! What has all this to do with power series and homogeneous polynomials? Recall from Section 4 that a polynomial $f(x_1, \ldots, x_n) \in \mathbf{Q}[x_1, \ldots, x_n]$ is *homogeneous of degree r* in case each of its constituent monomials, $bx_1^{m_1}x_2^{m_2} \ldots x_n^{m_n}$ (with $0 \neq b \in \mathbf{Q}$, and each m_i a nonnegative integer) satisfies $m_1 + m_2 + \ldots + m_n = r$. Apart from constant factors b, we claim that there are precisely ${}_nH_r$ such monomials. (This is practically obvious. Think of "x_i" as "i" in the earlier analysis. Our romantic's bouquet 1122245 is then associated with the homogeneous monomial $x_1^2x_2^3x_4x_5$ of degree $r = 7$ in $\mathbf{Q}[x_1, x_2, x_3, x_4, x_5]$. Convince yourself of the generality of such considerations.) Since the elementary symmetric polynomial $s_i \in \mathbf{Q}[x_1, \ldots, x_n]$ has $\binom{n}{i}$ constituent monomials, the temptation is irresistible to define *the i-th homogeneous polynomial* $h_i \in \mathbf{Q}[x_1, \ldots, x_n]$ to be the sum of all the ${}_nH_i$ homogeneous monomials of degree i (with constant factors $b = 1$) in $\mathbf{Q}[x_1, \ldots, x_n]$. For example, in $\mathbf{Q}[x_1, x_2, x_3]$, the first few h's are $h_1 = x_1 + x_2 + x_3$, $h_2 = x_1^2 + x_2^2 + x_3^2 + x_1x_2 + x_1x_3 + x_2x_3$, and $h_3 = x_1^3 + x_2^3 + x_3^3 + x_1^2x_2 + x_1^2x_3 + x_2^2x_1 + x_2^2x_3 + x_3^2x_1 + x_3^2x_2 + x_1x_2x_3$. From such meagre evidence, one may speculate that there is a relationship between the h_i on the one hand and the s_i (and, by Newton's formulas, the p_i) on the other. Indeed, there is, and we shall derive it next.

By passing to the multiplicative inverses of the polynomials in this section's key identity and then working in $F((x))$ as before, we have

$$\frac{1}{1 - s_1x + s_2x^2 + \ldots + (-1)^n s_n x^n}$$
$$= (1 - x_1x)^{-1}(1 - x_2x)^{-1} \ldots (1 - x_nx)^{-1}$$
$$= \left(1 + x_1x + x_1^2x^2 + \ldots\right)\left(1 + x_2x + x_2^2x^2 + \ldots\right)$$
$$\ldots \left(1 + x_nx + x_n^2x^2 + \ldots\right)$$
$$= 1 + h_1x + h_2x^2 + \ldots .$$

Cross-multiplying and equating corresponding coefficients produces the equations

$$h_k - s_1h_{k-1} + s_2h_{k-2} + \ldots + (-1)^{k-1}s_{k-1}h_1 + (-1)^k s_k = 0$$

indexed by k ($= 1, 2, 3, 4, \ldots$), where s_j is taken to be 0 whenever $j > n$. Evidently, one may solve for the h_i as expressions in terms of the s_i: no surprise, in view of the fundamental theorem in Section 5, since the h_i

happen to be symmetric polynomials. Also evidently, the symmetry of the displayed equations reveals that the s_i may be *similarly* expressed in terms of the h_i: quite a surprise, indicating that the theoretical importance of the h_i is every bit as important as the earlier roles of s_i and p_i. (Moral: *amor vincit omnia.*) Exercise 6 below pursues this theme. The reader should perform several of these solutions explicitly, verifying that $h_1 = s_1$, $h_2 = s_1^2 - s_2$, $s_2 = h_1^2 - h_2$, $h_3 = s_1^3 - 2s_1s_2 + s_3$, $s_3 = h_1^3 - 2h_1h_2 + h_3$, etc.

To close, we show that use of power series also *vincit omnia*, by rederiving the formula for ${}_n\mathrm{H}_r$. Indeed, take the analytic view of the argument at the beginning of the preceding paragraph, with F being $\mathbf{R}$ and $x_1, \ldots, x_n$ being nonzero real numbers. For real x satisfying $0 \leq |x| < \min\{|x_i|^{-1} : i = 1, \ldots, n\}$, by appealing again to facts established about power-series functions in calculus, we may justify the analytic interpretation of the assertion $(1 - x_1x)^{-1}(1 - x_2x)^{-1} \ldots (1 - x_nx)^{-1} = 1 + h_1(x_1, x_2, \ldots, x_n)x + h_2(x_1, x_2, \ldots, x_n)x^2 + \ldots$. The fortuitous choice $x_1 = x_2 = \ldots = x_n = 1$ reduces the right-hand side to

$$1 + ({}_n\mathrm{H}_1)x + ({}_n\mathrm{H}_2)x^2 + \ldots,$$

since $h_r(1, 1, \ldots, 1)$ just counts the ${}_n\mathrm{H}_r$ constituent monomials in the r-th homogeneous polynomial $h_r(x_1, \ldots, x_n)$. Moreover, the choice $x_1 = \ldots = x_n = 1$ converts the left-hand side to

$$(1-x)^{-1}(1-x)^{-1} \ldots (1-x)^{-1} = (1-x)^{-n}$$

which, since $|x| < 1$, is given by the binomial power series. Since the latter has nonzero radius of convergence (it's 1), "equating the corresponding coefficients" is a valid technique, whence ${}_n\mathrm{H}_r$ is the coefficient of x^r in

$$1 - n(-x) + \frac{(-n)(-n-1)}{2}(-x)^2 + \frac{(-n)(-n-1)(-n-2)}{3!}(-x)^3 + \ldots .$$

Therefore,

$$\begin{aligned} {}_n\mathrm{H}_r &= \frac{-n(-n-1)\ldots(-n-r+1)(-1)^r}{r!} \\ &= \frac{(-1)n(-1)(n+1)\ldots(-1)(n+r-1)(-1)^r}{r!} \\ &= \frac{n(n+1)\ldots(n+r-1)(-1)^r(-1)^r}{r!} = \frac{(n+r-1)!}{(n-1)!r!} \\ &= \binom{n+r-1}{r}, \end{aligned}$$

which completes the proof.

EXERCISES 2.7

1. Let F be a field, $F[[x]]$ the domain of formal power series, and $F((x))$ the field of formal Laurent series. Prove the following analogue of the fundamental theorem of arithmetic. For each nonzero $g \in F((x))$, there is exactly one way to express g in the form $x^n h$, where $n \in \mathbf{Z}$ and h is a nonzero formal power series whose multiplicative inverse also lies in $F[[x]]$.
2. Let $\mathbf{C}(x)$ be the quotient field of $\mathbf{C}[x]$. (As in Section 5, $\mathbf{C}(x)$ is called the field of rational functions over $\mathbf{C}$ in x.) Prove that $\mathbf{C}(x)$ is a proper subfield of $\mathbf{C}((x))$ by showing that the formal Laurent series $1 + x + (x^2/2!) + (x^3/3!) + \ldots$ is not a rational function. (*Hint*: Viewing the entities as functions of a complex variable x, recall (or believe) that the given series is just e^x, which is never 0. Use the fundamental theorem of algebra to reduce the task to showing that e^x is not constant. Then show it.)
3. Give an example of an infinite field of characteristic 2. Generalize.
4. (a) Define what ought to be meant by the derivative of a formal power series.

   (b) State – and prove – rules satisfied by the definition of differentiation which was given in (a), proceeding as much as possible by analogy with the treatment in Section 3.

   (c) Develop a "Maclaurin's formula" for formal power series analogous to that in #3(b) of Exercises 2.3. Is there also an analogue of Taylor's theorem for formal power series?
5. (a) Compute ${}_6H_1$, ${}_6H_2$, ${}_6H_3$, ${}_6H_6$, ${}_6H_7$, ${}_6H_8$, and ${}_6H_9$.

   (b) Which aspect of an appropriate selection process is being counted when one considers the sum ${}_nH_0 + {}_nH_1 + {}_nH_2 + \ldots + {}_nH_n$?

   (c) An incredibly wealthy oil merchant plans to buy four homes in a fashionable suburb. On the marketplace he/she finds six identical split-foyers, seven identical split-levels, six identical basement-ranchers, and four identical mobile homes. In how many ways may he/she purchase the set of four homes?

   (d) Continuing with (c), let x_1, x_2, x_3, x_4 correspond to (the selection of) a split-foyer, split-level, basement-rancher, mobile home, respectively. Which type of purchase is associated with the monomial $x_1{}^3x_4$? With $x_2{}^4$? How many such monomials represent admissible purchases?
6. (a) Prove that each symmetric polynomial in $\mathbf{Q}[x_1, \ldots, x_n]$ may be expressed as a sum of finitely many terms of the form $rh_1{}^{m_1} \ldots h_k{}^{m_k}$, where $r \in \mathbf{Q}$, $h_i(x_1, \ldots, x_n)$ is the i-th homogeneous polynomial defined in this section, and each m_i is a nonnegative integer. (*Hint*: use the fundamental theorem in Section 5.)

   (b) Express $\Sigma x_i{}^2 x_j$ in the manner described in (a). What, if anything, is special about the case $n = 2$?

3

There is much merit in constantly reminding oneself that without significant computational payoffs the concepts and bookkeeping devices of a mathematical theory are vapid and barren. Accordingly, one might be pardoned for wondering in what (if any) sense these pages extend the enterprise of equation solving, as typified by high-school's quadratic formula. This chapter is intended to quiet such worries by presenting the cubic formula (see Sections 1 and 2) and the quartic formula (in Section 4). Section 1 also relates some of the high drama, *dramatis personae* and Renaissance intrigue which figured in the discovery of such formulas. It is, however, in the remaining (optional) sections of the chapter that our subject truly comes of age. For instance, Section 5 explains how the Galois theory of modern algebra demonstrates that there

Cubics, Quartics, and Beyond

simply is no "quintic formula"; that is, *solutions of a certain type* (by radicals) *do not exist* for sufficiently general quintic (or higher-degree) polynomials over **Q**. Similarly, Section 3 explains how a linear-algebra analysis of the Cauchy–Kronecker–Steinitz construction (from Section 6 of Chapter 2) leads to the geometric payoff that *solutions of a certain type* (by ruler and compass) *do not exist* for the problem of trisecting sufficiently general angles. Of course, angles *can* be trisected (since $\theta/3$ measures some angle whenever $0 \leqslant \theta < 2\pi$), and *solutions* (not necessarily by radicals) *do exist* for arbitrary positive-degree complex polynomials. The latter assertion is the fundamental theorem of algebra, which has already seen *use* in #4 of Exercises 2.6 and #2 of Exercises 2.7; its *proof* will be the subject of Section 6.

1. Solution of the Cubic and the Story of Ferro-Cardan-Tartaglia

The reader will recall from the introduction that research on polynomial equations has been pursued by some of history's greatest mathematicians. Indeed, several of the results on these pages bear the names of their illustrious discoverers. However, the story behind Cardan's solution of the cubic is not particularly illustrious, for the (very human) actions of its principal characters cannot be termed "great." The pivotal event was the discovery, about 1505 by one Scipione del Ferro ("Scipio Ferro" in some accounts), of a formula with which one could express the roots of a cubic polynomial (over, say, **R**). Although the search for such an extension of antiquity's quadratic formula had occupied several centuries, Scipio chose not to publish the details of his new-found mathematical "holy grail." Indeed, Scipio confided his secret techniques to a few of his students, who subsequently gained fame and power through public demonstrations of their problem-solving prowess. Imagine the surprise of this clique some forty years later when an outsider, nicknamed Tartaglia (literally, "the stammerer"), began to win the public mathematical contests! Tartaglia, whose name derived from a speech impediment resulting from blows inflicted as a youngster when his hometown was laid waste by a French invasion in 1512, had stumbled onto the same solution as had Scipio—evidence for the view that "there are no great (wo)men, only great times." Naturally, Tartaglia was loath to reveal the secret of his new-found success, but agreed to confide this technique to a mathematician-physician named Cardan. What motive possessed Tartaglia? History speculates that Tartaglia, already the author of the first serious treatise on ballistics, hoped that publicity from so important an individual as Cardan would enable Tartaglia to curry favor, and employment, with the Spanish military. However, Tartaglia's trust was misplaced, for in 1545 Cardan broke his promise of secracy and provided a *complete* description of Tartaglia's technique in *Ars Magna* (literally "Great Art": math books had ritzy titles back then). Cardan included the details of his student Ferrari's solution of the quartic (an amazingly easy consequence of the cubic's solution: see Section 4). Tartaglia was incensed at Cardan's breach of faith, and rashly challenged Ferrari (who, like any good graduate student, sided with his advisor) to a public disputation in Milan in 1548. Predictably, the stammerer lost and returned to obscurity. For a time, Cardan's medical career soared, with scholarly contributions to the study of diseases such as typhus and syphilis. Cardan's interest in gambling also led to his writing a serious treatise on probability, fully a century before history's minion, Pascal. However, Cardan's personal life was filled with tragedy: the execution of one of Cardan's sons for the murder of his (the son's) unfaithful wife was followed by the humiliation of Cardan being publicly accused of heresy. We close the narrative of this fascinating, tawdry, human tale by asking the reader to supply his/her own moral(s). Was there a crime? If so, did it pay?

Before describing Cardan's solution, we shall motivate the initial step of it by giving an unusual proof of the quadratic formula. Suppose then that one is to find those x such that $ax^2 + bx + c = 0$. (Let's suppose throughout this section that all the polynomials are taken over $\mathbf{C}$; Exercise 5 will weaken this restriction considerably.) Equivalently, one is to find the roots of $f(x) = x^2 + (b/a)x + (c/a)$. Anyone who has graphed parabolas (or recalls the first derivative test of calculus) will perceive that $g(x) = f(x - (b/2a))$ is worth studying. Indeed, a computation gives $g(x) = x^2 - (D/4)$, where $D = (b^2 - 4ac)/a^2$, so that g's roots are $\sqrt{D}/2$ and $-\sqrt{D}/2$. According to the polynomial building method (1) in Section 1 of Chapter 2, the roots of $f(x)$ are then given by $r_1 = (-b/2a) + (\sqrt{D}/2)$ and $r_2 = (-b/2a) - (\sqrt{D}/2)$, a minor rewriting (and rederivation) of high school's quadratic formula. (For motivation of Section 2, note also that $(r_1 - r_2)^2 = (r_1 + r_2)^2 - 4r_1r_2 = (-b/a)^2 - 4(c/a) = D$.)

A corresponding reduction in the analysis of a cubic polynomial $f(x) = ax^3 + bx^2 + cx + d$ would involve $g(x) = f(x - k)$ for some $k \in \mathbf{C}$ with the property that the coefficient of x^2 in $g(x)$ is 0. To find such k, recall from Section 1 of Chapter 2 that, if r_1, r_2, r_3 are the roots of $f(x)$, then $r_1 + k$, $r_2 + k$, $r_3 + k$ are the roots of $g(x)$. Thus, by using the material on elementary symmetric functions in Section 4 of Chapter 2, we see that requiring the coefficient of x^2 in $g(x)$ to be 0 amounts to stipulating $0 = -0/a = s_1(r_1 + k, r_2 + k, r_3 + k) = (r_1 + r_2 + r_3) + 3k = s_1(r_1, r_2, r_3) + 3k = (-b/a) + 3k$, whence $k = b/3a$. In other words,

$g(x) = f(x - (b/3a))$ has no term in x^2. By dividing $g(x)$ by its leading coefficient, we don't change its roots (why?) and thus our task is reduced to finding the roots of $h(x) = x^3 + px + q \in \mathbf{C}[x]$.

There is no problem in case $p = 0$, for the roots of $h(x)$ are then just the third roots of $-q$, found using polar form as in Section 2 of Chapter 1. Similarly, the case $q = 0$ is transparent, for then $h(x) = x(x^2 + p)$ has 0 and the square roots of $-p$ as its roots. Accordingly, we may *suppose* during the rest of the derivation of Cardan's formula that *both p and q* are nonzero.

Now it is *our* turn to reveal the secret of Scipio and Tartaglia. The trick is to build a new polynomial, not merely by the methods (1) and (3) of Section 1 in Chapter 2, but by what might be termed a "weighted average" of these methods. Specifically, consider a new variable y such that $x = y - (p/3y)$. (This is backwards, but traditional. One should really describe y in terms of x, by solving $3y^2 - 3xy - p = 0$ for y, using the quadratic formula. The reader will note from that solution that $y \neq 0$ since $p \neq 0$, and so the backwards introduction of y, which involved division by y, is legitimate.) A computation reveals that $y^3h(x) = y^3h(y - (p/3y)) = y^6 + qy^3 - (p^3/27)$. Thus, if x is a root of h, we may solve for y^3 by the quadratic formula. Let $\alpha^3 \in \mathbf{C}$ be such a solution, and moreover let it be the one with a plus sign in front of the radical sign in the quadratic formula; that is,

$$\alpha^3 = \frac{-q + \sqrt{q^2 + (4p^3/27)}}{2}.$$

Of course, $\alpha \in \mathbf{C}$ is not uniquely determined (noting $\alpha^3 \neq 0$ since $p \neq 0$), and might as well be replaced by $\omega\alpha$ and $\omega^2\alpha$, where $\omega = (-1 + \sqrt{3}\, i)/2 = \operatorname{cis}(2\pi/3)$ is the primitive cubic root of 1 which was introduced in Chapter 1. This suggests that "all" *the roots of* $h(x) = x^3 + px + q$ *are given by* (our variant of Cardan's formula):

$$\alpha - \frac{p}{3\alpha}, \qquad \omega\alpha - \frac{p}{3\omega\alpha}, \qquad \text{and} \quad \omega^2\alpha - \frac{p}{3\omega^2\alpha}. \qquad \text{(Cardan's formula)}$$

(All the roots of $f(x)$ are then given by

$$\alpha - \frac{p}{3\alpha} - \frac{b}{3a}, \qquad \omega\alpha - \frac{p}{3\omega\alpha} - \frac{b}{3a}, \qquad \text{and} \quad \omega^2\alpha - \frac{p}{3\omega^2\alpha} - \frac{b}{3a}.)$$

We shall presently prove this suggestion. The weighted average construction won't be needed for the proof. It has served its purpose, namely facilitating the heuristic development of the putative list of roots.

First, let's put the formula to work by finding the roots r_1, r_2, r_3 of $f(x) = 2x^3 + 6x^2 + 8x + 2$. The above analysis diverts attention first to $g(x) = f(x - (b/3a)) = f(x - (6/2 \cdot 3)) = f(x - 1) = 2x^3 + 2x - 2$ and then to its monic associate $h(x) = x^3 + x - 1$, with $p = 1$ and $q = -1$.

(Does $h(x)$ look familiar? Glance back to Section 1 of Chapter 2.) To continue the analysis, we must produce $\alpha \in \mathbf{C}$ such that $\alpha^3 = (1 + \sqrt{1 + (4/27)}\,)/2$, which happens to be a real number, approximately 1.035758376. To ease the calculation, take α to be the real cubic root of this value of α^3. Then α is approximately 1.01178, and the roots of $h(x)$ (also of $g(x)$) are approximately

$$\alpha - \frac{p}{3\alpha} \doteq 0.68232,$$

$$\omega\alpha - \frac{p}{3\omega\alpha} \doteq -0.34116 + 1.1615i,$$

and

$$\omega^2\alpha - \frac{p}{3\omega^2\alpha} \doteq -0.34116 - 1.1615i.$$

Finally, the desired roots of $f(x) = g(x + 1)$ are approximated by

$$-0.31768, \qquad -1.34116 + 1.1615i, \quad \text{and} \quad -1.34116 - 1.1615i.$$

A word about the "approximate" solution of the preceding example. As the following proof will show, the formulas of Cardan are exact, not merely approximate. However, an approximate value of α (corresponding to an exact value α^3) may be all that tables/calculator/polar form can produce. The round-off error in describing α and the similar error in describing $\sqrt{3}$ (as part of ω) impose limitations on the numerical implementation of Cardan's method. Frequently, approximate solutions of a cubic are obtained more quickly by some of the techniques in Chapter 6. However, such numerical search procedures often work only if one knows *a priori* the nature of the roots being approximated, and such an analysis for cubics is available precisely because of Cardan's exact formulas: see Section 2 for details.

The proof that Cardan's formulas give all the roots of $h(x)$ is not nearly so profound as the intuitive insight which was needed to pass from $f(x)$ to $h(x)$. Indeed, having selected any $\alpha \in \mathbf{C}$ such that $\alpha^3 = (-q + \sqrt{q^2 + (4p^3/27)}\,)/2$, we need only to verify the factorization

$$x^3 + px + q = \left(x - \left(\alpha - \frac{p}{3\alpha}\right)\right)\left(x - \left(\omega\alpha - \frac{p}{3\omega\alpha}\right)\right)\left(x - \left(\omega^2\alpha - \frac{p}{3\omega^2\alpha}\right)\right).$$

Comparision of coefficients is uneventful (given #8 of Exercises 1.2). Indeed, the coefficient of x^2 in the putative factorization is

$$\frac{-(3\alpha^2 - p)(\omega^2 + \omega + 1)}{3\omega^2\alpha} = 0$$

(check). Similarly, the coefficient of x is

$$\alpha^2(\omega^3 + \omega + \omega^2) + \frac{p^2}{9\alpha^2}\left(\frac{1}{\omega^3} + \frac{1}{\omega} + \frac{1}{\omega^2}\right) - \frac{p}{3}\left(\frac{1}{\omega} + \omega + \frac{1}{\omega} + \omega + \frac{1}{\omega^2} + \omega^2\right)$$

$$= \frac{-p}{3}\, 3\left(\frac{1}{\omega} + \omega\right) = \frac{-p(1+\omega^2)}{\omega} = \frac{-p(-\omega)}{\omega} = p$$

(check that too!). Finally, the constant term simplifies as

$$-\alpha^3 - \frac{p^2}{9\alpha}\left(1 + \frac{1}{\omega} + \omega\right) + \frac{p\alpha}{3}\left(\frac{1}{\omega} + \omega + 1\right) + \frac{p^3}{27\alpha^3}$$

which, given the above choice of α, is

$$\frac{q - \sqrt{q^2 + \dfrac{4p^3}{27}}}{2} + \frac{p^3}{27}\left[\frac{2}{-q + \sqrt{q^2 + \dfrac{4p^3}{27}}}\right].$$

This, after rationalization of its second term, is

$$\frac{q - \sqrt{q^2 + \dfrac{4p^3}{27}}}{2} + \frac{2p^3}{27}\,\frac{\left(q + \sqrt{q^2 + \dfrac{4p^3}{27}}\right)}{4p^3/27} = \frac{q}{2} + \frac{q}{2} = q,$$

which completes the proof.

A final remark. It does not matter *which* square root of $q^2 + (4p^3/27)$ is used in determining the value of α^3, although *one such* must be fixed on, thus permitting the cancellation in the final step of the preceding proof.

EXERCISES 3.1

1. Use the methods of this section, approximating with a hand-held calculator as needed, to find all the roots of the following cubic polynomials: $2x^3 + (1+i)x^2 - 7$, $\sqrt{2}\,x^3 + ix^2 + \pi x - 1$, $8x^3 - 12ix^2 - 6x + i$, $x^3 + 4x^2 + 5x + 2$, $x^3 + x + 1$, $x^3 - 5x + 1$, $x^3 - 3x + 2$.
2. In the spirit of #1 of Exercises 1.4 (see also #1 of Exercises 1.5 and #1(a) of Exercises 2.3), follow the instructions of #1 for cubic polynomials of your own choice and creation.
3. If $f(x) = a_0x^n + a_1x^{n-1} + \ldots + a_n$ is an n-th-degree polynomial over a field F (such as **C**) in which $n \neq 0$, prove that the coefficient of x^{n-1} in $g(x) = f(x - (a_1/na_0))$ is 0. Using the theory of elementary symmetric functions, prove that a_1/na_0 is the only value of k in F such that $f(x-k)$ has no term in x^{n-1}.

4. In case either p or q is 0, is it either meaningful or valid to use Cardan's formula to describe the roots of $x^3 + px + q$?

*5. Check that the reduction to cubics of the form $x^3 + px + q$ and the subsequent analysis via Cardan's formula is valid in case the field of coefficients is (not just **C** but) any field F such that $\text{char}(F) \neq 2,3$. (*Hint*: What should ω be in this context? Apply the Cauchy–Kronecker–Steinitz construction in Section 6 of Chapter 2 to a suitable quadratic polynomial. Similarly, find suitable α.)

2. Discriminants and the Casus Irreducibilis

All the polynomials in this section will be taken over **R**. We begin by recalling the well-known, and useful, consequence of the quadratic formula that the nature of the roots r_1, r_2 of a quadratic polynomial $ax^2 + bx + c$ depends upon the *discriminant* $D = (b^2 - 4ac)/a^2 = (r_1 - r_2)^2$. Specifically, $D = 0$ if and only if $r_1 = r_2$ (in which case, the double root is a real number); $D > 0$ if and only if r_1 and r_2 are distinct real numbers; and $D < 0$ if and only if r_1 and r_2 are nonreal complex numbers (necessarily, conjugates of one another). There are times that a similar description of the qualitative nature of the roots of a cubic polynomial would be most opportune, for example in analyzing the discriminating cubic [16, (4), p. 185] to find real principal directions while seeking to classify the seventeen Euclidean quadric surfaces [4, 16]. Fortunately, such a criterion *is* available for cubics, and this section is devoted to its development.

Consider a real cubic polynomial $f(x) = ax^3 + bx^2 + cx + d$, with roots r_1, r_2, r_3 in **C**. By the final comment in Section 3 of Chapter 2 (namely, that the existence of a nonreal complex root z for a real polynomial entails its conjugate z^* being a root with the same multiplicity), there are three exclusive and exhaustive possibilities:

(1) There is a multiple root of $f(x)$, that is, $r_1 = r_2$ (which may or may not equal r_3) after suitable relabelling of the roots (in which case, r_1, r_2, and r_3 are each real numbers).

(2) There is exactly one real root of $f(x)$ (in which case, the other two roots are nonreal and conjugates of one another).

(3) The roots r_1, r_2, r_3 of $f(x)$ are real and distinct.

Note that passage from $f(x)$ to $g(x) = f(x - (b/3a))$ alters the roots (if $b \neq 0$) of the polynomial under consideration, but it does not change those roots' classification in terms of the possibilities (1), (2), (3). Moreover, passage to the monic associate $h(x)$ of $g(x)$ does not further alter the roots. Thus, in order to check which of (1), (2), (3) pertains to a given $f(x)$, attention may be focused instead on the corresponding $h(x)$. Writing

$h(x) = x^3 + px + q$ as in Section 1, we find it irresistible, given the above comment about the quadratic's discriminant D and the radical appearing in Section 1's description of α^3, to speculate that the pivotal factor in analyzing the roots of $h(x)$ should be the behavior of

$$\triangle = 4p^3 + 27q^2.$$

Indeed, this is so! *The main result of this section* is that (1) corresponds to $\triangle = 0$; (2) corresponds to $\triangle > 0$; and (3) corresponds to $\triangle < 0$.

Let's verify the preceding assertion for the degenerate cases in which either p or q vanishes. First, if $p = q = 0$, then $\triangle = 0$ and, as asserted, (1) applies since the roots of $h(x)$ are then all equal (to 0). Next, if $p = 0 \neq q$, then $\triangle = 27q^2 > 0$ and, as asserted, (2) applies since $h(x) = x^3 + q$ then has but one real root. Finally, if $p \neq 0 = q$, then $\triangle = 4p^3$ is positive (respectively, negative) if p is positive (respectively, negative); then (2) applies (respectively, (1) applies) since $h(x) = x(x^2 + p)$.

For the case in which neither p nor q is 0, let's first consider the assertion that (1) corresponds to $\triangle = 0$. One way to establish this is to prove, in case t_1, t_2, and t_3 denote the roots of $h(x)$, that $(t_1 - t_2)^2(t_1 - t_3)^2(t_2 - t_3)^2 = -\triangle$. (The left-hand side is, by our analysis of the quadratic's D, certainly what deserves to be called the cubic h's discriminant.) This expression for $-\triangle$ may be verified using elementary symmetric functions (as in Sections 4 and 5 of Chapter 2) or Cardan's formulas (from Section 1), but either calculation is rather tedious. Instead, we shall use the criterion for multiple roots developed in Section 3 of Chapter 2; namely, $h(x)$ has multiple roots if and only if $(h, h') \neq 1$. To compute this g.c.d. via Euclid's algorithm (see Section 3 of Chapter 1), we note that $h'(x) = 3x^2 + p$ and proceed to obtain $h(x) = (x/3)h'(x) + ((2px/3) + q)$. Accordingly, $(h, h') \neq 1$ if and only if $(2px/3) + q$ divides $h'(x)$; that is, by the factor theorem in Section 1 of Chapter 2, if and only if $h'(-3q/2p) = 0$. Since $4p^2h'(-3q/2p) = \triangle$ and $p \neq 0$, it follows that (1) obtains if and only if $\triangle = 0$, as asserted.

Next, we shall show that $h(x)$ has only one real root whenever $\triangle > 0$. Indeed (to revert to the notation in Section 1), $\triangle > 0$ implies $\alpha^3 \in \mathbf{R}$, in which α may be taken real, whence the root $r = \alpha - (p/3\alpha)$ is also real. Moreover, the other roots of $h(x)$ then fail to be real. To see this, suppose on the contrary that $\omega\alpha - (p/3\omega\alpha)$ is real. Then a simple computation (using $\omega^2 + \omega + 1 = 0$) yields that

$$-\frac{p}{3\omega\alpha} = \omega r + \frac{p}{3\alpha}\left(-1 - \frac{2}{\omega}\right) = \omega r - \frac{p}{3\alpha} + \frac{2p}{3\alpha}(1 + \omega)$$

$$= \frac{p}{3\alpha} + \omega\left(\alpha + \frac{p}{3\alpha}\right),$$

whence $\omega(\alpha + (p/3\alpha)) \in \mathbf{R}$. Since ω is not real, $\alpha + (p/3\alpha) = 0$, hence $p = -3\alpha^2$. By Cardan's formulas then, the roots of $h(x)$ are

$$r = 2\alpha,$$

$$\omega\alpha - \frac{p}{3\omega\alpha} = \omega\alpha - \frac{-3\alpha^2}{3\omega\alpha} = \omega\alpha + \frac{\alpha}{\omega}$$

$$= \frac{(\omega^2 + 1)\alpha}{\omega} = -\alpha, \qquad \text{and}$$

$$\omega^2\alpha - \frac{p}{3\omega^2\alpha} = \omega^2\alpha - \frac{-3\alpha^2}{3\omega^2\alpha} = \omega^2\alpha + \frac{\alpha}{\omega^2}$$

$$= \frac{(\omega^4 + 1)\alpha}{\omega^2} = -\alpha.$$

However, this contradicts the result of the preceding paragraph that the presence of a multiple root for $h(x)$ entails vanishing of the discriminant. Thus, if $\triangle > 0$, *only* one of h's roots is real.

It remains only to prove that if $h(x)$ has but one real root then $\triangle > 0$. (Why will this complete the proof of the main result of this section?) Indeed, if $h(x)$ has the real root r and the nonreal roots $a + bi$ and $a - bi$ (with $a \in \mathbf{R}$ and $0 \neq b \in \mathbf{R}$), then equating corresponding coefficients in

$$x^3 + px + q = h(x) = (x - r)(x - (a + bi))(x - (a - bi))$$

leads to $0 = -2a - r$, $p = a^2 + b^2 + 2ar$ and $q = -r(a^2 + b^2)$. Then substituting $r = -2a$ produces $p = b^2 - 3a^2$ and $q = 2a(a^2 + b^2)$, whence a straightforward calculation yields $\triangle = 4p^3 + 27q^2 = 4b^6 + 72b^4a^2 + 324b^2a^4$, which is indeed positive, as desired.

We should observe that the above result is very easy to apply. Consider last section's example $h(x) = x^3 + x - 1$. Its $\triangle$ is $4(1)^3 + 27(-1)^2 = 31$ which is positive, so that condition (2) obtains, and $h(x)$ has only one real root. But we already knew that, by virtue of the analysis using Cardan's formulas in Section 1. The virtue of computing $\triangle$ in this case, or for any other polynomial with positive $\triangle$, is that one thereby learns quickly that $h(x)$ has only one real root, so that any iterative root-approximation scheme which begins near a real root must converge to the "right" real root. In Section 4 of Chapter 6, we shall meet a cubic with three real roots (and negative $\triangle$) for which iteration of Newton's approximation scheme converges to the "wrong" root. As to the general question regarding the number of real roots for a given real polynomial, see Sturm's theorem in Section 4 of Chapter 5.

Finally, a word about this section's title. It is traditional to refer to case (3), in which a real cubic polynomial has three real roots, as being the *casus irreducibilis.* Its peculiarity is that Cardan's formulas for these real roots involve, through the expression for α^3, a nonreal (complex) square root of (the negative value) $\triangle$. Although such a back-door introduction of complex numbers troubled algebraists of times gone by, it is now known that there is *no* formula (of Cardan or anyone else) which uses only field operations and *real* radicals to express the roots of an irreducible cubic $h(x) \in \mathbf{Q}[x]$ such as $x^3 - 2x + 1$. (See [23, pp. 112–114] for a proof.) (How does one check whether a given rational polynomial is irreducible? See the algorithm in Section 3 of Chapter 4.) One should emphasize, in the spirit of Section 1, that solutions in the *casus irreducibilis* are "approximate" only in the sense that solutions to *any* cubic (or quadratic) may be approximate: see the comments in Section 1 regarding round-off errors resulting from use of trigonometric tables/calculator/polar form.

EXERCISES 3.2

1. By computing Δ (after reducing to suitable g and h), determine directly which of the cases (1), (2), (3) applies to each of the polynomials listed in #1 of Exercises 3.1. (Do *not* use Cardan's formula this time around.)
2. Do the analogue of #2 of Exercises 3.1.
3. Use a discriminant analysis to show that $x^3 - (4/3)x + (16/27)$ has a multiple root. What is its multiplicity? (*Hint*: Section 3 of Chapter 2.)
4. Suppose that $f(x) \in \mathbf{R}[x]$ is a cubic with a root of multiplicity 3. Show that the corresponding $h(x)$, built from $f(x)$ by the technique in Sections 1 and 2, is just x^3. (*Hint*: for some real numbers c and r, $f(x) = c(x - r)^3$.)
5. (a) Prove the assertion made in the text (using either method indicated there) that $-\Delta = (t_1 - t_2)^2(t_1 - t_3)^2(t_2 - t_3)^2$ whenever t_1, t_2, t_3 are the roots of a real cubic $x^3 + px + q$ (and, as usual, $\Delta = 4p^3 + 27q^2$).
   (b) If r_1, r_2, r_3 are the roots of a real cubic polynomial $ax^3 + bx^2 + cx + d$, express $(r_1 - r_2)^2(r_1 - r_3)^2(r_2 - r_3)^2$ in terms of a, b, c, d. (*Hint*: use (a).)

*3. How not to Trisect an Angle

In this section, we shall exploit certain standard facts about vector spaces which many readers will have met in undergraduate courses on linear algebra. The specific information that will be assumed will be described as it becomes necessary; for a detailed exposition, [12, Chapter 2] is more than adequate. By virtue of the "higher" perspective afforded by vector space theory, a satisfactory theoretical explanation for some of the calculations in Section 1 of Chapter 1 will be obtained. Another achievement of the theory will exhibit the naturality of the factor ring construction of Cauchy–Kronecker–Steinitz whereby roots were found in Section 6 of Chapter 2. By far the most important upshot of the theory will be a definitive, but negative, solution to a problem of ancient Greek geometry, from which this section draws its title. A reader whose interest in this section's material is confined to such geometric matters may choose instead to read the recent article [17], which makes no use of vector space theory.

Before getting to the higher algebra, let's notice a crucial dichotomy. If F is a subfield of a field K and if r is an element of K, there are two exhaustive, exclusive possibilities: either r is a root of some nonzero polynomial over F (in which case, r is said to be *algebraic over* F) or r is the root of no such polynomial (in which case, r is called *transcendental over* F). Although K serves merely as a convenient universe for the root testing in the preceding definitions, the role of F is fundamental. Precisely, each element r of any field F is algebraic over F (since r is a root of $x - r \in F[x]$), so that $\frac{2}{7}$ is algebraic over $\mathbf{Q}$, π is algebraic over $\mathbf{R}$, etc. Moreover (and this is what makes our enterprise of root seeking so challenging), certain elements beyond F may also be algebraic over F; for example, the irrational real number $\sqrt{3}$ is algebraic over $\mathbf{Q}$ (thanks to $x^2 - 3$), the nonreal complex number i is algebraic over $\mathbf{R}$ (thanks to $x^2 + 1$), etc. Instances of transcendence are just as easy to produce. If a field F is algebraically closed (in the sense of Section 6 of Chapter 2), for example $F = \mathbf{C}$ (as we shall prove in Section 6), then *each* field element beyond F is transcendental over F. For any field F, the element x (residing in the quotient field of $F[x]$) is transcendental over F. (Why? Recall the definition of equality of polynomials from Section 3 of Chapter 1.) In case $F = \mathbf{Q}$, the upshot of the nonconstructive cardinality analysis in #4 of Exercises 2.6 is that "*most*" *real numbers are transcendental over* $\mathbf{Q}$. The first specific real number which was shown to be transcendental over $\mathbf{Q}$ is the decimal expansion 0.1010010000001 . . . , in which the n-th string of 0's has $n!$ members (construction of Liouville, 1844). Other real numbers known to be transcendental over $\mathbf{Q}$ include π (theorem of Lindemann, 1882) and e (theorem of Hermite, 1873). Verifying transcendence over $\mathbf{Q}$ is typically much harder than verifying irrationality. Indeed, those readers who have seen easy proofs of the irrationality of π and e in advanced calculus courses will note the deeper nature (as, for example, in [13]) of the proofs that π and e are transcendental over $\mathbf{Q}$.

In order to develop a criterion to determine whether a given element is algebraic over a given field, we next introduce a notion of degree for field extensions. If F is a subfield of a field K, it is possible to view K as a vector space over F, by taking vector addition to be the given operation of addition in K and then taking scalar multiplication (of vectors in K by scalars in F) to be induced by the operation of multiplication in K. (The fact that such interpretations for vector addition and scalar multiplication actually satisfy the axioms for a vector space is a special case of K satisfying the domain axioms (i)–(vii) given in Section 1 of Chapter 1.) By a standard theorem, the F-(vector) space K has a *basis* $\{k_i\}$, that is, a subset of K such that each element of K may be expressed in precisely one way as a finite (linear combination) sum of the form $\sum r_i k_i$ arising from (uniquely determined) scalars $r_i \in F$. Concrete examples computed in Section 1 of Chapter 1 include: $\{1, \sqrt{2}\}$ is a **Q**-basis of $\mathbf{Q}(\sqrt{2})$, that is, a basis of the **Q**-space $\mathbf{Q}(\sqrt{2})$; and $\{1, \sqrt{2}, \sqrt{3}, \sqrt{6}\}$ is a **Q**-basis of $\mathbf{Q}(\sqrt{2}, \sqrt{3})$. Returning to the general situation, we let the *degree of K over F*, denoted by $[K : F]$, be the (cardinal) number of elements in any F-basis of K. ($[K : F]$ is well-defined by virtue of the standard theorem that any two bases of a given vector space have the same number of elements. That common number of elements is usually called the *dimension* of the vector space, and so the above definition amounts merely to letting $[K : F]$ denote the dimension of K as an F-space.) By the above remarks, $[\mathbf{Q}(\sqrt{2}) : \mathbf{Q}] = 2$ and $[\mathbf{Q}(\sqrt{2}, \sqrt{3}) : \mathbf{Q}] = 4$. In general, $[F : F] = 1$ since $\{1\}$ is an F-basis of F. Conversely, if $[K : F] = 1$, then $F = K$. (Proof sketch: By a standard theorem, each F-basis of F is a subset of some F-basis of K, since F is a vector subspace of the F-space K. How may $\{1\}$ be thus extended?)

Before developing those properties of degree which will figure in (pun intended) the geometric applications, let's develop the promised criterion for detecting algebraic elements. As usual, let F be a subfield of a field K. Consider an element α of K and, as in Section 1 of Chapter 1, let $F(\alpha)$ denote the subfield of K generated by F and α. Here's the *criterion*: α is algebraic over F if and only if $[F(\alpha) : F]$ is finite.

Its proof is most valuable. First, let $[F(\alpha) : F]$ be the finite (cardinal) number n. By the theorem recalled in the above parenthetical proof sketch, $W = \{r_0 + r_1\alpha + r_2\alpha^2 + \ldots + r_n\alpha^n \in F(\alpha) : r_i \in F \text{ for each } i\}$ is an F-vector subspace of K of dimension at most n. Since $\{1, \alpha, \alpha^2, \ldots, \alpha^n\}$ is therefore not an F-basis of W, there exist r_i, s_i in F such that $r_0 + r_1\alpha + \ldots + r_n\alpha^n = s_0 + s_1\alpha + \ldots + s_n\alpha^n$ and $r_j \neq s_j$ for at least one index j. Accordingly, $f(x) = (r_0 - s_0) + \ldots + (r_n - s_n)x^n \in F[x]$ is nonzero and $f(\alpha) = 0$, whence α is algebraic over F, as asserted. (For vector space veterans: speed up the preceding proof by using the fact that each set of $n + 1$ vectors in an n-dimensional vector space must be linearly independent.)

Conversely, if α is algebraic over F, combine the strategy of unique factorization with the second homomorphism property of evaluation in order to produce, as in the fourth paragraph of Section 6 of Chapter 2, an irreducible monic polynomial $p(x) \in F[x]$ such that $p(\alpha) = 0$. We assert that $p(x)$ coincides with $m(x)$, *the* monic polynomial of minimal degree in $F[x]$ such that $m(\alpha) = 0$. (By the well-ordering property, it is clear that there exists *a* minimal-degree monic polynomial in $F[x]$ having α as a root. Let $m(x)$ denote *any* such polynomial. Uniqueness of $m(x)$ will be a consequence of the following argument.) Indeed, the division algorithm yields $p(x) = q(x)m(x) + R(x)$, for certain $q(x)$ and $R(x)$ in $F[x]$ such that $\deg(R) < \deg(m)$. By the homomorphism properties of evaluation, $R(\alpha) = p(\alpha) - q(\alpha)m\alpha) = 0 - q(\alpha)0 = 0 - 0 = 0$ and so, by the minimality of m's degree, $R(x) = 0$. (A fussy detail: $R(x)$ is not guaranteed to be monic, but if $R(x)$ were nonzero, then its monic associate $T(x)$ would support the contradictory data $T(\alpha) = 0$ and $\deg(T) = \deg(R) < \deg(m)$.) Accordingly, $p(x) = q(x)m(x)$ and, since p was arranged irreducible and m was chosen with positive degree, $p(x)$ and $m(x)$ are associates. As $p(x)$ and $m(x)$ are each monic, $p(x) = m(x)$, as asserted. (It is customary to refer to $m(x)$, that is, $p(x)$, as the *minimum polynomial of* α over F. Don't confuse this unfortunate designation with matrix algebra's notion of minimum polynomial!)

Returning to our sheep, we shall complete the proof (for the case of algebraic α, as above) by showing that $[F(\alpha) : F] = \deg(p)$. Indeed, if $n = \deg(p)$, it is enough to show that $S = \{1, \alpha, \alpha^2, \ldots, \alpha^{n-1}\}$ is an F-basis of $F(\alpha)$. To this end, observe from the minimality of p's degree that S is evidently an F-basis of the vector space $V = \{r_0 + r_1\alpha + r_2\alpha^2 + \ldots + r_{n-1}\alpha^{n-1} \in F(\alpha) : r_i \in F$ for each $i\}$, so that our task may be completed by proving that $V = F(\alpha)$, that is, by proving V is a field. First things first: V is certainly closed under addition, by virtue of the distributive law; and closure under multiplication is a consequence of the fact that α^n, α^{n+1}, $\alpha^{n+2}, \ldots$ are each in V. (Why? Certainly α^n is in V, since the equation $p(\alpha) = 0$ allows you to solve for α^n in terms of 1, α, $\alpha^2, \ldots,$ and α^{n-1}. To see why higher powers of α are also in V, note that α^{n+1} is dispatched by trafficking with $\alpha p(\alpha) = 0$. Iterate.) Finally, and most importantly, verification of property (viii), existence of multiplicative inverses for nonzero elements, proceeds in V by suitably exploiting the irreducibility of $p(x)$. Similar arguments have already been given in the fourth application in Section 3 of Chapter 1; and in the sixth paragraph of Section 6 of Chapter 2. (Adapt those proofs to provide the necessary details here.) Thus V is a field, and we have completed the proof of the degree criterion for algebraic elements.

We pause to mention a particularly salutary benefit from the above proof. Recall that if α is a root of an irreducible n-th-degree polynomial $p(x) \in F[x]$, then $\{1, \alpha, \alpha^2, \ldots, \alpha^{n-1}\}$ is an F-basis of $F(\alpha)$; in other words, each element of $F(\alpha)$ is expressible as $r_0 + r_1\alpha + r_2\alpha^2 + \ldots + r_{n-1}\alpha^{n-1}$ for certain uniquely determined $r_i \in F$. Doesn't this construction ring a bell? In Section 6 of Chapter 2, the factor-ring construction of Cauchy–Kronecker–Steinitz produced a field L such that F is a subfield of L and L contains a root $\bar{x}$ of $p(x)$. According to its construction, $L = \{\bar{g} : g \in F[x]\}$, where $\bar{g} = \{g + qp \in F[x] : q \in F[x]\}$; the viewing of F as a subfield of L was accomplished by identifying elements $r \in F$ with the corresponding elements $\bar{r} \in L$. By using the definitions of addition and multiplication in L (which really just amounted to the homomorphism properties of overhead barring), it is easy to see that L coincides with $F(\bar{x})$. Indeed, each element of L is of the form $s_0 + s_1\bar{x} + s_2\bar{x}^2 + \ldots + s_m\bar{x}^m$ for suitable $s_i \in F$ and nonnegative integer m. More to the point, since an application of the division algorithm expresses any $g(x)$ as $q(x)p(x) +$

$R(x)$ for suitable polynomials $q(x)$ and $R(x)$ such that $\deg(R) < \deg(p) = n$, it follows that $\bar{g} = \bar{q}\bar{p} + \bar{R} = \bar{q}0 + \bar{R} = \bar{R}$. Thus, each element of L is expressible as $r_0 + r_1\bar{x} + r_2\bar{x}^2 + \ldots + r_{n-1}\bar{x}^{n-1}$ for suitable $r_i \in F$; by a degree argument (provide it), the coefficients r_i in such a description are uniquely determined. Hence $\{1, \bar{x}, \bar{x}^2, \ldots, \bar{x}^{n-1}\}$ is an F-basis of L. Accordingly, $F(\alpha)$ and L are *isomorphic* (have essentially the same structure) since $r_0 + r_1\alpha + \ldots + r_{n-1}\alpha^{n-1} \leftrightarrows r_0 + r_1\bar{x} + \ldots + r_{n-1}\bar{x}^{n-1}$ describes a one-to-one correspondence which preserves F and the operations of addition and multiplication. (Explain "preserves" and check the assertions, especially the one concerning multiplication.) Which brings us to the salutary benefit: if β is another root of the irreducible polynomial $p(x)$, then the field extensions $F(\alpha)$ and $F(\beta)$ of F are isomorphic, since each is isomorphic to the field L. By pursuing this result, one may prove the uniqueness (up to isomorphism) of a given polynomial's splitting field over a given field. (We shan't do so here, for reasons of space. Can you? Notice that whenever L is the splitting field of $f(x)$ over F and $\alpha \in L$ is a root of $f(x)$, then $f(x) = (x - \alpha)g(x)$, where L is the splitting field of $g(x)$ over $F(\alpha)$. Then use the well-ordering property.) One liberating aspect of isomorphism deserves to be noted; the fields $\mathbf{Q}(\sqrt[3]{2})$ and $\mathbf{Q}((\text{cis}\, 2\pi/3)\sqrt[3]{2})$ are isomorphic, since $\alpha = \sqrt[3]{2}$ and $\beta = (\text{cis}\, 2\pi/3)\alpha$ are roots of the irreducible polynomial $x^3 - 2 \in \mathbf{Q}[x]$, although only one of $\mathbf{Q}(\alpha)$ and $\mathbf{Q}(\beta)$ is a subfield of $\mathbf{R}$. (Poetically, complex fields can be (isomorphic to) real (fields).) Finally, a cautionary example: $\gamma = 0$ and $\delta = \sqrt{2}$ are each roots of $f(x) = x^3 - 2x$, but the fields $\mathbf{Q}(\gamma)$ and $\mathbf{Q}(\delta)$ are *not* isomorphic. What's wrong? (*Hint*: The proof that the mediating L is a field depended on $p(x)$ being irreducible. Is $f(x)$ irreducible?)

The proof sketch in the third paragraph of this section made use of the standard theorem that a vector subspace's dimension is at most the dimension of the ambient (bigger, given, universe) vector space. Accordingly, given fields $F \subset K \subset L$ with F a subfield of K and K a subfield of L, one has $[K : F] \leq [L : F]$, since K is a subspace of the F-space L. We can do even better: for the above tower of fields, $[L : F] = [L : K] \cdot [K : F]$. (Readers who are familiar with the arithmetic of cardinal numbers will notice that the following proof, although couched for the case of finite degree, carries over to the general situation.) For the proof, it is enough to show that if $\{t_1, t_2, \ldots, t_n\}$ is a K-basis of L and $\{k_1, k_2, \ldots, k_m\}$ is an F-basis of K, then $S = \{k_i t_j : i = 1, 2, \ldots, m;\ j = 1, 2, \ldots, n\}$ is an F-basis of L. Indeed, each element $\alpha \in L$ can be expressed as $\alpha = r_1 t_1 + \ldots + r_n t_n$ for suitable $r_j \in K$, and each r_j is of the form $r_j = s_{j1}k_1 + \ldots s_{jm}k_m$ for suitable $s_{ji} \in F$, whence $\alpha = \sum_{i,j} s_{ji}k_i t_j$. Accordingly, it remains only to establish that different linear combinations formed from S really are different; equivalently, to establish that if $\Sigma b_{ij}k_i t_j = 0$ and each $b_{ij} \in F$, then each $b_{ij} = 0$. Now we may infer from $\sum_j (\sum_i b_{ij}k_i)t_j = 0$ that $\sum_i b_{ij}k_i = 0$ for each j, since 0 can be expressed in only one way as a K-linear combination of $t_1, t_2, \ldots, t_n$. Then, because 0 can be expressed in only one way as an F-linear combination of $k_1, \ldots, k_m$, we next infer from $\sum_i b_{ij}k_i = 0$ that each $b_{ij} = 0$, which completes the proof.

Before indicating how the preceding result on the multiplicativity of degree for towers of fields is pertinent to geometric constructions, we pause to give six applications and to issue a warning. First, we reveal the structure of finite fields. Suppose that F is a finite field, with precisely q elements. What might q be? According to the material in Section 1 of Chapter 1, q might be prime (if, for example, F were $\mathbf{Z}_p$); q might be the composite number 4 (since we constructed the addition and multiplication tables for a field with 4 elements); but, by #10 of Exercises 1.1, q cannot be the composite number 6. The general fact is: q must be p^n, for a suitable prime integer p and positive integer n. (Notice that $4 = 2^2$ corresponds to $p = 2 = n$, but $6 = 2 \cdot 3$ is not a power of a prime.) For the proof, observe that char(F) $\neq 0$ since F is too small (that is, finite) to contain $\mathbf{Z}$ as a subdomain. Hence, char(F) is a prime p, and F contains (a copy of) $\mathbf{Z}_p$ as a subfield. Observe that the degree $n = [F : \mathbf{Z}_p]$ is finite, since a standard result in vector space theory is that any vector space (in this case, F) over any field K (for us, $\mathbf{Z}_p$) whose elements are expressible as not-necessarily-unique K-linear combinations of a finite set S (for us, F) has a (necessarily finite) K-basis which is a subset of S. Taking $\{b_1, \ldots, b_n\}$ as a $\mathbf{Z}_p$-basis of F, we see that each $\alpha \in F$ is uniquely expressible as $r_1b_1 + \ldots + r_nb_n$ for suitable $r_i \in \mathbf{Z}_p$. By making α correspond to the n-tuple $(r_1, \ldots, r_n)$, we achieve a one-to-one correspondence between F and the set of n-tuples with entries drawn from $\mathbf{Z}_p$. As the latter set evidently contains exactly p^n members, the proof that $q = p^n$ is complete.

One can do better still: for each prime integer p and each positive integer n, there is (up to isomorphism) exactly one field with precisely p^n elements. Indeed, we assert that if F is taken to be the splitting field over $\mathbf{Z}_p$ of $f(x) = x^{p^n} - x$, then F has precisely p^n elements. First, note that the roots of $f(x)$ are all simple, by the criterion in Section 3 of Chapter 2, since the g.c.d. $(f(x), f'(x)) = (f(x), p^nx^{p^n-1} - 1) = (f(x), 0 - 1) = 1$. Moreover, these p^n roots form a field K, since $(xy)^{p^n} = x^{p^n} \cdot y^{p^n}$ and $(x \pm y)^{p^n} = x^{p^n} \pm y^{p^n}$ in any field of characteristic p. (For the latter assertion, see the hint given for #8 of Exercises 1.3.) Since K is a subfield of F, the "minimum" subfields of K and F coincide (compare with #6(a) of Exercises 1.1), whence $\mathbf{Z}_p \subset K$. As F is the (splitting) field generated by $\mathbf{Z}_p$ and K, it then follows that $F = K$, whence F does indeed consist of p^n elements, as asserted. The proof that any other field L with precisely p^n elements must be isomorphic to the field F which was just constructed is a consequence of the isomorphic-uniqueness of splitting fields result. One need only check that L is a splitting field of $x^{p^n} - x$ over $\mathbf{Z}_p$. This, in turn, follows immediately from a result of Lagrange which is learned during any first course on group theory (the group in our case being the multiplicative structure on the set of nonzero elements of L). Rather than developing Lagrange's result, we purposely leave a *lacuna* (a classy word for "gap") so that we may return to more germane matters (leaving you with a small library exercise).

For the second of the promised six applications, let's examine the assertion that $S = \{1, \sqrt{2}, \sqrt{3}, \sqrt{6}\}$ is a $\mathbf{Q}$-basis of $\mathbf{Q}(\sqrt{2}, \sqrt{3})$. This example was mentioned earlier in the section and, in a sense, was treated adequately in Section 1 of Chapter 1. However the careful reader may have noticed that one detail remains: namely, showing that "different" $\mathbf{Q}$-linear combinations formed from S really are different. Although such verification of $\mathbf{Q}$-linear independence of S may proceed by reading between the lines in Section 1 of Chapter 1, we choose to argue as follows. Let $F = \mathbf{Q}$, $K = \mathbf{Q}(\sqrt{2})$, and $L = \mathbf{Q}(\sqrt{2}, \sqrt{3})$. By the preceding theoretical discussion, it is enough to prove that $\{1, \sqrt{2}\}$ is an F-basis of K and $\{1, \sqrt{3}\}$ is a K-basis of L. The first of these is immediate since $x^2 - 2$ is irreducible in $\mathbf{Q}[x]$. (Elaborate.) Finally, by virtue of the theorem that any generating set of a vector space must contain a basis, the assertion that $\{1, \sqrt{3}\}$ is a K-basis of L is almost evident. For if the assertion failed, $\{1\}$ or $\{\sqrt{3}\}$ would be a K-basis of L, so that $[L : K] = 1$, $K = L$, and $\sqrt{3} \in \mathbf{Q}(\sqrt{2})$, contradicting an explicit computation which was made in Section 1 of Chapter 1.

Another consequence of the computation which showed $\sqrt{3} \notin \mathbf{Q}(\sqrt{2})$ is the irrationality of $\sqrt{2} + \sqrt{3}$, more generally of expressions $a\sqrt{2} + b\sqrt{3}$ arising from nonzero rational numbers a and b. The reader may be interested in reading a far-reaching generalization of this fact in the recent article [18], which presupposes an acquaintance with Galois theory, as summarized in Section 5 below. A rather different proof of the irrationality of $\sqrt{2} + \sqrt{3}$ will issue from the rational root test, to be given in Section 1 of Chapter 4.

Our third application is a fast proof that $\mathbf{Q}(\sqrt{2}) \cap \mathbf{Q}(\sqrt{3}) = \mathbf{Q}$. (A direct proof was fashioned in Section 1 of Chapter 1.) Set $F = \mathbf{Q}$, $K = \mathbf{Q}(\sqrt{2}) \cap \mathbf{Q}(\sqrt{3})$, and $L = \mathbf{Q}(\sqrt{2})$, and observe that $2 = [L : F] = [L : K][K : F]$. If the assertion failed, then $[K : F] \neq 1$, whence $[K : F] = 2$, $[L : K] = 1$, $K = L$, and $\sqrt{2} \in \mathbf{Q}(\sqrt{3})$, the desired contradiction.

The fourth of our applications is, in actuality, several results arising from a common technique. To illustrate, we recover the motivating result in Section 6 of Chapter 2: if $f(x)$ is an irreducible cubic polynomial over $\mathbf{Q}$, then $f(x)$ remains irreducible over $\mathbf{Q}(\sqrt{n})$, where n is any squarefree integer exceeding 1. Indeed, thanks to the fundamental fact about degrees and the factor theorem, we see that reducibility of $f(x)$ over $\mathbf{Q}(\sqrt{n})$ would produce a root, α, of $f(x)$ in $\mathbf{Q}(\sqrt{n})$, so that $2 = [\mathbf{Q}(\sqrt{n}) : \mathbf{Q}] = [\mathbf{Q}(\sqrt{n}) :$

$\mathbf{Q}(\alpha)][\mathbf{Q}(\alpha):\mathbf{Q}]$. However, the above theory would then give $[\mathbf{Q}(\alpha):\mathbf{Q}] = \deg(f) = 3$, since $f(x)$ is an associate of the minimum polynomial of α over $\mathbf{Q}$, confronting us with the absurdity that 2 would be an integral multiple of 3. Another illustration of the technique (which would be tedious to obtain by more direct means) is: if $g(x)$ is an irreducible quadratic polynomial over $\mathbf{Q}$ (for instance, $g(x) = 2x^2 - 7x + 1$) and L is a field of characteristic 0 such that $[L:\mathbf{Q}] = 3$ (for instance, $L = \mathbf{Q}(\beta)$, where β is a root of $x^3 + 3x + 1$), then $g(x)$ remains irreducible over L. (Proof: As above, reducibility would lead to $\gamma \in L$ for some root γ of $g(x)$, whence $3 = [L:\mathbf{Q}(\gamma)][\mathbf{Q}(\gamma):\mathbf{Q}]$ and $[\mathbf{Q}(\gamma):\mathbf{Q}] = \deg(g) = 2$, producing the absurdity that 3 would be an integral multiple of 2.)

Our fifth application asserts that if $\alpha_1, \ldots, \alpha_n$ are each algebraic elements over a field F, then $[F(\alpha_1, \ldots, \alpha_n):F]$ is finite and, in fact, is at most $d_1 d_2 \ldots d_n$, where $d_i = [F(\alpha_i):F]$. Indeed, one computes $d = [F(\alpha_1, \ldots, \alpha_n):F]$ via the tower of fields

$$F \subset F(\alpha_1) \subset F(\alpha_1, \alpha_2) = (F(\alpha_1))(\alpha_2) \subset \ldots \subset F(\alpha_1, \ldots, \alpha_n)$$

as $d = [F(\alpha_1):F][F(\alpha_1, \alpha_2):F(\alpha_1)] \cdot \ldots \cdot [F(\alpha_1, \ldots, \alpha_n): F(\alpha_1, \ldots, \alpha_{n-1})] = e_1 e_2 \ldots e_n$, where e_i is the degree of the minimum polynomial of α_i over $F(\alpha_1, \ldots, \alpha_{i-1})$. By the very meaning of *minimum* polynomial, we have $e_i \leq \deg(f_i)$, where f_i is the minimum polynomial of α_i over F. Since $\deg(f_i) = d_i$, it follows that $e_i \leq d_i$ and $d \leq d_1 d_2 \ldots d_n$, as asserted.

One consequence of the preceding proof is particularly noteworthy. If L is "the" splitting field, over a field F, of an n-th-degree polynomial $f(x) \in F[x]$, then $[L:F] \leq n!$. Indeed, let $\alpha_1, \alpha_2, \ldots, \alpha_n$ be all the roots of $f(x)$ (in L) and observe, with the above notation, that $[L:F] = e_1 e_2 \ldots e_n$. As α_1 is a root of $f(x)$, we have $e_1 \leq \deg(f) = n$; as α_2 is a root of $f(x)/(x - \alpha_1)$ (which is a polynomial over $F(\alpha_1)$, by virtue of the factor theorem), we have $e_2 \leq \deg(f(x)/(x - \alpha_1)) = n - 1$; and, similarly (provide details), $e_3 \leq n - 2, \ldots, e_{n-1} \leq 2$ and $e_n = 1$. Thus, $[L:F] \leq n(n - 1) \cdot (n - 2) \cdot \ldots \cdot 2 \cdot 1 = n!$, as claimed. (The reader who is familiar with group theory, or whose interest was spurred by #4 of Exercises 2.5, will note that $n!$ is the size of the symmetric group of permutations of n objects. Relating degrees of certain splitting field extensions to sizes of related groups is part of the fundamental theorem of Galois theory: see Section 5 for details.)

The sixth application asserts that if F is a subfield of a field K such that $[K:F]$ is finite, then each element of K is algebraic over F. (Here's its easy proof. We show that $\alpha \in K$ is algebraic over F by checking the above criterion for algebraic elements, namely that $[F(\alpha):F]$ is finite, which is evident from the towerful computation (ouch) that $[K:F] = [K:F(\alpha)] \cdot [F(\alpha):F]$.) By combining the fifth and sixth applications, we see that if u and v are algebraic over a field F, then so are $u + v$, $u - v$, and uv. (Proof?

Consider $[F(u, v) : F]$.) Thus, one recovers that $\sqrt{2} + \sqrt{3}$ is algebraic over $\mathbf{Q}$ (no great accomplishment since $\sqrt{2} + \sqrt{3}$ is a root of $x^4 - 10x^2 + 1$) and, with just as little effort, $\sqrt[3]{7} + \sqrt[5]{9}$ is algebraic over $\mathbf{Q}$ (although, it is harder—isn't it?—to produce a nonzero rational polynomial with $\sqrt[3]{7} + \sqrt[5]{9}$ as a root).

Finally, the promised warning: the converse of the sixth application is false. Specifically, the datum that F is a subfield of a field K such that each element of K is algebraic over F does not guarantee that $[K : F]$ is finite. Indeed, consider $F = \mathbf{Q}$ and K "the" algebraic closure of F (in the sense described in Section 6 of Chapter 2). By the comments of the preceding paragraph, each element of K is indeed algebraic over F. Accordingly, it remains only to show, for each positive integer n, that $[K : F] \geq n$. (How will this prove that $[K : F]$ is infinite?) As with much of this section, the underlying reason is number-theoretic. Let p_n be the n-th positive prime integer, listed in increasing order (so that, for instance, $p_1 = 2$ and $p_8 = 19$), and set $\alpha_n = \sqrt[n]{p_n}$. Now α_n is the positive real root of the polynomial $f_n(x) = x^n - p_n \in \mathbf{Q}[x]$ and (by virtue of the strong version of Eisenstein's criterion, which will be given in Section 2 of Chapter 4) $f_n(x)$ is irreducible over $\mathbf{Q}$, thus $f_n(x)$ is the minimum polynomial of α_n over $\mathbf{Q}$, and hence $[K:F] \geqslant [\mathbf{Q}(\alpha_n):\mathbf{Q}] = \deg(f_n) = n$, as desired.

Enough algebraic strutting (for a while): we come next to the *geometric* applications of the (*algebraic* result on) multiplicativity of degrees in a tower of fields. The geometric problem which is to be addressed is: what may one construct in the Euclidean plane with the aid of ruler and compass, starting with data consisting of the perpendicular x- and y-axes and the length 1? (The ruler is unmarked and is not used for the purpose of measuring. It is to function solely as a straightedge, in order to draw the line which connects any two known points.) Certainly, all the elements of $\mathbf{Q}$ are obtainable under these conditions; for instance, the fact that similar triangles have corresponding sides proportional shows that the x-coordinate of P in Figure 4 is n/m. Additional points are obtained as points of intersection of either a pair of lines or a pair of circles or a line with a circle. (Thanks to our ruler, a line is available to us once two of its points have been obtained; similarly, a circle may be determined by either

FIGURE 4

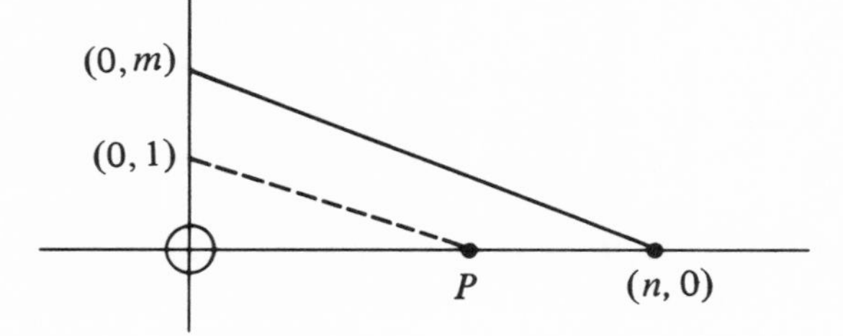

its center and radius, or its center and one of its points, or three of its points.) Observe that solving for the coordinates of such points of intersection involves only linear and/or quadratic equations. Thus, by virtue of the quadratic formula, whenever a *point* P is *constructible* (in the sense that P is the end product of a finite sequence of such intersection points), then the coordinates of P lie in a subfield F_n of $\mathbf{R}$, where $\mathbf{Q} = F_0 \subset F_1 \subset \ldots \subset F_n$, $F_i = F_{i-1}(\alpha_i)$ and $(\alpha_i)^2 \in F_{i-1}$. Focusing attention on coordinates is justified since (if you remember how to erect perpendiculars using ruler and compass) it is easy to see that a point (a, b) is constructible if and only if both $(a, 0)$ and $(0, b)$ are constructible. Accordingly, it is sensible to call a *real number* β *constructible* in case β is a coordinate of some constructible point. By use of diagrams such as Figure 4, it is straightforward to prove that the *set* $\mathcal{C}$ *of constructible numbers* is a subfield of $\mathbf{R}$. (Check.) Now, our geometric concerns depend on knowing how much larger $\mathcal{C}$ is than $\mathbf{Q}$. Certainly, $\mathcal{C}$ is somewhat larger than $\mathbf{Q}$ since, thanks to the theorem of Pythagoras, the irrational number $\sqrt{2}$ is in $\mathcal{C}$. However, in the sense of #4 of Exercises 2.6, $\mathcal{C}$ is much smaller than $\mathbf{R}$, as $\mathcal{C}$ is denumerable. This last assertion follows from the *main result on constructibility*: if the real number β is constructible (that is, if $\beta \in \mathcal{C}$), then β is algebraic over $\mathbf{Q}$ and $[\mathbf{Q}(\beta) : \mathbf{Q}] = 2^m$ for some nonnegative integer m (depending of β).

The proof of the result just stated follows from multiplicativity of degrees in a tower $\mathbf{Q} = F_0 \subset F_1 \subset \ldots \subset F_n$ such that $F_i = F_{i-1}(\alpha_i)$, $(\alpha_i)^2 \in F_{i-1}$, and $\beta \in F_n$. Indeed, the minimum polynomial of α_i over F_{i-1} has degree 1 or 2, whence $[F_i : F_{i-1}]$ is either 1 or 2, so that $[F_n : \mathbf{Q}] = [F_n : F_{n-1}] \ldots [F_1 : F_0]$ is a product of k 2s and $n - k$ 1s, for some $0 \leq k \leq n$. Since $[F_n : \mathbf{Q}(\beta)][\mathbf{Q}(\beta) : \mathbf{Q}] = [F_n : \mathbf{Q}] = 2^k$, the (uniqueness part of the) fundamental theorem of arithmetic implies $[\mathbf{Q}(\beta) : \mathbf{Q}] = 2^m$ for some $0 \leq m \leq k$, as desired.

As will be explained in Section 5, the main result on constructibility has a sharp converse. Without dwelling on that now, let's (finally!) get to this section's title. Which angles can be trisected by use of ruler and compass? Certainly 270° ($= 3\pi/2$ radians) can be: just construct a right angle. Similarly, 180° can be so trisected: take any interior angle of any equilateral triangle. However, 60° *cannot* be trisected by ruler and compass. Indeed, constructing an angle with radian measure θ (where $0 \leq \theta < 2\pi$) is equivalent to showing that both $\cos(\theta)$ and $\sin(\theta)$ are in $\mathcal{C}$, by our earlier observations about Cartesian coordinates. Thus, we need only show that $\beta = \cos(\pi/9)$ is not in $\mathcal{C}$. Now, recalling from Section 2 of Chapter 1 that $\cos(3\theta) = -3\cos(\theta) + 4\cos^3(\theta)$, we find that β is a root of $f(x) = 8x^3 - 6x - 1 \in \mathbf{Q}[x]$. Note that $f(x)$ has no rational roots (by the rational root test to be developed in Section 1 of Chapter 4) and so, by the usual combination of the factor theorem and the fundamental fact about degrees, we see that $f(x)$ is irreducible in $\mathbf{Q}[x]$. Therefore, $f(x)$ is the minimum polynomial of β over $\mathbf{Q}$, whence $[\mathbf{Q}(\beta) : \mathbf{Q}] = \deg(f) = 3$ which, by the fundamental theorem of arithmetic, is not an integral power of 2. Accordingly, β is not constructible, as claimed.

Exercise 9 below provides other geometric applications of the main result on constructibility. One deserving of mention (but whose treatment would take us too far afield) asserts that if $n \geq 3$, then the regular n-gon may be constructed by ruler and compass only if (and, by the promised sharp converse, if) n is the product of an integral power of 2 with finitely many distinct Fermat primes. (Recall that a prime p is called a *Fermat prime* if $p = 2^m + 1$ for some integer m, in which case m is necessarily an integral power of 2.) For instance, the values of n, $3 \leq n \leq 30$, such that the regular n-gon is constructible by ruler and compass are $n = 3$ (you knew that), 4 (you knew that, too), 5, 6, 8, 10, 12, 16, 17, 20, 24, and 30. (A discovery exercise for fans of $\phi(n)$, Euler's totient function: characterize the above twelve integers amongst the twenty-eight integers from 3 to 30, by using $\phi(n)$. Conjecture a more general result. Test it by trusting something over 30.)

EXERCISES 3.3

1. (a) Give an example of a complex number which is algebraic over **R** and transcendental over **Q**.
(b) A rectangular planar region, whose dimensions are $\sqrt{2}$ meters by π meters, is to be carpeted. Is it possible to accomplish this with nonoverlapping square fragments of carpet each of the same size? (*Hint*: If s is the length of such a fragment's side, produce positive integers n and m such that $ns = \sqrt{2}$ and $ms = \pi$. Is such an s transcendental or algebraic over **Q**?)
2. (a) Prove that $[\mathbf{Q}(\sqrt{2} + \sqrt{3}):\mathbf{Q}] = 4$.
(b) Let $\omega = (-1 + \sqrt{3}\, i)/2$, as usual. Compute $[\mathbf{Q}(\alpha):\mathbf{Q}]$ for each of the following values of α: $\sqrt[3]{2}$, $\omega\sqrt[3]{2}$, $\omega^2\sqrt[3]{2}$, $\sqrt{3}$, and $\sqrt{5}$.
3. Let F be a subfield of a field K such that each element of K is algebraic over F. Let S be a subdomain of K which contains F as a subsystem. Prove that S is a field. (*Hint*: In this instance, the relevant minimum polynomial has a nonzero constant term.) Give an example showing that the assertion fails without the "algebraic" hypothesis.
4. Let F be a subfield of a field K such that $[K:F] = 2$. Prove that K is the splitting field over F of some polynomial in $F[x]$. Give an example showing that the assertion fails if, instead, $[K:F] = 3$. (*Hint*: ω is not a real number.)
5. (a) Consider a tower of fields $F \subset K \subset L$. If $[L:F]$ is a prime integer, prove that either $K = F$ or $K = L$. (*Hint*: consider $[K:F]$ and $[L:K]$.)
(b) Let F be a subfield of a field containing an element α such that $[F(\alpha):F]$ is an odd integer. Prove that $F(\alpha^2) = F(\alpha)$. (*Hint*: consider $[F(\alpha):F(\alpha^2)]$.) Give such an example in which $F(\alpha^3) \neq F(\alpha)$. (*Hint*: consider ω.)
(c) Let $f(x)$ and $g(x)$ be irreducible polynomials over a field F, such that $m = \deg(f)$ is relatively prime to $n = \deg(g)$. (This means that 1 is the in-

teger g.c.d. of m and n.) Let $K = F(\alpha)$ and $L = F(\beta)$, where α and β are roots of $f(x)$ and $g(x)$, respectively, in an algebraic closure of F. Prove that $f(x)$ remains irreducible over L and $g(x)$ remains irreducible over K. (*Hint*: compute $[F(\alpha, \beta):F]$ using two different towers. Draw a number-theoretic conclusion, peeking, if necessary, at the first lemma in Section 1 of Chapter 4.)

6. (a) Prove that there is no field with precisely fifteen elements.
   (b) Let F be a field with four elements, constructed as in the text; that is, F is a splitting field over $\mathbf{Z}_2$ of $x^4 - x$. Explicitly exhibit an *isomorphism* (that is, a one-to-one correspondence preserving F and the operations of addition and multiplication) between F and the field with four elements which was given in Section 1 of Chapter 1.
   (c) Let p be a prime positive integer, m and n positive integers, and F and K finite fields with p^m and p^n elements, respectively. Prove that if F is a subfield of K, then n is an integral multiple of m. (*Hint*: consider the tower $\mathbf{Z}_p \subset F \subset K$.)
   (d) By analyzing the text's construction of finite fields, prove that each element, α, of the field $\mathbf{Z}_p$ satisfies $\alpha^p = \alpha$. (Number theory buffs: this is a fancy proof of the *little Fermat theorem*: $a^p \equiv a \pmod{p}$.) What can you conclude if $\alpha \neq 0$? Generalize to the case of fields with p^n elements.
   (e) Prove *Wilson's theorem*: for each positive prime integer p, $(p-1)! + 1$ is an integral multiple of p. (*Hint*: By (d), $\overline{1}, \overline{2}, \ldots, \overline{p-1}$ are all the roots of $x^{p-1} - 1 \in \mathbf{Z}_p[x]$. Exploit the blockbuster.)
   (f) If F is an algebraic closure of a finite field $\mathbf{Z}_p$, prove that $[F:\mathbf{Z}_p]$ is infinite.
7. Let F be a subfield of a field K. Prove that $L = \{\alpha \in K: \alpha \text{ is algebraic over } F\}$ is a subfield of K. (This subfield L is called the *algebraic closure of F in K*.) What can you conclude if $L = F$? What if $L = K$? Give an example in which L coincides with neither F nor K.
8. Consider a tower of fields $F \subset K \subset L$ such that each element of K is algebraic over F and each element of L is algebraic over K. Prove that each element of L is algebraic over F. (*Hint*: use the degree criterion for algebraic elements.)
9. (a) *Squaring the circle*: Prove that it is impossible to use ruler and compass (in the sense of the text) to construct (the real number which is the length of one side of) a square whose area is the same as the area of a circle whose radius is 1. (*Hint*: $\pi = (\sqrt{\pi})^2$.)
   (b) *Duplicating the cube*: Prove that it is impossible to use ruler and compass to construct a cube whose volume is twice the volume of a given cube with sides each of length 1. (*Hint*: $x^3 - 2$ is irreducible over $\mathbf{Q}$: see Sections 1 or 2 of Chapter 4.)
   (c) Prove that $(1080/17)^\circ$ can be trisected by ruler and compass. (*Hint*: what angle is produced by constructing a regular 17-gon?)
   (d) Prove that the regular pentagon (that is, regular 5-gon) can be constructed by ruler and compass, by verifying that 5 is a Fermat prime.

(e) Accomplish the analogue of (d) for regular 3-gons by showing that the minimum polynomial of $\cos(2\pi/3)$ over $\mathbf{Q}$ is of degree an integral power of 2. Can you treat (d) similarly, by producing the minimum polynomial of $\cos(2\pi/5)$ over $\mathbf{Q}$?

4. Solution of the Quartic: Methods of Ferrari and Descartes

In this brief section, we indicate how to obtain formulas for all the roots of a quartic polynomial $f(x) = ax^4 + bx^3 + cx^2 + dx + e$. We shall suppose that all the polynomials being considered are taken over $\mathbf{C}$, but as in #5 of Exercises 3.1, the analysis will hold for coefficients drawn from any field of characteristic unequal to 2 or 3. Indeed, Section 1 will be of more than passing relevance, for just as the quadratic formula was needed in Section 1's heuristic derivation of Cardan's formula for the cubic, the solution of a suitably devised cubic will be the key to developing a quartic formula.

There are, in fact, two standard ways of suitably devising cubics. The first one that we shall discuss was discovered in 1637 by Descartes, the French philosopher-mathematician of "*Cogito, egro sum*" fame who was a principal architect of analytic geometry (hence, the term "Cartesian coordinates") and whose exposition of a mind-body dualism has provoked much theological and secular thought. Descartes's procedure begins with a reduction analogous to that in Section 1. Specifically, using #3 of Exercises 3.1, observe that $g(x) = f(x - (b/4a))$ is a quartic with no term in x^3, and so attention is diverted to the monic associate, $h(x)$, of $g(x)$. For example, if $f(x) = 2x^4 - 8x^3 + \sqrt{2}\, x^2 - ix + 7$, (the reader should verify that) $g(x) = f(x+1) = 2x^4 + (\sqrt{2} - 12)x^2 + (-16 + 2\sqrt{2} - i)x + (\sqrt{2} - i)$ and $h(x) = \frac{1}{2} g(x) = x^4 + ((1/\sqrt{2}) - 6)x^2 + (-8 + \sqrt{2} - (i/2))x + ((1/\sqrt{2}) - (i/2))$. As in Section 1 of Chapter 2, the roots r_1, r_2, r_3, r_4 of $f(x)$ are given by $r_j = t_j - (b/4a)$, where t_1, t_2, t_3, t_4 are the roots of $h(x)$. It will be convenient to write $h(x) = x^4 + px^2 + qx + r$ in the following treatment.

One degenerate case needs individual care. If $q = 0$, then the t_j may be found after two applications of the quadratic formula (solve first for the $(t_j)^2$), precisely as in #7(e) of Exercises 1.2. Accordingly, we may henceforth assume that $q \neq 0$.

Here is the critical step: by the son of blockbuster (Section 1 of Chapter 2) or the fundamental theorem of algebra (whose proof, you'll recall, awaits in Section 6), $h(x)$ *is expressible as the product of two monic quadratics.* Finding the roots of these quadratics is a simple matter thanks to the quadratic formula, and *those* roots are precisely the t_j that are being sought. The catch is that we need to find the quadratics' coefficients (otherwise, to *what* would you apply the quadratic formula?). Specifically, we are to solve for m, n, v, and w, given that

$$x^4 + px^2 + qx + r = (x^2 + mx + n)(x^2 + vx + w).$$

Equating corresponding coefficients leads to the set of equations

$$\begin{aligned} v + m &= 0, \\ w + mv + n &= p, \\ mw + nv &= q, \\ nw &= r. \end{aligned}$$

As $v = -m$ from the first of these equations, the equations simplify to

$$\begin{aligned} w - m^2 + n &= p, \\ m(w - n) &= q, \\ nw &= r. \end{aligned}$$

Substituting $w = m^2 + p - n$, we reduce to the equations

$$m(m^2 + p - 2n) = q \quad \text{and} \quad n(m^2 + p - n) = r.$$

Since we are in the case $q \neq 0$, m is also nonzero, whence $n = (m^3 + pm - q)/2m$. Substituting this value of n into the expression for r and cross-multiplying by $4m^2$ shows that $\beta = m^2$ *is a root of the cubic polynomial*

$$x^3 + 2px^2 + (p^2 - 4r)x - q^2.$$

Any root of this polynomial (and Section 1 supplies *three* of them) is a satisfactory value of β; *either* square root of the selected β may be taken as m; and then computing $n = (m^3 + pm - q)/2m$, $w = m^2 + p - n$, and $v = -m$ completes the (time-consuming but uneventful) process.

An example is in order. To find the roots of $f(x) = 2x^4 - 4x^3 - 10x^2 + 20x - 6$, attend to $g(x) = f(x + \frac{1}{2}) = 2x^4 - 13x^2 + 8x + \frac{9}{8}$ and $h(x) = x^4 - \frac{13}{2}x^2 + 4x + \frac{9}{16}$, with $p = -\frac{13}{2}$, $q = 4$, and $r = \frac{9}{16}$. Solving for m, n, v, and w as above, we have $\beta = m^2$ is a root of $x^3 - 13x^2 + 40x - 16$. Sooner or later, the reader will discover that 4 is a root of this cubic. (Sooner, using the rational root test to be developed in Section 1 of Chapter 4; later, using the techniques of this chapter's Section 1.) Taking $m^2 = \beta = 4$ and choosing $m = 2$ (what would happen if we were to choose $m = -2$?), we then find $n = (2^3 + (-13/2)2 - 4)/4 = -9/4$, $w = 2^2 + (-13/2) + 9/4 = -1/4$, and $v = -2$. This achieves the factorization

$$h(x) = \left(x^2 + 2x - \frac{9}{4}\right)\left(x^2 - 2x - \frac{1}{4}\right);$$

consequently, two applications of the quadratic formula reveal that the t_j are $-1 + (\sqrt{13}/2)$, $-1 - (\sqrt{13}/2)$, $1 + (\sqrt{5}/2)$, and $1 - (\sqrt{5}/2)$. Thus, we finally are able to identify the roots $r_j = t_j + \frac{1}{2}$ of $f(x)$ as $(-1 + \sqrt{13})/2$, $(-1 - \sqrt{13})/2$, $(3 + \sqrt{5})/2$, and $(3 - \sqrt{5})/2$.

The reader will recall from Section 1 that Ferrari's method for solving the quartic appeared in 1545 in Cardan's *Ars Magna.* Rather than describe *this* great art for a general quartic, let's merely illustrate how one would approach the preceding example with Ferrari's technique. The first step is to pass to the monic associate, with the same roots as the original polynomial. (Note the gain in this method: there's no need to remove the coefficient of x^3.) In our example, attention is then focused on $x^4 - 2x^3 - 5x^2 + 10x - 3$. The *critical step for Ferrari's method* is the determination of complex numbers a and b so that $(x^4 - 2x^3 - 5x^2 + 10x - 3) + (ax + b)^2$ is the square of some monic quadratic. Our task is then to solve for a, b, c, and d such that

$$x^4 - 2x^3 + (-5 + a^2)x^2 + (10 + 2ab)x + (-3 + b^2) = (x^2 + cx + d)^2.$$

Equating corresponding coefficients leads to a set of equations which yields, after several substitutions of the type considered for Descartes's method, the following data: $c = -1$, $a^2 = 6 + 2d$, and $(2d + 10)^2 = 4a^2b^2 = 4(6 + 2d)\cdot(d^2 + 3)$; in particular, d is a root of the cubic polynomial $2x^3 + 5x^2 - 4x - 7$ (check) which one traditionally calls the *resolvent cubic* of the given quartic. Sooner or later, the reader will discover that -1 is a root of this resolvent cubic. Taking $d = -1$ (what would happen if we had seized a different root of the resolvent?), we solve for all the missing coefficients thus: $c = -1$, $a^2 = 6 + 2d = 4$, $a = 2$ (what if we were to take $a = -2$?), and $b = (2cd - 10)/2a = -2$. Accordingly, the critical step amounts to

$$x^4 - 2x^3 - 5x^2 + 10x - 3 + (2x - 2)^2 = (x^2 - x - 1)^2,$$

whence *the given polynomial is a difference of squares, factoring as*

$$x^4 - 2x^3 - 5x^2 + 10x - 3 = (x^2 - 3x + 1)(x^2 + x - 3).$$

Applying the quadratic formula to each of the factors on the right-hand side, we recover the roots of the given quartic: $(3 + \sqrt{5})/2$, $(3 - \sqrt{5})/2$, $(-1 + \sqrt{13})/2$, and $(-1 - \sqrt{13})/2$. Notice that Descartes's method gave the same roots (of course!), but made use of a different auxiliary cubic. We leave to the reader to determine his/her preference regarding Descartes's method *vis-à-vis* that of Ferrari. Our general feeling is that the former's advantage lies in its treatment paralleling that of the cubic case, while the latter's advantage is (*ars magna*, indeed!) that it tends to be quicker.

EXERCISES 3.4

1. Using the method of Descartes (after suitably dealing with the terms in x^3), identify the cubic polynomials to which one is led while analyzing the following three quartics: $x^4 - 4x^3 + 18x^2 - 36x + 81$, $x^4 - (1+i)x^3 + (1+i)x^2 - (1+i)x + i$, $2x^4 + 6x^3 + 22x^2 + 54x + 36$.
2. Repeat #1 using the method of Ferrari.
3. Do the analogue of #2 of Exercises 3.1.

*4. In the spirit of Section 2, devise a discriminant theory for quartics. Show, in particular, that a quartic's discriminant is 64 times the discriminant of the resolvent cubic.

*5. On Quintics and Formulas: a Mathematical Success (Hi)story

Scientific progress is possible only by a careful balance between perseverance and skepticism. The former should not be underestimated: for example, the cubic and quartic formulas described above in Sections 1 and 4 were only obtained several centuries after mathematicians had become aware of the quadratic formula. Following more than two hundred additional years of searching for a similar quintic formula, a healthy skepticism won the day and, in 1799, an Italian physician-mathematician, Ruffini (not to be confused with Cardan!), proved that the additional searching had been in vain. A careful statement of Ruffini's result (that the general polynomial of degree at least five cannot be solved by radicals) will be given below. Various aspects of Ruffini's proof lacked total rigor, and it remained for the Norwegian mathematician, Abel (after whom commutative groups are dubbed "Abelian"), in 1824 to provide a flawless demonstration of Ruffini's result. Within 10 years, "modern algebra" came of age, through the efforts of the French mathematician Galois whose "Galois theory" featured a penetrating interaction between groups and fields. Galois theory's applications to the theory of equations are many and varied—we shall examine some of these below—including a most satisfactory proof of the Ruffini–Abel result. First, however, we shall indulge in a little history concerning the lives of Abel and Galois.

While still a schoolboy, Abel was reading the works of the masters, including Euler (whose expertise at computation ranks among the greatest in history), Lagrange (who interpolated his way into our hearts in Section 2 of Chapter 2), and Newton (whose work is herein met in Sections 4 and 7 of Chapter 2 and Section 4 of Chapter 6). Indeed, Abel's first attempt at rigorizing Ruffini's argument dates from that period. When Abel finally succeeded in proving Ruffini's result, he found, as is often the case, clay feet in (some of) the ivory towers: the incomparable Gauss (about whom, more in the next section) chose not to spend time reading the manuscript of the young outsider, Abel, and Abel eventually had to pay by himself for the initial printing and distribution of his work. By that time, Abel's father, a minister, had died, reportedly of alcoholic excesses, and the impoverished Abel decided to travel throughout Europe in order to foster professional contacts and popularize his own work. Fortunately, in Berlin, Abel was befriended in 1826 by Crelle, who published some of Abel's papers in the newly-founded *Journal für die reine und angewandte Mathematik* ("Crelle's Journal"). Besides his work on algebra (including Abelian groups), Abel did important foundational work at that time on the subject of elliptic functions (including a nonalgebraic solution of quintics) in competition with Jacobi. Regrettably, the young artist's fortunes then went into rapid decline. Weakened by tuberculosis, he contracted an enervating cold during a Christmas sleighing trip with his fiancée, and died of complications shortly thereafter, early in 1829 at the age of twenty-six. Ironically, two days after Abel's death, Crelle managed to secure for Abel what had long been his goal, a professorship at the University of Berlin. In 1830, Abel was posthumously honored as a corecipient, with Jacobi, of the *Grand Prix*.

Galois's life story is, if possible, even more poignant and tragic than that of Abel. Galois received virtually no professional recognition during his lifetime, and lived to be only twenty, dying in 1832 of wounds suffered in a pistol duel. As a youngster, Galois was impatient with the required formalities of traditional schoolwork, and instead read the works of the masters, including Lagrange and (by then) Abel. His distaste for the discipline of his schools was matched by an equally unorthodox, abbrviated mode of expression, with the result that Galois twice failed the entrance examination of the *Ecole Polytechnique.* Even moreso than Abel, Galois was the unappreciated outsider: a manuscript which Galois wrote at age seventeen was misplaced by the great Cauchy; two years later, another manuscript was lost following the death of a secretary of the *Académie des Sciences*; and a third manuscript was rejected for publication upon the advice of the referee, Poisson. Having been raised in a liberal household (his father was a mayor), Galois felt these slights keenly, and became involved in the political uprisings in Paris in 1830. For his idealistic championing of a government based on merit, ruled by the untraditional and the young, Galois paid the price: six months in jail. Subsequently, since his political activities did not abate, Galois was inveigled into a romantic entanglement, leading to his duel with (what some historians believe to be) an *agent provocateur.* Fortunately, on the eve of that fatal duel, Galois wrote a letter to a friend in which he recorded results on elliptic integrals (shades of Abel) and the Galois theory of equations (about which, more shortly). The letter was published posthumously in 1832, and Galois's fame increased with the passing years. Galois has been honored in the notation for finite fields (whose structure he revealed, and which we studied in Section 3): the field with p^n elements is denoted $GF(p^n)$, the GF standing for Galois field.

It is remarkable that Abel, Galois, Schreier (see "refinement" later in this section), and Eisenstein (see Sections 1 and 2 of Chapter 4) each died before reaching the age of thirty. As the authors can attest, such is not the fate of all mathematicians!

We shall next describe the fundamental theorem in Galois's approach to the theory of equations. Let $f(x)$ be an n-th-degree polynomial over a field F, and let L be "the" splitting field of f over F; that is, $L = F(r_1, \ldots, r_n)$, where $r_1, \ldots, r_n$ are all the roots of $f(x)$ (in L, listed according to their multiplicities). An *automorphism* (*of* L *over* F) is, in the terminology of Section 3, an isomorphism of L with itself, that is, a one-to-one correspondence $L \to L$ which preserves F and the operations of addition and multiplication. Evidently, any composite of automorphisms is itself an automorphism, and so the set of automorphisms forms a group, called the *Galois group of* L *over* F (loosely speaking, the "Galois group of $f(x)$"), and denoted by $\mathrm{Gal}(L/F)$. Of course, $\mathrm{Gal}(L/F)$ is finite, since any automorphism induces a permutation of $\{r_1, \ldots, r_n\}$ (why?), and hence of *order* (that is, cardinality) at most $n!$. (Indeed, by the theorem of Lagrange alluded to in Section 3—have you done the small library exercise yet?—the order of $\mathrm{Gal}(L/F)$ is a divisor of $n!$.) To be able to assert anything further, we assume also that each irreducible polynomial in $F[x]$ which divides $f(x)$ has no multiple roots. (This assumption holds automatically if $\mathrm{char}(F) = 0$, which is the case of classical interest. Indeed, one need only show that an irreducible $p(x)$ satisfies $(p(x), p'(x)) = 1$, according to the criterion in Section 3 of Chapter 2. This obtains by virtue of condition (**) in Section 3 of Chapter 1, since the assumption that $\mathrm{char}(F) = 0$ guarantees that $\deg(p'(x)) = \deg(p(x)) - 1$ and $p(x) \nmid p'(x)$.) In this case, we say nowadays that L/F is a *Galois field extension* and that $f(x)$ is a *Galois polynomial* (*over* F). For this context, the *fundamental theorem of Galois theory* asserts the following. There is a one-to-one correspondence between the set of subgroups of $\mathrm{Gal}(L/F)$ and the set of fields intermediate between F and L. (Precisely, K is such an *intermediate field* if $F \subset K \subset L$ and K is a subfield of L. We assume that the reader can provide the definition of the term "subgroup.") In this correspondence, a subgroup H of $\mathrm{Gal}(L/F)$ is associated with the intermediate field $H' = \{\alpha \in L : \sigma(\alpha) = \alpha$ for each $\sigma \in H\}$; and an intermediate field K is associated to the subgroup $K' = \mathrm{Gal}(L/K) = \{\sigma \in G : \sigma(\alpha) = \alpha$ for each $\alpha \in K\}$. For each intermediate field K, the extension L/K is a Galois field extension. (Moreover, for readers familiar with group theory: K/F is a Galois field extension if and only if K' is a normal subgroup of Gal(L/F), in which case Gal(K/F) is group-isomorphic to the factor group Gal(L/F)$/K'$. Given intermediate fields $K \subset M$, the group-theoretic index $[K' : M']$ coincides with the (Section 3's) dimension $[M : K]$. In particular, by passing to $K = F$ and $M = L$, one obtains the following.) For any Galois field extension L/F, the order of Gal(L/F) coincides with the (Section 3's) dimension $[L : F]$.

Before discussing quintics and related matters, we pause to describe three results in order to indicate that Galois theory does, indeed, apply to our subject's concerns. First, we sharpen Section 3's criterion for constructibility as follows: a real number β is constructible (that is, $\beta \in \mathcal{C}$) if and only if there exists a Galois field extension $L/\mathbf{Q}$ such that $\beta \in L$ and $[L : \mathbf{Q}] = 2^m$ for some nonnegative integer m (depending on β). A sketch of the proof will be instructive. For the "only if" half, begin with a constructible β, and obtain, as in Section 3, a tower $\mathbf{Q} = F_0 \subset \ldots \subset F_n$ such that $F_i = F_{i-1}(\alpha_i)$, $\alpha_i^2 \in F_{i-1}$, and $\beta \in F_n$. Now, the smallest splitting field E over $\mathbf{Q}$ which contains F_n is known to be the field generated by the fields $\tau(F_n)$, as τ ranges over the isomorphisms of F_n with copies inside a preassigned algebraic closure of F_n. For each such τ, one has a tower $\mathbf{Q} = \tau(F_0) \subset \ldots \subset \tau(F_n)$ such that $\tau(F_i) = \tau(F_{i-1})(\tau(\alpha_i))$ and $\tau(\alpha_i)^2 \in \tau(F_{i-1})$. Moreover, there are only finitely many such τ. (Why? Recall the blockbuster in Section 1 of Chapter 2.) Accordingly, by adjoining the $\tau(\alpha_i)$'s one at a time in an appropriate order, one may obtain E from $\mathbf{Q}$ *via* a tower of 2-dimensional extensions. An application of Section 3's result on multiplicativity of degrees in towers reveals that $[E : \mathbf{Q}]$ is a power of 2, and completes the proof of the "only if" half. To prove the converse, we shall require some group theory, in concert with the fundamental theorem of Galois theory. Indeed, starting with $\beta \in L$, where $L/\mathbf{Q}$ is Galois and $[L : \mathbf{Q}] = 2^m$, we have that Gal($L/\mathbf{Q}$) has order 2^m. By virtue of group theory's result that any p-group is solvable, one obtains a subgroup tower $G_0 = Gal(L/\mathbf{Q}) \supset G_1 \supset \ldots \supset G_m = \{\text{identity map}\}$ such that each group-theoretic index $[G_i : G_{i+1}]$ equals 2 (whence G_{i+1} is necessarily a normal subgroup of G_i). An(other) application of the fundamental theorem produces the tower of fields $\mathbf{Q} = G_0' \subset G_1' \subset \ldots \subset G_m' = L$ such that each dimension $[G_{i+1}' : G_i']$ equals 2. Hence, $G_{i+1}' = G_i'(\gamma_{i+1})$, where $\gamma_{i+1}^2 \in G_i'$. (Why? Use the quadratic formula.) As in Section 3, obtaining γ_{i+1} from G_i' amounts to intersecting an appropriate line and circle (provide details), so that L's elements, in particular β, are each constructible by ruler and compass, completing the proof.

A second noteworthy use of Galois theory concerns the reälization of abstract groups as Galois groups. Specifically, one can show that, if G is *any* finite group and K is any field, then there exists a Galois field extension L/F such that K is a subfield of F and G is group-isomorphic to Gal(L/F). (A remark for advanced readers: the extension to certain possibly infinite groups called *profinite* groups is given in [27].) One offshoot of the Galois-theoretic proof of the cited result is a new (non-algorithmic) proof of the fundamental theorem on symmetric polynomials, which saw important service in Sections 4 and 5 of Chapter 2. A related question which has received much attention during the past century (and for which several affirmative instances have been settled) asks whether each finite group G is isomorphic to Gal($L/\mathbf{Q}$) for some appropriate Galois field extension $L/\mathbf{Q}$. As with most good questions, this has led to unanticipated research on areas both old and new, through the efforts of twentieth-century giants such as E. Artin, D. Hilbert, E. Noether, and J. Safarevic.

A third use of Galois theory appears in [18]. We mentioned this pretty article in Section 3 while discussing the irrationality of $\sqrt{2} + \sqrt{3}$; we mention it again since it is now somewhat more accessible to the reader.

Next, we turn to the central application of Galois theory, the topic of *solvability by radicals.* (The definition to be given below is one of several possible choices. The recent article [8] compares various historically prominent, but inequivalent, competing definitions of this term.) For simplicity, we shall consider only the case of polynomials over fields of characteristic 0. Let $f(x)$ be a polynomial over a field F of characteristic 0, and let L be a splitting field of $f(x)$ over F. We shall say that $f(x)$ is *solvable by radicals* (*over* F) in case there exists a tower of fields $F = F_0 \subset \dots \subset F_n$ such that $F_i = F_{i-1}(\alpha_i)$, $\alpha_i^{m_i} \in F_{i-1}$ for appropriate nonnegative integers m_i, and L is a subfield of F_n. (The analogy between this condition and the above characterization of constructibility is evident.) According to the quadratic formula, any quadratic polynomial $ax^2 + bx + c$ is solvable by radicals: it suffices to take $n = 1$, $m_1 = 2$, and $\alpha_1^{m_1} = b^2 - 4ac$. Similarly, the material in Section 1 reveals that each cubic polynomial (without loss of generality, of the form) $x^3 + px + q$ is solvable by radicals: take $n = 3$, with $\alpha_1^2 = -3$, $\alpha_2^2 = q^2 + (4p^3/27)$, and $\alpha_3^3 = (-q + \alpha_2)/2$. We leave to the reader the task of verifying (by use of either the method of Descartes or the method of Ferrari) that the material in Section 4 shows that each quartic is solvable by radicals.

Regrettably, the pattern does not continue for higher degrees. Specifically, let $n \geq 5$ and consider $n + 1$ independent, commuting indeterminates $a_1, a_2, \dots, a_n, x$ over a field F of characteristic 0. The *theorem of Ruffini–Abel* asserts that the *general* (*monic*) polynomial *of degree* $n \geq 5$, namely $f_n(x) = x^n + a_1x^{n-1} + \dots + a_n \in F(a_1, \dots, a_n)[x]$, is not solvable by radicals over the field $F(a_1, \dots, a_n)$. Historically, there was a controversy whether the general polynomial was specific enough, and so a need arose to produce a specific rational quintic which is not solvable by radicals over $\mathbf{Q}$. One such quintic is $x^5 - 6x + 3$. Indeed, no root of this polynomial lies in the end result of a tower obtained by successive adjunctions of radicals; in particular, none of the roots of $x^5 - 6x + 3$ is constructible.

In order to describe how the Galois theory allows one to prove the assertions made in the preceding paragraph, a group-theoretic digression is needed. Familiarity with the notions of "normal subgroup" and "factor group" will be presupposed. First, we say that a group G is *cyclic* in case there is at least one element $\alpha \in G$ with the property that the smallest subgroup of G which contains α is G itself. Apart from isomorphic copies, the only cyclic groups are $\{0\}$, $\mathbf{Z}$, and the structures $\mathbf{Z}_n$ encountered in Section 1 of Chapter 1. (In each case, the group operation is addition. What are the appropriate values of α?) Next, we introduce the building blocks of finite group theory. A group $G \neq \{1\}$ is called *simple* in case $\{1\}$ and G (which are, in the nature of things, normal subgroups of G) are the *only* normal subgroups of G. The reader should verify that $\mathbf{Z}_2$ and $\mathbf{Z}_3$ are simple groups, whereas $\mathbf{Z}_4$ is not simple. Indeed, the Abelian simple groups are precisely the groups $\mathbf{Z}_p$ corresponding to prime p: these are the so-called "simple simple groups." As we shall see, the key to proving the theorem of Ruffini–Abel lies in discerning the role of nonabelian simple groups. To see how the "building blocks" build, we state the *Jordan–Hölder theorem*. For each finite group G, there is a *composition series* $G = G_0 \supset \ldots \supset G_m = \{1\}$, that is, a finite tower of subgroups with the property that each G_{i+1} is maximal among the proper normal subgroups of G_i (equivalently, with the property that each G_{i+1} is normal in G_i, with the corresponding *composition factor* G_i/G_{i+1} being a simple group). Moreover, although G may have more than one composition series, the *length m* does not vary; and the set of composition factors is (apart from ordering) the same for any composition series of G. (For practice in constructing composition series, see Exercise 7 below.) Finally (and prophetically) we say that a finite group G is *solvable* in case each of its composition factors is a simple simple group. Any finite Abelian group is solvable; as we shall see, so are certain important nonabelian ones.

What has all this group theory to do the the theory of equations? As before, let L be the splitting field over a field F (of characteristic 0) of a polynomial $f(x)$. Let G be the Galois group of $f(x)$ over F; that is, $G = \text{Gal}(L/F)$. It is quite rare that G is Abelian: indeed, a deep theorem of Kronecker asserts, in case $F = \mathbf{Q}$, that G is Abelian if and only if L is a subfield of $\mathbf{Q}(\tau)$, for τ some primitive n-th root of 1, in the sense of Section 3 in Chapter 1. However, G often turns out to be solvable. Indeed, the *main theorem* of the subject (due, of course, to Galois) states that $f(x)$ is solvable by radicals over F if and only if G is a solvable group.

To apply Galois's theorem, first let S_n be the n-th *symmetric group,* that is, the group of permutations of n given (distinct) objects. (What is the group operation of S_n? What is the order of S_n?) In the spirit of #4 of Exercises 2.5, observe that each element of S_n may be written as a (finite) product of *transpositions,* that is, of permutations moving only two of the n objects. The n-th *alternating group, A_n,* consists of those permutations in S_n which may be expressed as the product of an even number of transpositions. It can be shown that A_n is a proper normal subgroup of S_n, with $[S_n : A_n] = 2$. To test whether S_n is solvable, that is, whether the tower $S_n \supset A_n \supset \{1\}$ may be refined to a composition series with simple simple composition factors, one needs to know the following facts. For $n = 2$, 3, and 4, A_n is solvable (try proving this); for $n \geq 5$, A_n is simple (and not solvable). By appealing to Schreier refinement theory, one infers: S_2, S_3, and S_4 are each solvable, but S_n fails to be solvable whenever $n \geq 5$.

The reader may have already guessed how Galois's theorem leads to the Ruffini–Abel result. Specifically, if $f_n(x)$ is the general polynomial of degree $n \geq 1$, then it is known that S_n is (isomorphic to) the Galois group of $f_n(x)$ over the field $K = F(a_1, a_2, \ldots, a_n)$. Accordingly, if $n \geq 5$, the above information about S_n combines with Galois's theorem to show that $f_n(x)$ is not solvable by radicals over K. By the same token, f_2, f_3, and f_4 *are* solvable by radicals over K. In fact, with due notice of characteristics, the latter result may be obtained via a computational approach to Galois's theorem with the aid of the *resolvents* of Lagrange (who, you see, did much more than just interpolate), the result being a rederivation of Cardan's formula for cubics (Section 1) and Ferrari's quartic formula (Section 4). For details, see [25, pp. 178–182], which exploits appropriate composition series for S_3 and S_4; see also Exercise 7(b) below.

It remains only to justify the assertion that $x^5 - 6x + 3$ is not solvable by radicals over $\mathbf{Q}$. Of course, this results from the computation that its Galois group is S_5, which isn't solvable. In a more positive vein, what *can* we say about the roots of $x^5 - 6x + 3$? First, there are exactly five of them in $\mathbf{C}$, by the fundamental theorem of algebra (see Section 6). There are no multiple roots, since $(x^5 - 6x + 3, 5x^4 - 6) = 1$. None of the roots is rational (see either the rational root test or the weak version of Eisenstein's criterion in Section 1 of Chapter 4). Indeed, a careful reading of the foregoing reveals that none of the roots is constructible. (It may help to notice that $x^5 - 6x + 3$ is irreducible in $\mathbf{Q}[x]$: see either Eisenstein's criterion in Section 2 of Chapter 4 or the factoring algorithm in Section 3 of Chapter 4.) As for the real roots, 0 is not a root; according to either Descartes's rule of signs (Section 3 of Chapter 5) or Budan's theorem (Section 6 of Chapter 5), there is exactly one negative root; and, by Sturm's theorem (Sections 4 and 5 of Chapter 5), precisely two positive roots are present. The three real roots may be "isolated" in the intervals $(-2, -1)$, $(0, 1)$, and $(1, 2)$, by the use of the intermediate-value theorem familiar from calculus, a theme which is effectively pursued in Sections 1 and 2 of Chapter 5. Approximations (to any desired accuracy) to the three real and two nonreal (but complex) roots may be effected by the techniques in Chapter 6: see especially Sections 4, 5, and 8 therein.

A closing thought for this most abstract of sections. Theory and practice are, indeed, inseparable. The discussion of just *one* example in the preceding paragraph has served to motivate *each* of this book's remaining sections!

EXERCISES 3.5

1. Let n be a positive integer. Prove that the n-th symmetric group S_n is Abelian if and only if $n \leqslant 2$.
2. Each of the following rational polynomials has a Galois group which is isomorphic to a group which you have met: $x^2 - 3$, $x^3 - 2$, $x^3 - 3$, $x^2 + x + 1$, $x^4 + x^2 + 1$, and $x^3 - 3x + 2$. Identify those grôups. (For the last of these, you may wish to use Cardan's formula.)
3. (a) Use the fundamental theorem of Galois theory to show that if L/F is a Galois field extension, then there are only finitely many fields intermediate between F and L. (*Hint*: think of the groups involved.)
   (b) If F is a subfield of a finite field L, prove that the field extension L/F is Galois. (*Hint*: recall from Section 3's "small library exercise" that $GF(p^n)$ is the splitting field of $x^{p^n} - x$ over $GF(p)$.)
   (c) Let $\tau \in \mathbf{C}$ satisfy $\tau^n = 1$ for some positive integer n, that is, let τ be an n-th root of 1. Let $L = \mathbf{Q}(\tau)$. Prove that $L/\mathbf{Q}$ is a Galois field extension if τ is a primitive n-th root of 1. What happens if τ is not a primitive n-th root of 1?
4. Let F be a subfield of $\mathbf{C}$ which contains ω, a primitive n-th root of 1, and let $\alpha \in F$. Prove that the Galois group of $x^n - \alpha$ over F is Abelian. (*Hint*: Show that the roots are of the form $\omega^i\beta$, $i = 0, \ldots, n-1$, so that each automorphism σ is determined by $\sigma(\beta)$.)
5. Let F be a field, and let $f(x) \in F[x]$ be a polynomial with no multiple roots. Assume that G, the Galois group of $f(x)$ over F, is *transitive*; that is, for any roots r, s of $f(x)$, there exists a suitable $\sigma \in G$ such that $\sigma(r) = s$. Prove that $f(x)$ is irreducible in $F[x]$. (*Remark.* Conversely, if f is irreducible and has no multiple roots, then G is transitive. The proof of this converse is trickier, and is reminiscent of Section 3's (sketch of the) proof of "uniqueness" of splitting fields. Try it!)
6. Show that the particular quintic $x^5 + 3x^4 + 3x^3 + x^2 - x - 1$ *is* solvable by radicals over $\mathbf{Q}$. (*Hint*: find a rational root. If necessary, peek at Section 1 of Chapter 4.)
7. (a) Let G be the set of ordered pairs (a, b) with $a \in \mathbf{Z}_2$ and $b \in \mathbf{Z}_3$. Show that componentwise addition is a group operation for G. Display two distinct composition series for G.
   (b) Prove that the alternating group A_4 is solvable by first finding a non-cyclic normal subgroup V, of order 4. (*Remark.* V is the famous *four-group* of Klein. In German, "Vier" means "four.")
   (c) For which value(s) of n is A_n both simple and solvable?

*6. The Fundamental Theorem of Algebra: a Needle for Every Haystack

It is possible to view the construction of bigger and bigger algebraic structures as a response to the search for roots of polynomials with coefficients in smaller algebraic structures. For instance, passage from the system of nonnegative integers to **Z** is motivated by the search for roots of polynomials such as $x + 3 - 2$ (practical applications: describe a debt of 1 dollar, determine the time of occurrence of a felony under investigation, . . .). Similarly, the need to solve practical problems diverts attention from **Z** to **Q** (example: the root of $2x - 3$ describes an equitable share when two thieves divide a booty of three dollars) and from **Q** to **R** (example: according to the theorem of Pythagoras, the positive root of $x^2 - 2$ describes the length of fence needed for one edge of a suitable triangular garden plot). As we mentioned when introducing **C**, various scientific considerations (of great practical importance: ask any engineer!) compel the study of the roots of polynomials such as $x^2 + 1$ and, consequently, a shift of attention from **R** to **C**. It is truly remarkable that no further expansion of algebraic structures is warranted by searches for roots, in view of the *fundamental theorem of algebra*: for each polynomial $f(x) \in \mathbf{C}[x]$ of degree $n \geq 1$, one can write $f(x) = c(x - \alpha_1)(x - \alpha_2) \ldots (x - \alpha_n)$, for suitable complex numbers $c, \alpha_1, \alpha_2, \ldots, \alpha_n$, possibly listed with repetition. To rephrase the result using the jargon used in stating the son of blockbuster in Sections 1 and 6 of Chapter 2: each nonzero complex polynomial has all its roots in **C**. Other paraphrases noted in Section 6 of Chapter 2 state that **C** is the algebraic closure of **R**; and that **C** is algebraically closed. We have already put the fundamental theorem of algebra to work (see, for example, #2 in Exercises 2.7). Another application, in Section 1 of Chapter 5, will be the cataloguing of the irreducible real polynomials, one consequence of which is the specialization of the material in Section 4 of Chapter 1 to the theory of partial fractions as typically used in courses on integral calculus.

The first rigorous proof of the fundamental theorem of algebra was published by Gauss, one of the greatest mathematicians of all time, in his dissertation in 1799. (A historical aside: that year also marked the discovery of the Rosetta stone by Napoleon's soldiers in Egypt, thus permitting translation of hieroglyphics of tremendous archeological importance. For a recent bicentennial survey describing Gauss's scientific accomplishments, including his collaboration on magnetism with the physicist, Weber, see the very readable article [22].) Prior to 1799, the fundamental theorem had been carefully stated and used, most notably by the physicist-mathematician, d'Alembert. Gauss's powerful insight is indicated by the fact that he obtained, over the course of many years, several different proofs of the fundamental theorem of algebra. Inevitably, his methods involved techniques from mathematical areas besides algebra alone. Even today, techniques from analysis (and, frequently, topology) are needed to prove the result. More advanced readers will know that a short proof is possible, early in a course on complex variables, with the aid of Liouville's theorem. The proof which we have chosen to present needs only calculus's intermediate-value theorem for *continuous* functions. More particularly, we shall need the following corollary of it: *each real polynomial of odd degree has* (*at least*) *one real root*. A proof of this corollary, together with a most satisfactory sharpening of differential calculus's results for *polynomial* functions, will be given in the first two sections of Chapter 5.

Our proof of the fundamental theorem of algebra begins with a pair of reductions. First, we claim that it is enough to prove that each complex polynomial of positive degree has at least one complex root. The proof that this seemingly weaker conclusion suffices proceeds, as in the fourth paragraph of Section 6 in Chapter 2, via appeals to the factor theorem, the fundamental fact about degrees, and the well-ordering property.

Second, we claim that it suffices to restrict attention to (finding roots of) real polynomials. To prove this, we revisit a technique encountered in the final paragraph of Section 2 of Chapter 1. To wit: for each $f \in \mathbf{C}[x]$, let f^* be the polynomial obtained from f by conjugating f's coefficients. Specifically, if $f(x) = \sum_j (r_j + is_j)x^j$ for some real numbers r_j and s_j, then $f^*(x) = \sum_j (r_j - is_j)x^j$. (Compare the passage from f to f^* with the method of reduction modulo p in Section 1 of Chapter 2. Apart from noting the formal analogy, the review should be helpful, since the latter method will recur in Section 2 of Chapter 4.) Set $g(x) = f(x)f^*(x)$. Now, a typical coefficient of g is

$$\sum_{j=0}^{k} (r_j + is_j)(r_{k-j} + is_{k-j})$$

which is readily seen to be a real number. (Expand and check.) If the result is supposed known for real polynomials, $g(\alpha) = 0$ for some $\alpha \in \mathbf{C}$. Thus, by the second fundamental (homomorphism) property of evaluation, $0 = f(\alpha)f^*(\alpha)$, whence either $f(\alpha) = 0$ or $f^*(\alpha) = 0$. If the former holds, our task has been completed. In the other case, $f^*(\alpha) = 0$, argue precisely as in Section 2 of Chapter 1, and get $f(\alpha^*) = (f^*(\alpha))^* = 0^* = 0$, producing α^* as the desired root of $f(x)$.

The above reductions have reduced our task to proving that each real polynomial $f(x)$ of degree $n \geq 1$ has at least one complex root. By the fundamental theorem of arithmetic, $n = 2^v m$, for some nonnegative integer v and some odd positive integer m. Note that the case $v = 0$ has been asserted above. (This amounts to the case of odd n.) The remainder of the proof will proceed by *induction on v*. (The technique of mathematical induction, which is equivalent to the well-ordering property, is making its first appearance in these pages, but it is a tool which is surely familiar to most readers. It will recur in Section 3 of Chapter 5.)

A final reduction. By using the strategy of unique factorization and the second fundamental property of evaluation, one may reduce, precisely as at the close of the fourth paragraph of Section 6 in Chapter 2, to the case in which $f(x)$ is irreducible in $\mathbf{R}[x]$.

A third appeal to Section 6 in Chapter 2 (a.k.a. the son of blockbuster or the theorem of Cauchy–Kronecker–Steinitz or the existence of splitting fields) produces n roots $\alpha_1, \alpha_2, \ldots, \alpha_n$ of $f(x)$ in some field L of which $\mathbf{C}$ is a subfield. (Incidentally, the α_j are distinct. The reason was given in Section 5. Can you find—or reproduce— it?) Now, for each integer t, consider the polynomial

$$g_t = \prod_{j<k} (x - \alpha_j - \alpha_k - t\alpha_j\alpha_k)$$

in $L[x]$. Observe that $\deg(g_t) = n(n-1)/2 = 2^{v-1}p$, for some odd positive integer p. Moreover, each coefficient of g_t is of the form $h(\alpha_1, \ldots, \alpha_n)$ for a suitable symmetric polynomial $h \in \mathbf{R}[x_1, \ldots, x_n]$, and so the final result in Section 5 of Chapter 2 may be used to show that $g_t \in \mathbf{R}[x]$. By induction (that is, since $\deg(g_t)$ is divisible by a smaller power of 2 than is $\deg(f)$), at least one of the roots of g_t is complex. In other words, $\alpha_j + \alpha_k + t\alpha_j\alpha_k \in \mathbf{C}$ for some j, k depending on t. Pick such a pair (j, k) for each t. Now, as t runs through $(n(n-1)/2) + 1$ distinct values (thank goodness $\mathbf{Z}$ is infinite!) and the corresponding pairs (j, k) are recorded, some pair must pop up at least twice. (Arguments like this are said to refer to the *pigeonhole principle.* Why is the principle true?) In other words, there exist distinct integers t_1 and t_2 and distinct indexes j and k such that both $\alpha_j + \alpha_k + t_1\alpha_j\alpha_k$ and $\alpha_j + \alpha_k + t_2\alpha_j\alpha_k$ are complex numbers. Subtraction, followed by division by (the nonzero complex number) $t_1 - t_2$, reveals that $\alpha_j\alpha_k \in \mathbf{C}$. Hence, $\alpha_j + \alpha_k \in \mathbf{C}$ as well. (Why?)

To complete the proof, it suffices to establish that α_j and α_k are each complex. To this end, let $P(x) = x^2 - (\alpha_j + \alpha_k)x + \alpha_j\alpha_k \in \mathbf{C}[x]$. The roots of $P(x)$ may, of course, be obtained by the quadratic formula. As $\mathbf{C}$ is closed under square roots—see Section 2 of Chapter 1—the roots of $P(x)$ therefore all belong to $\mathbf{C}$. However, α_j and α_k *are* the roots of $P(x)$, since $P(x)$ factors as $(x - \alpha_j)(x - \alpha_k)$. This completes the proof.

This will be the only section of the book which lists no exercises. Instead, we ask the reader to contemplate the makeup of the foregoing proof. He/she will doubtlessly be impressed with Gauss's theorem (and the review will be helpful). In Section 2 of Chapter 4, we shall meet with more of Gauss's work, the beautiful theory of content.

4

Polynomials over the Rational Numbers

This chapter continues the application of the material in Chapter 2 to more specific contexts. Whereas much of the specificity of Chapter 3 arose from our dwelling on polynomials of "low" *degree,* this chapter's precision will derive from a focussing on familiar *domains* from which the relevant polynomial's *coefficients* are to be drawn. What could be more familiar than the rational field **Q** or its subdomain **Z** of integers? (We hope that's a rhetorical question!) Accordingly, we attend to $\mathbf{Z}[x]$ and $\mathbf{Q}[x]$ in this chapter. (Don't worry: $\mathbf{R}[x]$ and $\mathbf{C}[x]$ will have their turn, in subsequent chapters.) Our main concern in Section 1 is to develop the rational root test, a quick, effective way of finding all the rational roots of a given polynomial in $\mathbf{Q}[x]$. Of course, finding roots, whether rational or not, is important–think of the factor theorem–it just happens that the rational roots are particularly accessible. Moreover, specific information about rational roots was needed earlier (in, for example, Section 1 of Chapter 2 and Section 3 of Chapter 3), and we shall presently have the means to produce such information with ease. Unlike the dividends gleaned from reduction modulo p in Section 1 of Chapter 2, this chapter's methods apply to nonintegral rational roots as easily as to integral roots. In #3(c) of Exercises 4.1, the reader will find that the outstanding promissory notes are all paid; that is, all the earlier chapters' claims regarding rational/integral roots are settled there. Section 1 also contains a weak form of Eisenstein's sufficient condition that a polynomial in $\mathbf{Q}[x]$ not have a rational root. (We have not seen this weak version and its treatment elsewhere.) The usual (strong) version of Eisenstein's criterion concerns irreducibility. It, together with a celebrated lemma of Gauss, forms part of the algebraic riot in (the optional) Section 2, where proofs, making effective use of earlier theory (such as homomorphism properties of reduction modulo p), are somewhat neater than is customary for texts at this level. The more general concerns of factorization in Section 2 (as opposed to Section 1's root hunts) lead to (the optional) Section 3's algorithm for complete factorization in $\mathbf{Q}[x]$. For that, first brush up on Lagrange interpolation.

A final word about notation, before proceeding. The earlier chapters' skirmishes with rational roots make it plausible that number theory will figure heavily in such roots' analyses. In fact, that's so. It will be convenient to adapt the notation in Section 3 of Chapter 1 so that, when appropriate, symbols such as $(a_1, \ldots, a_n)$ will refer to the integer g.c.d. of the finite set of integers a_1, $\ldots, a_n$; and, when appropriate, $a|b$ will indicate that an integer b is an integral multiple of an integer a. Context will surely determine whether such factorization symbols pertain to $\mathbf{Z}$ or (as they used to, and still may) to $\mathbf{Q}[x]$.

1. Rational Roots (Candidates and Survivors) and a Result of Eisenstein

Before getting to the rational root test and the other items mentioned above, we must dispatch three tasks. The first of these is the *divisibility lemma*: whenever nonzero integers a, b, and c satisfy $a \mid bc$ and $(a, b) = 1$, then $a \mid c$. Its easy proof is left to the reader. (Here's a hint: either use uniqueness of factorization into products of primes—the fundamental theorem of arithmetic—or adapt the argument involving linear combinations which was used in Section 3 of Chapter 1 to show that (**)$\Rightarrow$(*).)

Secondly, each nonzero rational number q may be uniquely expressed in *lowest terms* in the sense that $q = a/b$, where a and b are nonzero, relatively prime integers such that b is positive. Surely, such information is not news to any reader ($60/-50 = -6/5$ is the stuff of elementary school) but how is such an assertion to be proved? Naturally, because there is "method to [our] madness," the answer involves the preceding paragraph's divisibility lemma. The reader is encouraged to fashion a proof for himself/herself. (If need be, peek at the second paragraph of Section 2 below. That will answer more than just the question here, and you might be seduced into reading an optional section! Intrigued? Develop a lowest-terms result in $\mathbf{Q}[x]$, and then generalize it.)

Finally, we next lessen the task to be confronted by the rational root test. To wit: given any nonzero rational polynomial $f(x) \in \mathbf{Q}[x]$, there exists a nonzero integral polynomial $g(x) \in \mathbf{Z}[x]$ such that $f(x)$ and $g(x)$ have the same roots (rational or otherwise, in, say, an algebraic closure $\overline{\mathbf{Q}}$ of $\mathbf{Q}$). One way to produce such $g(x)$ is to multiply $f(x)$ by a common multiple of the denominators in the lowest-terms descriptions of $f(x)$'s nonzero coefficients; then $g(x) = cf(x)$ for a suitable (nonzero) constant $c \in \mathbf{Z}$. For instance, if $f(x) = (-2/27)x^3 + (70/105)x - 9$, then c may be 27 (or any positive integral multiple thereof), so that $-2x^3 + 18x - 243$ is a satisfactory $g(x)$. (Does the $cf(x)$ construction of clearing denominators look familiar? Peek again at the second paragraph of Section 2 below. Are you seduced yet?) The verification that $f(x)$ and $g(x) = cf(x)$ have the same roots boils down to the fact that $\mathbf{Z}$ is a domain since, by the second homomorphism property of evaluation (in Section 1 of Chapter 2), each $\alpha \in \overline{\mathbf{Q}}$ satisfies $g(\alpha) = c \cdot f(\alpha)$.

Now for the rational root test. Our datum is a rational polynomial $f(x) = a_0x^n + a_1x^{n-1} + \ldots + a_n$ of degree $n \geq 1$. Our task is to produce a set S of finitely many rational numbers such that each rational root α of $f(x)$ satisfies $\alpha \in S$. (To be sure, some members of S may fail to be roots of $f(x)$. Each candidate $\beta \in S$ must be tested further, either by brute-force evaluation of $f(\beta)$ or by synthetic division, as in Section 1 of Chapter 6 below. Most readers will have encountered the candidate-survivor syndrome before: think of the first and second derivative tests for extrema in calculus.) The comments of the preceding two paragraphs allow us to assume that each $a_i \in \mathbf{Z}$, and to express a given rational root α of $f(x)$ in lowest terms: $\alpha = r/s$ where $(r, s) = 1$. The *rational root test asserts* that $r \mid a_n$ and $s \mid a_0$, so that S may be built by knowing the lists of integral divisors of a_n and of a_0. (As for how one knows *those*, consult any reputable text on number theory such as [21]. Factorizations into products of primes in $\mathbf{Z}$ will need to be executed effectively. For this, see the "sieve of Eratosthenes.")

To prove the assertions of the rational root test, observe that

$$0 = s^n \cdot 0 = s^n \cdot f(\alpha) = a_0r^n + a_1r^{n-1}s + \ldots + a_{n-1}rs^{n-1} + a_ns^n.$$

In particular, $r(a_0r^{n-1} + a_1r^{n-2}s + \ldots + a_{n-1}s^{n-1}) = -a_ns^n$, whence r divides the product of a_n and $-s^n$. However, r and $-s^n$ are relatively prime (why?), and so the above divisibility lemma (more "method to madness"!) guarantees that $r \mid a_n$, as asserted. To show that $s \mid a_0$, observe that $s(a_1 r^{n-1} + a_2 r^{n-2}s + \ldots + a_n s^{n-1}) = -a_0 r^n$ and argue similarly.

We offer five applications of the rational root test. First, let's find the rational roots, if there are any, of the polynomial $f(x) = \frac{2}{3}x^5 - 2x^4 + \frac{40}{3}x^3 - \frac{32}{3}x^2 + 40x + \frac{200}{3}$. To that end, clear denominators via multiplication by $c = 3$ and consider, instead, the polynomial $g(x) = cf(x) = 2x^5 - 6x^4 + 40x^3 - 32x^2 + 120x + 200$, with $n = 5$, $a_0 = 2$, and $a_n = 200$. Now, the possible values for r are the integral divisors of a_n, namely $1, -1, 2, -2, 4, -4, 8, -8, 5, -5, 25, -25, 10, -10, 50, -50, 20, -20, 100, -100, 40, -40, 200$, and -200. The possible values for s are the positive integral divisors of a_0, namely 1 and 2. Accordingly, there are thirty candidates formed by all "possible" values of r/s, namely $1, -1, 2, -2, 4, -4, 8, -8, 5, -5, 25, -25, 10, -10, 50, -50, 20, -20, 100, -100, 40, -40, 200, -200, \frac{1}{2}, -\frac{1}{2}, \frac{5}{2}, -\frac{5}{2}, \frac{25}{2}$, and $-\frac{25}{2}$. Test each of these thirty candidates by evaluating each $f(r/s)$. (Of course, you may stop if/when you have found $5 = \deg(f)$ roots. To further ease the burden, note that considering $\frac{1}{2}g(x)$ would have produced only eighteen candidates r/s.) One finds (doesn't one?) that -1 is the only rational root of $f(x)$. How would the material in Section 4 of Chapter 3 then be employed to find all the five roots of $f(x)$?

For a second application, let's find the rational roots of $f(x) = 2x^4 - x^3 - 53x^2 + 25x + 75$. Here, $n = 4$, $a_0 = 2$, $a_n = 75$, and one may check (do so) that there are twelve candidates for r, two candidates for s, and, as a result, twenty-four candidates for r/s. In fact, four of these *are* roots (which ones are they?), so that $f(x)$ factors as $f(x) = 2(x - 5)(x - \frac{3}{2})(x + 5)(x + 1)$. Observe that time would be saved, after noting that $f(5) = 0$, by dividing to get $f(x) = (2x^3 + 9x^2 - 8x - 15)(x - 5)$, for, by homing in on the quotient's coefficients, subsequent root searches proceed with $r \in \{1, -1, 3, -3, 5, -5, 15, -15\}$ and $s \in \{1, 2\}$. There would then be only sixteen additional candidates for r/s, as opposed to twenty-three additional candidates based on the list generated from $f(x)$'s coefficients.

For our third application, consider $f(x) = x^3 - 36x + 72$. Here, $n = 3$, $a_0 = 1$, $a_n = 72$, $r \in \{1, -1, 2, -2, 4, -4, 8, -8, 3, -3, 9, -9, 6, -6, 18, -18, 12, -12, 36, -36, 24, -24, 72, -72\}$, and $s = 1$. This illustrates the general principle that *if a rational polynomial is monic, then each of its rational roots must be an integer.* (Proof? Observe that $1 \leq s$ and $s \mid 1$. Hence $s = 1$.) Unfortunately, the reader will discover (eventually!) that, for the above example, none of the twenty-four candidates is a root. One moral, of great applicability in science, may be drawn: somewhat depressing examples *occasionally* reveal important general principles!

The preceding principle may be put to work for another interesting application: if $p_1, \ldots, p_k$ are distinct positive prime integers and n is any integer exceeding 1, then no root of $f(x) = x^n - p_1 p_2 \ldots p_k$ is rational. (How many roots are there? If necessary, see Section 2 of Chapter 1 or Section 6 of Chapter 3.) In particular, the positive (real) number $\sqrt[n]{p_1 p_2 \ldots p_k}$ is irrational. As a special case, we recover the "pesky fact" from Section 1 of Chapter 1 that $\sqrt{2}$ is irrational; and, more generally, that $\sqrt{m}$ is irrational whenever m is a squarefree integer exceeding 1. To prove the general statement, note first that each rational root α of $f(x)$ must be an integer other than 1 and -1, whence by relabelling if necessary, $p_1 \mid \alpha$. Let q be the quotient $\alpha \div p_1$. As $f(\alpha) = 0$, we then have $p_1^n q^n = p_1 p_2 \ldots p_k$, a contradiction to the uniqueness assertion in the fundamental theorem of arithmetic. Indeed, the prime p_1 appears to at least the n-th power in the prime-power factorization of the left-hand side, while only the first power of p_1 divides the right-hand side. This contradiction establishes that $f(x)$ has no rational roots, as desired.

For our fifth application of the rational root test, consider $\alpha = \sqrt{2} + \sqrt{3}$. It was remarked in Section 3 of Chapter 3 that α is irrational, essentially because $\sqrt{2}$ and $\sqrt{3}$ are each irrational (see the preceding paragraph!) and $\mathbf{Q}(\sqrt{2}) \cap \mathbf{Q}(\sqrt{3}) = \mathbf{Q}$ (a fact proved twice already, in Section 1 of Chapter 1 and Section 3 of Chapter 3). Here's another way to see that α is irrational. Since $\alpha^2 = 2 + 2\sqrt{2}\sqrt{3} + 3 = 5 + 2\sqrt{6}$, solving for $\sqrt{6}$ and then squaring both sides reveals that $24 = (\alpha^2 - 5)^2$, whence α is a root of the polynomial $f(x) = x^4 - 10x^2 + 1$. (Thus, you've seen the hat from which *that* particular rabbit was pulled in Section 1 of Chapter 2. For more of this hat trick, see #5 in Exercises 4.1 below.) However, the rational root test reveals that 1 and -1 are the only possible rational roots of $f(x)$, while $-1 < 1 < \sqrt{2} < \alpha$. Thus α is, indeed, irrational, as claimed. A modest generalization of the fact that $\mathbf{Q}(\sqrt{2}) \cap \mathbf{Q}(\sqrt{3}) = \mathbf{Q}$ will be given in the next-to-the-last paragraph of this section; for a far-reaching extension, see [18].

Because a time-consuming application of the rational root test often yields no rational roots, it is of interest to develop a quicker test, of wide utility, which often identifies such situations. We've seen such tests in Section 1 of Chapter 2 (recall, especially, #8 of Exercises 2.1) via reduction modulo the *specific* primes 2 and 3. One test which works for an *arbitrary* prime is the *weak version of Eisenstein's criterion*, which we now proceed to state. An integral polynomial $f(x) = a_0 x^n + a_1 x^{n-1} + \ldots + a_n$ of degree $n \geq 2$ has no rational roots *if* there exists a positive prime integer p with the following three properties: $p \nmid a_0$, $p \mid a_i$ whenever $1 \leq i \leq n$, and $p^2 \nmid a_n$. For its indirect proof, express a rational root α of $f(x)$ in lowest terms as $\alpha = r/s$, for suitable relatively prime integers r and s. As in the proof of the rational root test, we have that $0 = s^n f(\alpha)$ leads to

$$0 = a_0 r^n + a_1 r^{n-1} s + \ldots + a_{n-1} r s^{n-1} + a_n s^n.$$

Solving for a_0r^n and applying the integer analogue of the linear combination principle from Section 3 of Chapter 1, we infer that $p \mid a_0r^n$. (Which hypothesis on p was used?) Since p is prime and $p \nmid a_0$, we next invoke the integer analogue of (*) from Section 3 of Chapter 1 in order to conclude successively that $p \mid r^n$ and and $p \mid r$. Accordingly, $p^n \mid r^n$ and, in fact, p^2 then divides each of the terms a_0r^n, $a_1r^{n-1}s, \ldots, a_{n-1}rs^{n-1}$. (Why?) Thus, solving for a_ns^n in the equation displayed above and then applying the linear combination principle reveals that $p^2 \mid a_ns^n$. Since $p^2 \nmid a_n$, uniqueness of factorization (the fundamental theorem of arithmetic, again) gives $p \mid s^n$ and, as above using (*), $p \mid s$. Therefore, $p \mid (r, s) = 1$, whence $p = 1$, the desired contradiction, to complete the proof.

Four applications of the weak version of Eisenstein's criterion are given next. First, the integral polynomial $x^8 + 294x^6 + 7x^4 - 98x + 28$ has no rational roots. Indeed, select $p = 7$ and verify the conditions required by the criterion: $7 \nmid 1$, $7 \mid 294$, $7 \mid 7$, $7 \mid -98$, $7 \mid 28$, 7 divides each of the (undisplayed) coefficients which equal 0, and $7^2 = 49 \nmid 28$. Contrast this approach with an application of the rational root test, which in this case turns up twelve hapless candidates. (Question: What about the earlier "depressing" polynomial $x^3 - 36x + 72$: does the weak version of Eisenstein's criterion dispatch it so easily?)

Our second application concerns an old friend, treated differently four paragraphs ago, namely the polynomial $x^n - p_1p_2 \cdots p_k$, where $n \geq 2$, and the p_i's are distinct positive primes. Verifying that the conditions of the weak version of Eisenstein's criterion hold in this case, with any of the p_i's playing the role of p, just amounts to (what else?) uniqueness of factorization into products of positive primes in **Z**. (Check.)

The third application is to the polynomial $f(x) = 1 + x + (x^2/2!) + \ldots + (x^p/p!)$, where p is any prime. We'll show that $f(x)$ has no rational roots. (For more detailed information about the roots of $f(x)$, see the discussion involving a noncanonical Sturm sequence in Section 4 of Chapter 5.) The analysis begins by shifting attention to $g(x) = p!f(x)$, an integral polynomial whose roots coincide with those of $f(x)$. (Why do $g(x)$ and $f(x)$ have the same roots? If necessary, review the reasoning in the third paragraph of this section.) However, the analysis is no sooner begun than it ends, for the weak version of Eisenstein's criterion applies to $g(x)$. (Check carefully. Why is the $p^2 \nmid a_n$ condition satisfied by $g(x)$?)

Besides the last paragraph's technique of altering a given polynomial via multiplication by a constant factor prior to applying the weak version of Eisenstein's criterion, there is another way to bring about such a useful alteration. Specifically, given a rational polynomial $f(x)$ and rational numbers a and b such that $a \neq 0$, then $f(x)$ has a rational root if and only if the polynomial $g(x) = f(ax + b)$ has a rational root. The proof of this assertion is straightforward: each complex number α satisfies $g(\alpha)$

$= f(a\alpha + b)$ and $f(\alpha) = g((\alpha - b)/a)$, and it now is only a matter of noticing that the rationality of any of α, $a\alpha + b$, and $(\alpha - b)/a$ entails the rationality of all three. For instance (here is the promised fourth application) consider $f(x) = x^4 + 4x + 1$. As it stands, $f(x)$ cannot be treated by the weak version of Eisenstein's criterion, since no prime p satisfies $p \mid a_n = 1$. However, the change of variable effected by $a = b = 1$ diverts attention to the polynomial $g(x) = f(x + 1) = x^4 + 4x^3 + 6x^2 + 8x + 6$ to which the weak version of Eisenstein's criterion applies, with the aid of the prime 2. Accordingly, $g(x)$ has no rational roots and, by the above general observation, therefore neither does $f(x)$.

We next prove a fact which was alluded to earlier in this section (and which was actually mentioned in Section 1 of Chapter 1), namely that $\mathbf{Q}(\sqrt{m}) \cap \mathbf{Q}(\sqrt{n}) = \mathbf{Q}$ whenever m and n are distinct squarefree integers, each exceeding 1. For its indirect proof, suppose instead that some irrational (real) number may be expressed both as $a + b\sqrt{m}$ and as $c + d\sqrt{n}$, for suitable rational numbers a, b, c, and d. Then, both b and d are nonzero, so that solving for $\sqrt{m}$ and $\sqrt{n}$ reveals $\sqrt{m} \in \mathbf{Q}(\sqrt{n})$ and $\sqrt{n} \in \mathbf{Q}(\sqrt{m})$. (Note that the equivalent fact that $\mathbf{Q}(\sqrt{m}) = \mathbf{Q}(\sqrt{n})$ is a fast consequence of the material on degrees of field extensions in the optional third section of Chapter 3.) Now, as $\sqrt{m} \in \mathbf{Q}(\sqrt{n})$, there exist rational numbers α and β such that $\sqrt{m} = \alpha + \beta\sqrt{n}$, and so, by squaring both sides, we have $m + 0\sqrt{n} = \alpha^2 + \beta^2 n + 2\alpha\beta\sqrt{n}$. Since $\sqrt{n} \notin \mathbf{Q}$ (because $x^2 - n$ has no rational roots), we infer $m = \alpha^2 + \beta^2 n$ and $0 = 2\alpha\beta$. As $\mathbf{Q}$ is a domain, the last of these conditions guarantees that either $\alpha = 0$ or $\beta = 0$. If $\beta = 0$, then the first condition reduces to $m = \alpha^2$, contradicting the hypothesis that m is squarefree. (Explain.) Thus, $\alpha = 0$, and so $m = \beta^2 n$. Put differently: β is a root of the rational polynomial $f(x) = nx^2 - m$. One way to complete the analysis is to express β in lowest terms, clear denominators in the equation $m = \beta^2 n$, and apply the divisibility lemma. We leave this approach for the interested reader to carry out, and shall instead apply the weak version of Eisenstein's criterion to $f(x)$. The problem is that it *can't* be applied because $f(x)$ *has* a rational root, β. Therefore, any prime p that divides m fails to satisfy at least one of the conditions $p \nmid n$ and $p^2 \nmid -m$. However, $p^2 \nmid -m$ since m is squarefree, and so $p \mid n$. As no square of a prime divides m (because that's what it means to say that m is squarefree), we infer that $m \mid n$. (Explain.) Similarly, the condition $\sqrt{n} \in \mathbf{Q}(\sqrt{m})$ allows one to deduce that $n \mid m$, from which one then has $m = n$ (why?), the desired contradiction.

To close this section, we indicate an interesting consequence of the preceding result. Let $\overline{\mathbf{Q}}$ be the algebraic closure of $\mathbf{Q}$, in the sense of Section 6 of Chapter 2. (We know that was an optional section, but this is our last chance this section for seduction!) The field extension $\mathbf{Q} \subset \overline{\mathbf{Q}}$ is an example of an *infinite-dimensional Galois extension.* Finite-dimensional Galois extensions were surveyed in Section 5 of Chapter 3—yes, another optional section. (For information about the infinite-dimensional case, see the comments about [27] in the annotated list of references. That the extension $\mathbf{Q} \subset \overline{\mathbf{Q}}$ *is* infinite-dimensional, in the sense that $[\overline{\mathbf{Q}} : \mathbf{Q}]$ is not finite, was proved in Section 3 of Chapter 3—yes, also optional.) One consequence of the fundamental theorem of Galois theory as stated in Section 5 of Chapter 3, is that whenever $\mathbf{Q} \subset F$ is a finite-dimensional Galois field extension, there are only finitely many fields intermediate between $\mathbf{Q}$ and F. (The reason is that such fields correspond to subgroups of the Galois group, and $\mathrm{Gal}(F/\mathbf{Q})$, being finite, has only finitely many subgroups.) Note, however, that the situation in this regard is quite different for the infinite-dimensional case. Indeed, $\mathbf{Q} \subset \overline{\mathbf{Q}}$ has infinitely many distinct intermediate fields: by the material in the preceding paragraph, it suffices to cite the fields of the form $\mathbf{Q}(\sqrt{p}\,)$, where p traverses the infinite (thanks again, Euclid!) set of positive primes.

EXERCISES 4.1

1. Find an integral polynomial whose roots coincide with the roots of the rational polynomial $(3/2)x^3 + (15/8)x^2 - (21/4)x + (3/4)$.
2. Observe that $4 \mid 12$, $4 \nmid 6$, $4 \nmid 2$, and $6 \cdot 2 = 12$. Why don't these facts serve to contradict the divisibility lemma?
3. Using the rational root test,
   (a) Find all the rational roots of $2x^9 - 9x^2 - 3$.
   (b) Prove that neither $x^3 + 3x - 2$ nor $2x^4 - 8x^3 + 1$ has any rational roots.
   (c) Check the following polynomials, each of which has been mentioned earlier, for rational roots: $x^3 + x - 1$, $x^4 + x^3 - 6x + 7$, $x^n + nx + 1$ (where n is an odd integer exceeding 1), $x^3 - 2$, $8x^3 - 6x - 1$, and $x^5 - 6x + 3$.
4. Provide two different explanations for the fact that the real number $\sqrt[3]{6}$ is irrational.
5. (a) Find a (in fact, *the*) monic quartic rational polynomial having $\sqrt{7} - \sqrt{2}$ as a root.
   (b) Does the polynomial in (a) have any rational roots?
   (c) Verify that, if m and n are positive integers, then $\sqrt{m} + \sqrt{n}$ is a root of $x^4 - 2(m + n)x^2 + (m - n)^2$. Generalize to find a monic quartic rational polynomial having $\alpha\sqrt{m} + \beta\sqrt{n}$ as a root, for any given rational numbers α and β.
6. Use the weak version of Eisenstein's criterion to show that neither $2x^4 + 6x^3 + 18x^2 - 21$ nor $2x^5 - 6x^3 - 21x^2 + 9x - 15$ has a rational root.

7. Prove that for each integer k, the polynomial $x^4 + 4k + 1$ has no rational root. (*Hint*: replace x by $x + 1$ and apply the weak version of Eisenstein's criterion.)
8. How was the condition $n \geqslant 2$ used in the text's proof of the weak version of Eisenstein's criterion? Is the result valid without that assumption?
9. If m and n are distinct, squarefree integers exceeding 1 and if α and β are nonzero rational numbers, prove that $\alpha\sqrt{m} + \beta\sqrt{n}$ is irrational.

*2. Content, the Lemma of Gauss, and Eisenstein Revisited

Consider the rational polynomial $f(x) = \frac{7}{3}x^2 + \frac{7}{2}x - \frac{21}{5}$. By passing to a common denominator, we may rewrite $f(x)$ as $(70x^2 + 105x - 126)/30$; then factoring the integer g.c.d. $(70, 105, -126) = 7$ from each coefficient in the numerator yields $f(x)$ as the product of $\frac{7}{30}$ and $10x^2 + 15x - 18$. Observe that $\frac{7}{30}$ is a positive rational number and $10x^2 + 15x - 18$ is a *primitive* polynomial, that is, an integral polynomial the set of whose coefficients has integer g.c.d. $= 1$: $(10, 15, -18) = 1$. From so simple an example comes the idea for the first step in a program which ultimately will reduce questions concerning factorizations in $\mathbf{Q}[x]$ to related questions couched in $\mathbf{Z}[x]$. To wit: each nonzero rational polynomial $f(x)$ may be expressed in exactly one way as the product $f(x) = c(f) \cdot f^*(x)$ where $c(f)$ is a positive rational number (called the *content* of $f(x)$) and $f^*(x)$ is a primitive polynomial. Of course, for the particular $f(x)$ which was considered above, $c(f) = \frac{7}{30}$ and $f^*(x) = 10x^2 + 15x - 18$.

Before indicating how the aforementioned program is to be accomplished, we pause to give the instructive proof that each nonzero rational polynomial has a unique description as a product of the specified type. The "existence" half of the proof follows the lines of the example's analysis. Indeed, if $f(x) = (b_0/c_0)x^n + (b_1/c_1)x^{n-1} + \ldots + (b_n/c_n)$, where the b_i and c_i are integers and each $c_i > 0$, let $k = c_0c_1 \ldots c_n$, and note that $f(x) = (a_0x^n + \ldots + a_n)/c$, where $a_i = b_i \cdot \prod_{j \neq i} c_j$. Thus, setting $D = (a_0, a_1, \ldots, a_n)$, $e_i = a_i/D$, and $p(x) = e_0x^n + e_1x^{n-1} + \ldots + e_n$ leads to $f(x) = (D/k)p(x)$. Evidently, D/k is positive, since both D and k are positive. Moreover, $p(x)$ is primitive, that is, $(e_0, e_1, \ldots, e_n) = 1$, by the integer analogue of #4(c) of Exercises 1.3, thus completing the existence proof. As for uniqueness, suppose that $f(x) = (a/b)g(x) = (c/d)h(x)$, where a and b are relatively prime positive integers, c and d are also relatively prime positive integers, and $g(x)$ and $h(x)$ are each primitive. What is the g.c.d. of the set of coefficients of the polynomial $bdf(x)$? Since that polynomial may be expressed as $adg(x)$ and $g(x)$ is primitive, another appeal to the integer analogue of #4(c) of Exercises 1.3 shows that the g.c.d. in question is ad. By the same token, $bdf(x) = bch(x)$, whence the g.c.d. in question is bc, and we have $ad = bc$. Section 1's divisibility lemma now comes to the rescue: as $a \mid bc$ and $(a, b) = 1$, we deduce that $a \mid c$. Similar reasoning yields $c \mid a$, whence $a = c$ (why?). Cancelling this common (nonzero) value from the members of the equation $ad = bc$ produces $b = d$. Finally, $g(x) = (b/a)f(x) = (d/c)f(x) = h(x)$, which completes the proof.

Which brings us to the program's main technical result, *Gauss's lemma*: if $f(x)$ and $g(x)$ are any nonzero rational polynomials, then $c(fg) = c(f) \cdot c(g)$. For a proof, write $f(x) = c(f)f^*(x)$ and $g(x) = c(g)g^*(x)$ for suitable primitive polynomials $f^*(x)$ and $g^*(x)$, so that $f(x)g(x)$ is the product of the positive rational number $c(f)c(g)$ and the integral polynomial $f^*(x) \cdot g^*(x)$. By the "uniqueness" half of the preceding result, the problem comes down to showing that $f^*(x)g^*(x)$ is primitive, that is, to showing that *the product of any two primitive polynomials is primitive.* Accordingly, we change notation and now assume that $f(x)$ and $g(x)$ are each primitive. The proof will be indirect. If $f(x)g(x)$ isn't primitive, then the fundamental theorem of arithmetic supplies a positive prime integer p such that each coefficient of $f(x)g(x)$ is an integral multiple of p.(Any prime divisor of $c(fg)$ will do.) Accordingly, $\overline{f(x)g(x)} = 0$ where, as in Section 1 of Chapter 2, "overhead-barring" denotes reduction modulo p. Then, by the second fundamental (homomorphism) property of reduction modulo p, the product $\bar{f}(x) \cdot \bar{g}(x)$ is 0 in $\mathbf{Z}_p[x]$. However, neither of the factors, $\bar{f}(x)$ and $\bar{g}(x)$, is 0 in $\mathbf{Z}_p[x]$; otherwise, each coefficient of one of f, g would be an integral multiple of p and so, by the preceding result, p would divide the content of either f or g, whence $p \mid 1$, an absurdity. Consequently, $\mathbf{Z}_p[x]$ fails to satisfy condition (vi)′ in Section 1 of Chapter 1; that is, $\mathbf{Z}_p[x]$ is not a domain. However, primeness of p guarantees that $\mathbf{Z}_p$ is a field and, in particular, a domain: recall the proof in Section 3 of Chapter 1 via **Z**'s g.c.d. algorithm. Moreover, the algebraic structure $S[x]$ is a domain for *any* domain S: recall the proof in Section 3 of Chapter 1 via the fundamental fact about degrees. In particular, $\mathbf{Z}_p[x]$ really is a domain (independently of whatever internal, provisional other assumptions were made in the body of this proof), and so the condition $\overline{f(x)g(x)} = 0$ has led to a contradiction. Thus $f(x)g(x)$ is primitive, and the proof is complete.

We shall complete the program soon, but we pause next to indicate the power of Gauss's lemma by using it to produce a new proof of Section 1's rational root test. As before, we suppose that one root of the integral polynomial $f(x) = a_0x^n + a_1x^{n-1} + \ldots + a_n$ is the rational number r/s, expressed in lowest terms as a quotient of the relatively prime integers r and s. The new proof begins by applying the factor theorem: $f(x) = q(x) \cdot (x - (r/s))$ for a suitable rational polynomial $q(x)$. Subjecting $q(x)$ to this section's first result leads to $q(x) = c(q) \cdot q^*(x)$ with, in particular, $q^*(x)$

$= d_0x^{n-1} + \ldots + d_{n-1}$ being primitive. Consequently, $f(x) = (c(q)/s)q^*(x) \cdot (sx - r)$. As the polynomials $q^*(x)$ and $sx - r$ are each primitive, Gauss's lemma guarantees that their product is also primitive. Taking $s > 0$ without loss of generality, we then have $c(f) = c(q)/s$ (why?). However, $c(f) \in \mathbf{Z}$ (why?), and so equating the corresponding leading coefficients and also the corresponding constant terms in the equation $f(x) = c(f)q^*(x)(sx - r)$ produces the equations $c(f) \cdot d_0 \cdot s = a_0$ and $c(f) \cdot d_{n-1} \cdot (-r) = a_n$, in which each displayed factor is an integer. Accordingly, $s \mid a_0$ and $r \mid a_n$, to complete the proof.

If you answered the second "why?" in the preceding paragraph, you've already made the basic observation that the content of an integral polynomial must be an integer. This has the following important consequence. If an integral polynomial $f(x)$ factors nontrivially in $\mathbf{Q}[x]$, that is, if $f(x) = g(x)h(x)$ for suitable rational polynomials $g(x)$ and $h(x)$ each of which has positive degree, then there is a corresponding factorization "in" $\mathbf{Z}[x]$: namely $f(x) = G(x)H(x)$, where $G(x)$ and $H(x)$ are integral polynomials such that $\deg(G) = \deg(g)$ and $\deg(H) = \deg(h)$. (Sketch of proof: Write $g(x) = c(g)g^*(x)$ and $h(x) = c(h)h^*(x)$ for suitable primitives $g^*(x)$ and $h^*(x)$. By Gauss's lemma, $c(g) \cdot c(h) = c(f)$ which, by the above observation, is an integer. Thus, setting $G(x) = c(f) \cdot g^*(x)$ and $H(x) = h^*(x)$ suffices.) One upshot is *the program's key fact*: in order to obtain the complete factorization in $\mathbf{Q}[x]$ for a given rational polynomial $f(x)$, write $f(x) = c(f)f^*(x)$, where $f^*(x)$ is an effectively computable primitive polynomial, and proceed to factor $f^*(x)$ as far as possible using integral polynomials with positive degrees. (Sketch of proof of "one upshot": If $f(x)$ is the product of rational polynomials $g(x)$ and $h(x)$, factor $g(x)$ and $h(x)$ as in the above sketch of proof and conclude that $f^*(x) = g^*(x) \cdot h^*(x)$.) What supplies the program's punch is the fact that the above command, "proceed to factor $f^*(x)$," may be obeyed algorithmically: see Section 3 for details.

To pursue the above techniques, we first introduce the following definition. A nonzero integral polynomial $f(x)$ is said to be *irreducible in* $\mathbf{Z}[x]$ if and only if, whenever integral polynomials $g(x)$ and $h(x)$ satisfy $f(x) = g(x)h(x)$, then precisely one of $g(x)$, $h(x)$ is either 1 or -1. (To predict the definition of "irreducible in $R[x]$" corresponding to an arbitrary domain R, note that 1 and -1 are the only integral polynomials which possess multiplicative inverses in $\mathbf{Z}[x]$.) By means of the fundamental fact about degrees, one may show (do so!) that an integer p is prime (*qua* integer) if and only if p is irreducible in $\mathbf{Z}[x]$. Moreover, to catalogue the nonconstant irreducible polynomials, first notice, via $f(x) = c(f)f^*(x)$, that an integral polynomial of positive degree which is irreducible in $\mathbf{Z}[x]$ must be primitive; and then, arguing as in last paragraph's sketch of proof, observe that a primitive polynomial of positive degree is irreducible in $\mathbf{Z}[x]$ if and only if it is prime in $\mathbf{Q}[x]$ (in the sense defined in Section 3 of

Chapter 1). We remark that virtually identical reasoning and conclusions serve to classify the polynomials which are irreducible in $R[x]$, where R is any UFD (unique factorization domain). The principal upshot, namely that $R[x]$ is a UFD if (and only if) R is a UFD, readily supplies examples, such as $\mathbf{Z}[x]$ and $\mathbf{Q}[x, y]$, of UFDs which are not Euclidean domains.

Such highbrow algebra serves perfectly to remind the reader of an important class of UFDs which was treated in Section 3 of Chapter 1, namely the domains $F[x]$ arising from fields F. The case in which F is the finite field $\mathbf{Z}_p$ will next be exploited as we shall prove the *strong version of Eisenstein's criterion.* As in Section 1, the data consist of an integral polynomial $f(x) = a_0x^n + a_1x^{n-1} + \ldots + a_n$ of degree $n \geq 2$ and a positive prime integer p such that $p \nmid a_0$, $p \mid a_i$ whenever $1 \leq i \leq n$, and $p^2 \nmid a_n$. The strong conclusion is that $f(x)$ is irreducible (that is, prime) in $\mathbf{Q}[x]$. Here is its indirect proof. After supposing that $f(x)$ is reducible in $\mathbf{Q}[x]$, apply the "important consequence" established two paragraphs ago, to get $f(x) = g(x)h(x)$, where $g(x)$ and $h(x)$ are integral polynomials with positive degrees. As in the proof of Gauss's lemma, apply the homomorphism property of reduction modulo p, whence $\bar{f}(x) = \bar{g}(x)\bar{h}(x)$. Since $p \mid a_i$ for $1 \leq i \leq n$, note that $\bar{f}(x) = \bar{a}_0x^n$. Moreover, the condition $p \nmid a_0$ guarantees that $\bar{a}_0 \neq 0 \in \mathbf{Z}_p$, so that $\bar{f}(x)$ is associated to x^n. Since x is a prime in $\mathbf{Z}_p[x]$, unique factorization then assures that $\bar{g}(x)$ and $\bar{h}(x)$ are each (nonzero) associates of appropriate powers of x. The exponents on those "appropriate powers" can't be 0 since, if, for example, $\deg(\bar{g}) = 0$, then $n = \deg(\bar{f}) = \deg(\bar{g}) + \deg(\bar{h}) = \deg(\bar{h}) \leq \deg(h) < \deg(h) + \deg(g) = \deg(f) = n$, a contradiction. Thus, $\bar{g}(x)$ and $\bar{h}(x)$ each have constant term 0 $(= \bar{0} \in \mathbf{Z}_p)$, whence p divides the constant terms of both $g(x)$ and $h(x)$. Accordingly, p^2 divides the product of those two constant terms. But that product is just a_n and we supposed that $p^2 \nmid a_n$. Therefore, the assumption of reducibility has led to a contradiction, which completes the proof of the strong version of Eisenstein's criterion.

To find applications, one need look no farther than Section 1. Thus, for instance, one infers irreducibility in $\mathbf{Q}[x]$ for $x^4 + 4x + 1$; ditto for $1 + x + (x^2/2!) + \ldots + (x^p/p!)$, where p is an arbitrary positive prime integer. Another application, which was promised in Section 3 of Chapter 3, guarantees that for each positive integer n, there exists an n-th-degree polynomial which is irreducible in $\mathbf{Q}[x]$. In fact, besides the example thereof which was described in Chapter 3, Eisenstein's criterion also establishes the (easier?) example $x^n + px + p$. Additional applications of Eisenstein's criterion are given in Exercises 4, 5, and 6 below.

As in Section 1, the strong version's applicability may be extended with the help of suitable changes of variable. Specifically, for any rational polynomial $f(x)$, the following three statements are equivalent: $f(x)$ is irreducible in $\mathbf{Q}[x]$; $g(x) = f(ax + b)$ is irreducible in $\mathbf{Q}[x]$ for each pair of rational numbers a and b such that $a \neq 0$; and $g(x) = f(ax + b)$ is irreducible in $\mathbf{Q}[x]$ for at least one pair of rational numbers a and b such that $a \neq 0$. (Sketch of proof: If $f(x)$ is irreducible, then $f(x) = f_1(x)f_2(x)$ for certain rational polynomials $f_1(x)$ and $f_2(x)$ with positive degrees. Then the second homomorphism property of evaluation gives, for all rationals a and b such that $a \neq 0$, that $g(x) = f(ax + b) = g_1(x)g_2(x)$, where $g_i(x) = f_i(ax + b)$. As $\deg(g_i) = \deg(f_i)$ (check this), $g(x)$ inherits reducibility in $\mathbf{Q}[x]$ from $f(x)$. The other cases are treated similarly.) By way of application, we next establish irreducibility in $\mathbf{Q}[x]$ for $f(x) = 1 + x + x^2 + \ldots + x^{p-1}$, where p is any prime. Indeed, since we have $f(x) = (x^p - 1)/(x - 1)$, focus instead on

$$h(x) = f(x+1) = \frac{(x+1)^p - 1}{x}$$

$$= x^{p-1} + a_1 x^{p-2} + \ldots + a_{p-1},$$

where $a_i = \binom{p}{i}$. By the hint in #8 of Exercises 1.3, each a_i is an integral multiple of p. Since $a_{p-1} = p$ is not divisible by p^2, the strong version of Eisenstein's criterion shows that $h(x)$ is irreducible in $\mathbf{Q}[x]$, and so an application of the foregoing principle completes the demonstration.

We close with an admonition. There exist polynomials whose irreducibility in $\mathbf{Q}[x]$ is not decided by the above method of combining a change of variables with Eisenstein's criterion. For instance, $f(x) = x^3 + x - 1$ is irreducible in $\mathbf{Q}[x]$ (why?) but, for any choice of rational numbers a and b such that $a \neq 0$, the strong version of Eisenstein's criterion fails to apply to the (irreducible) polynomial $g(x) = f(ax + b) = a^3x^3 + 3a^2bx^2 + (3ab^2 + a)x + (b^3 + b - 1)$.

EXERCISES 4.2

1. Determine which of the following integral polynomials is primitive: $2x^4 - 6x^3 + 15x^2 + 10$, $10x^4 + 15x^3 - 6x^2 + 2$, and $3x^9 + 9x^3 - 6$.
2. (a) Let $f(x)$ be the polynomial in #1 of Exercises 4.1. Write $f(x)$ explicitly as the product $c(f) \cdot f^*(x)$, where $c(f)$ denotes the content of $f(x)$ and where $f^*(x)$ is primitive.

   (b) For each of the polynomials in #1 above, reduce the problem of factoring it completely in $\mathbf{Q}[x]$ to the corresponding factoring problem in $\mathbf{Z}[x]$ for a related primitive polynomial.
3. Which of the following integral polynomials is irreducible in $\mathbf{Z}[x]$: 0, 1, −1, 2, −2, 3, −3, 4, −4, $2x - 6$, $x - 3$, $x^3 - x + 1$, $2x^3 - 2x + 2$?
4. Use the strong version of Eisenstein's criterion to deduce irreducibility in $\mathbf{Q}[x]$ for each of the polynomials in #6 and #7 of Exercises 4.1. By testing for primitivity, determine which (if any) of these polynomials is irreducible in $\mathbf{Z}[x]$.

5. Consider positive integers $m \leqslant n$. Use Eisenstein's criterion to show that $x^n + (1 + x)^m + (1 - x)^m$ is irreducible in $\mathbf{Q}[x]$. (*Hint*: Expand and combine like terms, treating the cases $m < n$ and $m = n$ separately.)
6. For each prime integer $p \geqslant 3$, prove that $x^p + px + 1$ is irreducible in $\mathbf{Z}[x]$. (*Hint*: replace x by $x - 1$ and use Eisenstein's criterion.) What if $p = 2$?
7. Let F be an infinite field and $f(x) \in F[x]$ an irreducible n-th-degree polynomial over F. Prove that $F[x]$ contains infinitely many pairwise-nonassociated irreducible n-th-degree polynomials. (*Hint*: when does $f(x + b_1) = f(x + b_2)$, for field elements b_1 and b_2?)

*3. The Factoring Algorithm of Kronecker

As announced, this section presents Kronecker's algorithm whereby one may effectively obtain the complete factorization in $\mathbf{Q}[x]$ of any given rational polynomial. Before proceeding further, the reader is urged to review the identity theorem (in Section 1 of Chapter 2) and Lagrange interpolation (Section 2 of Chapter 2). Because the value of recurring useful *techniques* (rather than just their *results*) must be appreciated by any working scientist, we close the section by using the two aforementioned results from Chapter 2 in order to also study the possible factorizations of an interesting family of polynomials. For both topics in this section, familiarity with the preceding section is essential.

Here is the data which Kronecker's algorithm addresses: a rational polynomial $f(x)$ of degree $n \geqslant 1$. As in Section 2, we may write $f(x)$ as $f(x) = c(f)f^*(x)$, where the effectively computable positive rational number $c(f)$ is the content of $f(x)$ and where $f^*(x)$ is a (necessarily integral) primitive polynomial. By appealing to the corollaries of Gauss's lemma in Section 2, we see that $f(x)$ is reducible in $\mathbf{Q}[x]$ if and only if there exists an integral (in fact, primitive) polynomial $g(x)$, of positive degree less than n, such that $g(x) \mid f^*(x)$, with quotient in $\mathbf{Z}[x]$. Indeed, one may restrict the hunt to g's of degree at most $[\frac{n}{2}]$ (where $[d]$ denotes the largest integer less than or equal to d). The justification for this reduction derives from the fundamental fact about degrees (in Section 3 of Chapter 1), since $f(x)$ cannot be the product of two polynomials *each* of which has degree exceeding $[\frac{n}{2}]$. Accordingly, we need only concoct a procedure which settles the reducibility/irreducibility question for an *integral* polynomial $f(x)$ of degree $n \geq 1$; by the above reasoning, such an $f(x)$ is reducible in $\mathbf{Q}[x]$ if and only if there exists an *integral* polynomial $g(x)$ such that $1 \leq \deg(g) \leq [\frac{n}{2}]$ and $g(x) \mid f(x)$.

To the hunt. Observe that *if* one had an integral polynomial $g(x)$ such that $g(x) \mid f(x)$ in $\mathbf{Q}[x]$, then the second homomorphism property of evaluation (in Section 1 of Chapter 2) would show that $g(m) \mid f(m)$ in $\mathbf{Z}$, for each integer m. We have already seen earlier in this section that knowledge of an integer's integral divisors is frequently useful, and that theme is about to be rejoined. A special case: the divisors of 0 aren't especially fascinating; so, what if $f(m) = 0$? The factor theorem would then give $f(x) = (x - m)q(x)$ and, thanks to the unique factorization result in Section 3 of Chapter 1, attention would be diverted from $f(x)$ to $q(x)$, of degree $n - 1$.

To the rejoining. Let $N = [\frac{n}{2}] + 1$ and consider N distinct integers $a_1, \ldots, a_N$ such that each evaluation $f(a_i)$ is nonzero. For each i, construct the (finite) set $\{d_{ij} : j = 1, \ldots, t_i\}$ of distinct integral divisors of $f(a_i)$. (For example, if $f(a_i) = 2$, you'd be considering the set $\{1, -1, 2, -2\}$.) Corresponding to each selection $b_1, \ldots, b_N$ such that $b_i \in \{d_{ij} : j = 1, \ldots, t_i\}$, use Lagrange interpolation to construct *the* rational polynomial $h(x)$ of degree at most $[\frac{n}{2}]$ such that $h(a_i) = b_i$ for each i. If $h(x) \in \mathbf{Z}[x]$ and has positive degree, divide polynomials to check whether $h \mid f$. If the response is affirmative, then $h(x)$ is a satisfactory $g(x)$, and the further factoring (if any) of $f(x)$ is treated by similar, separate analyses of $h(x)$ and $f(x)/h(x)$. If *no* choice of $b_1, \ldots, b_N$ leads to an $h(x)$ for which the check whether $h \mid f$ has an affirmative response, then $f(x)$ is irreducible in $\mathbf{Q}[x]$.

Here is a typical illustration of the above algorithm. Presented with the task of factoring $f(x) = \frac{2}{7}x^4 + \frac{8}{7}$ in $\mathbf{Q}[x]$, observe that $c(f) = \frac{2}{7}$ and consider, instead, ths problem of factoring the primitive polynomial $f^*(x) = x^4 + 4$. In this case, $n = 4$, so that $N = [\frac{n}{2}] + 1 = 3$. With an eye to generating reasonably short lists of divisors, select $a_1 = 0$, $a_2 = 1$, and $a_3 = -1$. Then $f(a_1) = 4$ whose set of integral divisors is $\{1, -1, 2, -2, 4, -4\}$, and $f(a_2) = 5 = f(a_3)$ with the set(s) of integral divisors $\{1, -1, 5, -5\}$. Consequently, there are $6 \cdot 4 \cdot 4 = 96$ possible selections of (b_1, b_2, b_3). Let's begin sifting through these 96 candidates. The selection $b_1 = 1$, $b_2 = -1$, $b_3 = 5$ leads to the unique rational polynomial $g(x)$ of degree at most 2 such that $g(0) = 1$, $g(1) = -1$, and $g(-1) = 5$. By Lagrange interpolation, the corresponding candidate is

$$g(x) = \frac{1(x-1)(x+1)}{(0-1)(0+1)} + \frac{(-1)(x-0)(x+1)}{(1-0)(1+1)} + \frac{5(x-0)(x-1)}{(-1-0)(-1-1)}$$

$= x^2 - 3x + 1$ which, although integral, does not divide $f^*(x)$. (Expletive deleted.) So: one down, at most ninety-five to go! Next try the selection $(b_1, b_2, b_3) = (2, 5, 1)$, and check that the corresponding $g(x) = x^2 + 2x + 2$. This time, following the dictates of the algorithm leads to success, as division of $f^*(x)$ by the newly acquired $g(x)$ reveals that $x^4 + 4 = (x^2 - 2x + 2)(x^2 + 2x + 2)$. Moreover, $x^2 - 2x + 2$ and $x^2 + 2x + 2$ are each irreducible in $\mathbf{Q}[x]$: either repeat the algorithm for them or (more sensibly, for quadratics) notice (by either the quadratic formula or Section 1's rational root test) that they have no rational roots. Therefore, the desired complete factorization is given by

$$\frac{2}{7}x^4 + \frac{8}{7} = \frac{2}{7}(x^2 - 2x + 2)(x^2 + 2x + 2).$$

Exercise 1 below will provide the reader with another opportunity to apply Kronecker's algorithm. He/she may also wish to gnaw on the somewhat more taxing example, $(x^5 + x^4 + x^2 + x + 2)/16$, in the meantime.

Finally, we shall address the following problem. If n is a positive integer and $a_1, \ldots, a_n$ are distinct integers, determine the complete factorization in $\mathbf{Q}[x]$ of

$$f(x) = (x - a_1)(x - a_2) \ldots (x - a_n) + 1.$$

A straightforward calculation reveals *reducibility in two cases*:

(i) $n = 2$, $a_1 - a_2 = 2$: then $f(x) = (x - a_2 - 1)^2$; and

(ii) $n = 4$ and the a_i are consecutive, say $a_2 = a_1 - 3$, $a_3 = a_1 - 2$, and $a_4 = a_1 - 1$: then $f(x) = ((x - a_1)(x - a_1 + 3) + 1)^2$, and the quadratic inside the outer parentheses cannot be factored further in $\mathbf{Q}[x]$.

The following shows how one might sensibly hope to have discovered (i) and (ii) and shows, moreover, that $f(x)$ *is irreducible* in $\mathbf{Q}[x]$ *in all other cases.*

Consider the situation when $f(x)$ happens to be reducible in $\mathbf{Q}[x]$. Then, by the results in Section 2, $f(x) = g(x)h(x)$ for certain (primitive) integral polynomials, $g(x)$ and $h(x)$, with positive degree(s). Observe, for each $i = 1, \ldots, n$, that $g(a_i)h(a_i) = f(a_i) = 1$ (why?) and so, since $g(a_i)$ and $h(a_i)$ are integers, one necessarily has $g(a_i) = h(a_i)$ $(= \pm 1)$. Accordingly, the identity theorem guarantees that $g(x) = h(x)$ (explain), whence $f(x) = g(x)^2$. If $r = \deg(g)$, then comparing degrees establishes that $n = 2\deg(g) = 2r$ (explain), an even integer. Next, we claim that $g(a_i) = 1$ for precisely half of the indexes i (in which case, $g(a_j) = -1$ for the other r indexes). Indeed, if $g(x)$ takes on the value 1 more often than asserted, relabel so that $g(a_i) = 1$ for $i = 1, \ldots, r, \ldots, s$, where $r < s \leq n$. Then the identity theorem and Lagrange interpolation conspire, as in #4(b) of Exercises 2.2, to show that $g(x) = (x - a_1) \ldots (x - a_r) \ldots (x - a_s) + 1$, *the* monic rational polynomial of degree at most s which takes the value 1 at each of $a_1, \ldots, a_r, \ldots,$ and a_s. One then has the desired contradiction: $s = \deg(g) = r$. Moreover, $g(a_i)$ takes on the value 1 for *at least* r of the indexes i since, otherwise, one proves as above that $g(a_j)$ is -1 for *too many* indexes j. This establishes the claim. After we relabel so that $g(a_1) = \ldots = g(a_r) = 1$ and $g(a_{r+1}) = \ldots = g(a_n) = -1$, the identity theorem/Lagrange interpolation axis yields two expressions for $g(x)$:

$$(x - a_1) \ldots (x - a_r) + 1 = g(x) = (x - a_{r+1}) \ldots (x - a_n) - 1$$

which may be exploited via the second homomorphism property of evaluation. Indeed, by successively substituting $a_1, \ldots, a_r$ for x and simplifying, we have

$$(a_1 - a_{r+1}) \ldots (a_1 - a_n) = 2,$$

$$\vdots$$

$$(a_r - a_{r+1}) \ldots (a_r - a_n) = 2,$$

each of which equations expresses 2 as a product of r distinct integers. Now, apart from the order of factors the *only* such expressions are 2, $2 \cdot 1$, $(-2) \cdot (-1)$, and $(-2) \cdot (-1) \cdot 1$, the longest of which displays three factors. Therefore, $r \leq 3$. We next proceed to treat the cases $r = 1$, $r = 2$, and $r = 3$ separately.

Case $r = 1$ (so that $n = 2$): in this case, only one representation of 2 was gleaned above, it states that $a_1 - a_2 = 2$ (possibly after relabelling), and this situation was designated as instance (i) above.

Case $r = 2$ (so that $n = 4$): in this case, the above representations of 2 assert that $(a_1 - a_3)(a_1 - a_4) = 2 = (a_2 - a_3)(a_2 - a_4)$. One way that this could happen would entail $a_1 - a_3 = 2$, $a_1 - a_4 = 1$, $a_2 - a_3 = -1$, and $a_2 - a_4 = -2$, the situation of consecutive integers noted as instance (ii) above. (The reader is urged to check inconsistency of the set of four equations $a_1 - a_3 = 2$, $a_1 - a_4 = 1$, $a_2 - a_3 = -2$, and $a_2 - a_4 = -1$. Apart from relabelling, there are no additional subcases.) To explain the magic factorization announced earlier in (ii), you need only observe that

$$g(x) = (x - a_1)(x - a_2) + 1 = (x - a_1)(x - a_1 + 3) + 1.$$

Incidentally, regardless of the choice of a_i, this quadratic $g(x)$ is irreducible in $\mathbf{Q}[x]$ because its discriminant is the constant (fill in this space, dear reader).

(Final) *case* $r = 3$ (so that $n = 6$): the above necessary condition for reducibility asserts that, by permuting the three factors in the equation $2 = (-2) \cdot (-1) \cdot 1$, one has

$$2 = (a_1 - a_4)(a_1 - a_5)(a_1 - a_6) = (a_2 - a_4)(a_2 - a_5)(a_2 - a_6)$$
$$= (a_3 - a_4)(a_3 - a_5)(a_3 - a_6).$$

Without any loss of generality (that is, by relabelling) take $a_1 - a_4 = -2$, $a_1 - a_5 = -1$, and $a_1 - a_6 = 1$. Since $(a_2 - a_4) - (a_2 - a_5) = a_5 - a_4 = (a_1 + 1) - (a_1 + 2) = -1$, the ordered triple $(a_2 - a_4, a_2 - a_5, a_2 - a_6)$ cannot be any of the possibilities $(-2, 1, -1)$, $(1, -1, -2)$, $(1, -2, -1)$, $(-1, 1, -2)$, $(-1, -2, 1)$. Therefore, $(a_2 - a_4, a_2 - a_5, a_2 - a_6) = (-2, -1, 1)$ and, in particular, $a_2 - a_4 = a_1 - a_4$, forcing $a_1 = a_2$, the desired contradiction. Thus, $f(x)$ is irreducible when $n = 6$, and the analysis has been completed.

EXERCISES 4.3

1. Factor $f(x) = 5x^5 + 16x^4 + 6x^3 - 4x^2 + x + 1$ in $\mathbf{Q}[x]$ as the product of a cubic and a quadratic. Verify that you have obtained the complete factorization of $f(x)$ in $\mathbf{Q}[x]$.
2. Factor rational polynomials of your own choosing (for example, ones met earlier in this chapter), by means of this section's algorithm. In particular, reestablish that each of the polynomials in #6 of Exercises 4.1 and #6 of Exercises 4.2 is irreducible in $\mathbf{Q}[x]$.
3. Let n be a positive integer, and let $a_1, a_2, \ldots, a_n$ be distinct integers. Discuss the factorization in $\mathbf{Q}[x]$ of the polynomial $(x - a_1)(x - a_2) \ldots (x - a_n) - 1$. (*Hint*: as in the text, reduce to the case of even n. But then consider the polynomial function's limit as $n \to \infty$, peeking at Section 1 of Chapter 5, if necessary.)

5

Polynomials over the Real Numbers

This chapter comes to grips with the problem of analyzing the roots of a real polynomial. (As is shown at the start of Section 1, it is no more difficult, in principle, to treat the case of a complex polynomial.) By virtue of the fundamental theorem of algebra, the roots are either real numbers or nonreal complex numbers. Besides reexamining the multiplicities in such a catalogue, Section 1 obtains some preliminary results on the number of real roots by means of two tools which are familiar from study of the calculus, namely, limit computations (which have already been met in considering Bernoulli iteration in Section 4 of Chapter 2) and the intermediate-value theorem for continuous functions. The aforementioned preliminary results deal with signs of polynomials' coefficients—at once a remembrance of #8 in Exercises 2.1 concerning integral roots and a powerful motivation for the remainder of this chapter—and are sharp enough to close the gap which was (intentionally and explicitly) left in the proof of the fundamental theorem of algebra in Section 6 of Chapter 3. Sharper results are obtained in Section 2, with the help of appropriate strengthenings of the intermediate-value theorem and Rolle's theorem (another result familiar from calculus). In certain easy cases (when the roots of a given polynomial's derivative are known), these lead to the *separation* or isolation of the real roots into nonoverlapping open intervals, each containing at most one real root. (Such separation is the goal of any theoretical analysis of real roots and must, in many cases, be accomplished before one can sensibly interpret the results of Chapter 6's numerical root-approximation techniques.) The study of the sequence of signs of a polynomial's coefficients leads, in Section 3, to the notion of *variations* and thereby to Descartes's rule of signs, a rough, but often useful, estimate on the number of real roots. Finally, Section 4 presents *the* answer, namely the notion of *Sturm sequences* and the theorem of Sturm, which produces the precise number of real roots of *any* given real polynomial and then algorithmically isolates those roots. The organic nature of our subject is reinforced by the basic role of variations in Section 4. Even more pleasing in this regard is the explanation why Sturm's isolation technique is algorithmic: everything comes down to Euclid's g.c.d. algorithm (from Section 3 of Chapter 1). Hence, as was the case in each earlier chapter, one may regard the thrust of this chapter as a celebration of the division algorithm and, more generally, of the consequences of the fundamental fact about degrees. In (the optional) Sections 5 and 6, a proof is given for Sturm's theorem, and related theorems of Budan and Vincent are sketched.

1. The Nature of the Real Roots and of the Unreal Roots

As promised, we shall begin by indicating how, in theory, one may reduce the study of the roots of a complex polynomial to the special case involving a real polynomial. Consider, then, $f(x) \in \mathbf{C}[x]$. Let $f^*(x)$ be the polynomial obtained from f by conjugating f's coefficients. For example, if $f(x) = x^2 + (1 - i)x - i$, then $f^*(x) = x^2 + (1 + i)x + i$. It was shown in the fourth paragraph of Section 6 of Chapter 3 that each coefficient of the product $g(x) = f(x)f^*(x)$ is a real number; and, moreover, that (counting multiplicities) the roots of $g(x)$ are described by merely listing the roots of $f(x)$ and the conjugates of the roots of $f(x)$. Consequently, if we are (somehow) given the roots of $g(x)$, we may then obtain the roots of $f(x)$ by testing a "few" possibilities. For example, if $f(x)$ is the above complex quadratic, then $g(x) = x^4 + 2x^3 + 2x^2 + 2x + 1$ (check), whose roots are found (either by the factor theorem, two applications of the rational root test and the quadratic formula, or by one of the methods in Section 4 of Chapter 3) to be -1, -1, i, and $-i$. Accordingly, the preceding discussion reveals that -1 is a root of $f(x)$, and so is either i or $-i$. By evaluation, the former possibility is shown to hold. Of course, the quadratic formula applies directly to this example, giving (once again) that -1 and i are the roots of $f(x)$. Note also that the procedure of passing from $f(x)$ to the corresponding $g(x)$ is applicable to complex polynomials of arbitrary degree, not just to quadratics, so that the material in Chapter 6 combines well with the present discussion in order to *numerically approximate the roots of complex polynomials.* Practice in applying the ideas of this paragraph will be afforded by Exercises 1 and 2 below. Apart from #2, *each polynomial henceforth considered in this chapter will be assumed to have real coefficients.*

The basic result on cataloguing the roots of a (real) polynomial $f(x)$ was given as the final remark in Section 3 of Chapter 2. Recall that it stated that whenever a complex number z is a root of $f(x)$ with multiplicity m, then the conjugate z^* is also a root of $f(x)$ with multiplicity m. We pause to offer an alternate, direct proof of this important assertion. First, notice that it is enough to prove, in case $z = a + bi$ for real numbers a and b, that $f(x)$ is divisible by $h(x) = (x - a)^2 + b^2$, since $h(x)$ is the product of the (whoops, complex) polynomials $x - z$ and $x - z^*$. (Strictly speaking, this would only dispatch the case $m = 1$. Fill in the details in case $m > 1$ by appropriate appeals to unique factorization and the factor theorem.) For the proof, apply the division algorithm in $\mathbf{R}[x]$ to obtain $f(x) = q(x) \cdot h(x) + Ax + B$, for a suitable polynomial $q(x)$ and real numbers A and B. Since $h(z) = 0$, the homomorphism properties of evaluation yield that $0 = f(z) = 0 + Az + B = (Aa + B) + Abi$, whence $Aa + B = 0 = Ab$. Now, without loss of generality, $b \neq 0$ (else, there is nothing to prove!), so that $A = 0$ (by the far reaches of property (vi)′ in Section 1 of Chapter 1), and then $B = -Aa = 0$ as well. Accordingly, $f(x) = q(x)h(x)$, whence $f(z^*) = q(z^*)h(z^*) = q(z^*) \cdot 0 = 0$, which completes the proof.

The preceding discussion, when coupled with unique factorization, shows that each nonzero real polynomial $f(x)$ can be expressed in the form $f(x) = a_0(x - r_1)^{m_1} \ldots (x - r_k)^{m_k}F(x)$, where a_0 is the leading coefficient of $f(x)$; $r_1, \ldots, r_k$ are the distinct real roots of $f(x)$ with corresponding multiplicities $m_1, \ldots, m_k$; and the real polynomial $F(x)$ is positive when evaluated at each real number. Specifically, $F(x)$ is the product of the $h(x)$'s considered above; that is, $F(x)$ is the product of the polynomials $(x - a)^2 + b^2$ arising from conjugate pairs of nonreal roots, $a + bi$ and $a - bi$, of $f(x)$. This paragraph's observation will lead to a quick proof of a key lemma in Section 2. Another application of it is given next.

We next proceed to characterize the irreducible polynomials in $\mathbf{R}[x]$. As observed in Section 3 of Chapter 1, attention may be restricted to monic polynomials, since associates of irreducibles are irreducible. Of course, each monic linear (degree-1) polynomial is irreducible: that's true over *any* field. Over $\mathbf{C}$, the story ends there, by virtue of the fundamental theorem of algebra. As (some of) you saw in (the optional) Sections 2 and 3 of Chapter 4, the situation is much more complicated over $\mathbf{Q}$. We next show that, over $\mathbf{R}$, the story hasn't yet ended, but neither is it complicated! Specifically, the monic irreducible real polynomials with degree exceeding 1 are the quadratics $x^2 + bx + c$ with discriminant $b^2 - 4c < 0$. Indeed, the discussion in the preceding paragraph leads to the description of the "extra" irreducibles as those quadratics of the form $(x - a)^2 + b^2 = x^2 - 2ax + (a^2 + b^2)$ with $a \in \mathbf{R}$ and $0 \neq b \in \mathbf{R}$. Such polynomials have discriminant $(-2a)^2 - 4(a^2 + b^2) = -4b^2$, which is indeed negative. Moreover, real quadratics with negative discriminant have no real roots (by the quadratic formula), and hence must be irreducible (why?), completing the proof of the "real story." Note, incidentally, that calculus's approach to the denominators used in partial fractions has now been reconciled with the treatment in Section 4 of Chapter 1.

Next, we intensify our attention on the view of polynomial as function. Recall the *intermediate-value theorem*: if $g(x)$ is a continuous real-valued function defined on the closed interval $[a, b]$ such that $g(a)$ and $g(b)$ are nonzero and of opposite sign, then $g(c) = 0$ for some c such that $a < c < b$. (Put differently, this location principle for roots of the continuous function $g(x)$ states: the graph of $g(x)$ crosses the x-axis between a and b in case one of the points $(a, g(a))$, $(b, g(b))$ lies above the x-axis and the other lies below the x-axis.) Of course, the intermediate-value theorem *is* applicable to polynomials, since each polynomial function, being differentiable, is necessarily continuous. An immediate corollary of these comments asserts that if a (real) polynomial $f(x)$ has no roots in $[a, b]$, then $f(d)$ is either always positive or always negative as d varies over $[a, b]$. For an important class of "rootless" polynomials of the type just mentioned, see property (4) of Sturm sequences in Section 4 below. Later in this section, we shall present other corollaries of the intermediate-value theorem, which will guarantee the presence of real roots in certain cases. These must await the introduction of limit techniques for detecting sign changes amongst a polynomial's values, to which we now turn.

Consider a nonzero n-th-degree polynomial $f(x) = a_0x^n + \ldots + a_n$. The main facts needed below may be summarized as follows:

$$\lim_{x \to \infty} f(x) = \begin{cases} \infty, & \text{if } a_0 > 0, \\ -\infty, & \text{if } a_0 < 0, \end{cases}$$

and

$$\lim_{x\to-\infty} f(x) = \begin{cases} \infty, & \text{if } a_0 > 0 \text{ and } n \text{ is even,} \\ -\infty, & \text{if } a_0 < 0 \text{ and } n \text{ is even,} \\ -\infty, & \text{if } a_0 > 0 \text{ and } n \text{ is odd,} \\ \infty, & \text{if } a_0 < 0 \text{ and } n \text{ is odd.} \end{cases}$$

In particular, these facts assert that for each real number d whose absolute value is sufficiently large, $f(d)$ is nonzero and has the same sign as its "first term," a_0d^n. As sign (change)s are all that need concern us while applying the intermediate-value theorem, we shall use the symbols $f(\infty)$ and $f(-\infty)$ to denote the sign of the type of infinity which was asserted to be the corresponding limit of $f(x)$. For example, $f(\infty) = -$ if $a_0 < 0$; $f(-\infty) = +$ if $a_0 > 0$ and n is even; etc.

It is rather easy to establish the above limit assertions. Indeed, it is enough to show that the limit (as x approaches the infinity of your choice) of $f(x)/a_0x^n$ is 1 (why does this suffice?) or, equivalently, that the limit of $g(x) = (f(x) - a_0x^n)/a_0x^n$ is 0. Now, $g(x) = (a_1/a_0)(1/x) + (a_2/a_0) \cdot (1/x)^2 + \ldots + (a_n/a_0)(1/x)^n$ and, since the limit of $1/x$ is 0, standard limit theorems yield the limit of $g(x)$ to be $(a_1/a_0)0 + (a_2/a_0)0^2 + \ldots + (a_n/a_0)0^n = 0$, as desired. (The reader may also wish to evaluate the limit of $g(x)$ with the aid of $n - 1$ applications of L'Hospital's rule.)

Next, we turn to the promised corollaries which guarantee the presence of real roots. The first of these asserts that if $f(x)$ is a (real) polynomial of odd degree, then $f(x)$ has at least one real root. (*Don't* use the fundamental theorem of algebra to prove this! That would be circular reasoning, as the result to be proved now was already employed in the proof given earlier for the fundamental theorem of algebra.) To prove this via the intermediate-value theorem, you only need to observe that, because $\deg(f)$ is odd, the signs $f(-\infty)$ and $f(\infty)$ are not the same. More precisely put: if you select positive b and negative a with sufficiently large absolute value, so that the sign of $f(a)$ is $f(-\infty)$ and the sign of $f(b)$ is $f(\infty)$, then you're guaranteed the existence of a root of $f(x)$ in the open interval (a, b).

Finally, consider the case of a real polynomial $f(x) = a_0x^n + \ldots + a_n$ whose leading coefficient, a_0, and constant term, a_n, are nonzero and of opposite sign. Then we assert that $f(x)$ is guaranteed to have at least one *positive* root (but needn't have a negative root). Indeed, the presence of a positive root follows from the intermediate-value theorem, since $f(\infty)$ is the sign of a_0 while $f(0)$ *is* a_n. (Question: if n were even, must $f(x)$ have a *negative* root?)

Although the result in the preceding paragraph will be sharpened early in Section 2, it is already strong enough to show, for example, that $x^4 - x - 2$ and $x^2 - 7$ each have at least one positive root. In fact, an easy consequence of Descartes's rule of signs (Section 3) will be that these two polynomials each have precisely one positive root and precisely one negative root.

EXERCISES 5.1

1. Find all the roots of $f(x) = 2x^5 - 4x^4 + 4x^3 - 8x^2 + 2x - 4$, and hence obtain the complete factorization of $f(x)$ in $\mathbf{R}[x]$. (*Hint*: use the rational root test (Section 1 of Chapter 4) and either method in Section 4 of Chapter 3 for the analysis of the resulting quartic quotient.)
2. Let $r_1, \ldots, r_4$ denote the roots of the complex quartic $f(x)$ listed in #6 of Exercises 2.3. Use the "$f(x)f^*(x)$ trick" to obtain a real octic (eighth-degree polynomial) with the roots $r_1, r_1^*, \ldots, r_4, r_4^*$.
3. Compute the signs defined in the text as $f(\infty)$ and $f(-\infty)$ for each of the following values of $f(x)$: $2x^4 - \pi x^3 - \sqrt{2}$, $-\pi x^4 + \sqrt{2}x^3 + 2x + 7$, $\sqrt{2}x^5 + \pi x^2 - 7x - 2$, and $-7x^5 + 2x^4 - \sqrt{2}x^3 - \pi$.
4. Using the results in this section concerning parity of degree and comparison of the signs of leading and constant coefficients, show that each of the polynomials listed in #3 has at least one real root. Use the intermediate-value theorem to isolate the discerned real roots, with the aid of open intervals of finite length.

2. Rolle's Theorem and Applications

The reader will recall from his/her study of the calculus that the derivative $f'(x)$ may be interpreted as the slope of the tangent line to the graph of the differentiable function $f(x)$. One familiar consequence of such a geometric, nonformal view of $f'(x)$ is the theorem of Rolle. The most important result in this section is a strengthening of Rolle's theorem for the special case of a polynomial function. Several applications are given, most importantly the isolation of roots in certain instances. These results issue from a corresponding strengthening of the intermediate-value theorem, which we shall call the *parity lemma.*

The parity lemma asserts that if $f(x)$ is a real polynomial function considered on a closed interval $[a, b]$ and if neither a nor b is a root of $f(x)$, then the number of roots, counting multiplicities, of $f(x)$ in the open interval (a, b) is even or odd according as to whether $f(a)$ and $f(b)$ do or do not have the same sign. For a proof, use Section 1's observation that $f(x) = (x - r_1) \ldots (x - r_k)(x - s_1) \ldots (x - s_m)(x - t_1) \ldots (x - t_n)F(x)$, where (always listing with multiplicity) $r_1, \ldots, r_k$ are the roots of $f(x)$ in (a, b); $s_1, \ldots, s_m$ are the roots of $f(x)$ which are less than a; $t_1, \ldots, t_n$ are the roots of $f(x)$ which are greater than b; and the polynomial $F(x)$ is positive when evaluated at each real number. Since each of the fractions $(a - s_i)/(b - s_i)$ and $(a - t_j)/(b - t_j)$ is positive, the sign of $f(a)/f(b)$ is the same as the sign of $(a - r_1) \ldots (a - r_k)/(b - r_1) \ldots (b - r_k)$. As each fraction $(a - r_u)/(b - r_u)$ is negative, $f(a)/f(b)$ therefore shares the sign of $(-1)^k$, from which the assertion is immediate. For some applications of the partity lemma, see Exercise 1 below.

One consequence of the parity lemma deserves to be noted at this point. On the one hand, it will be seen to extend the final result in Section 1; and, on the other hand, it amounts to a substantial fragment of Descartes's rule of signs. (For the other fragment, see Section 3). Specifically, a (real) polynomial $f(x) = a_0x^n + \ldots + a_n$, with $a_0 \neq 0 \neq a_n$, has an even or an odd number of positive roots, counting multiplicities, according as to whether a_0 and a_n do or do not have the same sign. (Proof sketch: as in Section 1, note that $f(0) = a_n$, while $f(\infty)$ is the sign of a_0.) You may cut your teeth, or whatever, on the above fragment by considering Exercise 2 below.

The idea of playing off $f(a)$ against $f(b)$ two paragraphs ago may suggest playing off $f'(x)$ against $f(x)$, i.e., considering the fraction $f'(x)/f(x)$. And indeed, such whimsy is as familiar to us as logarithmic differentiation, and has already paid off, in Section 4 of Chapter 2, with a proof of Newton's formulas. We next present another payoff of this whimsy, an immediate consequence of which will be the desired extension of Rolle's theorem. Specifically, if d is a real root of the polynomial $f(x)$, then $\lim_{x \to d^-}(f'(x)/f(x)) = -\infty$ and $\lim_{x \to d^+}(f'(x)/f(x)) = \infty$; in particular, if the positive number ϵ is sufficiently small, then $f'(d - \epsilon)/f(d - \epsilon)$ is negative and $f'(d + \epsilon)/f(d + \epsilon)$ is positive. For a proof, suppose that the multiplicity of d as a root of $f(x)$ is $m \geq 1$, so that $f(x) = (x - d)^m g(x)$, for some polynomial $g(x)$ such that $g(d) \neq 0$. After differentiating $f(x)$ with the aid of rules (a)–(e) in Section 3 of Chapter 2, simplify to get

$$\frac{f'(x)}{f(x)} = \frac{g'(x)}{g(x)} + \frac{m}{x - d}.$$

Since $g(d) \neq 0$ and $g'(x)/g(x)$ is a continuous function, $g'(x)/g(x)$ is bounded (in absolute value) "near" d. However, $\lim_{x \to d^-}(m/(x - d)) = -\infty$ and $\lim_{x \to d^+}(m/(x - d)) = \infty$, from which the desired assertion follows easily.

Which leads us to *Rolle's theorem*: if real roots a and b of a (real) polynomial $f(x)$ are *consecutive roots*, in the sense that $a < b$ and the open interval (a, b) contains no roots of $f(x)$, then there is at least one—and counting multiplicities, precisely an odd number—of roots of $f'(x)$ in (a, b). For a proof, observe, via the preceding result, that $f'(a + \epsilon)/f(a + \epsilon) > 0$ and $f'(b - \epsilon)/f(b - \epsilon) < 0$ for sufficiently small $\epsilon > 0$. By consecutivity of a and b, the parity lemma reveals that $f(a + \epsilon)$ and $f(b - \epsilon)$ have the same sign, whence $f'(a+\epsilon)$ and $f'(b-\epsilon)$ must have opposite signs. Accordingly, another application of the parity lemma, this time to the polynomial $f'(x)$, shows that, counting multiplicities, the number of roots of $f'(x)$ in the open interval $(a+\epsilon, b-\epsilon)$ is odd. However, by the blockbuster, one may choose ϵ so that neither $(a, a+\epsilon]$ nor $[b-\epsilon, b)$ contains a root of $f'(x)$, from which Rolle's theorem rolls.

Notice that Rolle's theorem may be rephrased as the assertion that there can be at most one (*not* counting multiplicities) root of a (real) polynomial $f(x)$ which is strictly between two consecutive roots of $f'(x)$. Accordingly, *if* the distinct real roots $r_1 < \ldots < r_k$ of $f'(x)$ are known, then isolation of $f(x)$'s real roots is effected by the intervals $(-\infty, r_1)$, $[r_1, r_2), \ldots, [r_{k-1}, r_k)$, $[r_k, \infty)$, in the sense that each of these intervals contains *at most one* of $f(x)$'s roots. To see which (if any) of these *do* contain a root of $f(x)$, check to see if any endpoint r_j is a root and then use the parity lemma to treat the open intervals $(-\infty, r_1)$, $(r_1, r_2), \ldots, (r_{k-1}, r_k)$, (r_k, ∞). For example, if $f(x) = 2x^3 - 3x^2 - 36x - 90$, then $f'(x) = 6x^2 - 6x - 36 = 6(x+2)(x-3)$ has the roots $r_1 = -2$ and $r_2 = 3$. Thus, by the preceding remarks, the intervals $(-\infty, -2)$, $[-2, 3)$, and $[3, \infty)$ each contain at most one root of $f(x)$. Straightforward computations give $f(-\infty) = -$, $f(-2) = -46 < 0$, $f(3) = -171 < 0$, and $f(\infty) = +$, whence isolation is effected: namely, in this case $f(x)$ has but one real root, and that root is in the interval $(3, \infty)$. By similar reasoning, one may show (and *you* should) that the polynomial $2x^3 - 3x^2 - 36x - 9$ has three real roots, one each in the intervals $(-\infty, -2)$, $(-2, 3)$, and $(3, \infty)$. For additional illustrations of this paragraph's isolation technique, see Exercise 3 below.

We next introduce a very useful inequality. To wit: if the (nonzero, real) polynomial $f(x)$ has precisely m real roots, then $f'(x)$ has *at least* $m-1$ real roots, counting multiplicities in both cases. For a proof, let $r_1 < \ldots < r_k$ be the distinct real roots of $f(x)$, with corresponding multiplicities $m_1, \ldots, m_k$ (so that $m_1 + \ldots + m_k = m$). Now the discussion in Section 3 of Chapter 2 establishes (since $\text{char}(\mathbf{R}) = 0$) that each r_i is a root of $f'(x)$ with multiplicity $m_i - 1$. (If $m_i = 1$, this means that the corresponding r_i is not a root of $f'(x)$). Moreover, by Rolle's theorem, $f'(x)$ is guaranteed to have at least one root in each of the $k-1$ open intervals $(r_1, r_2), \ldots, (r_{k-1}, r_k)$. Therefore, the number of real roots of $f'(x)$, counting multiplicities, is at least $(m_1 - 1) + \ldots + (m_k - 1) + (k-1) = m - 1$, completing the proof.

Giving equal time to the nonreal roots, we note the following equivalent formulation of the above result: if the (nonzero, real) polynomial $f(x)$ has precisely M nonreal (complex) roots, then $f'(x)$ has *at most* M nonreal roots, counting multiplicities in both cases. (In particular, if each root of $f(x)$ is real, then each root of $f'(x)$ is also real, an observation that will be useful in dealing with Exercise 4(a) below.) For a proof, let $n = \deg(f)$ and observe that $m + M = n$, by the fundamental theorem of algebra. Similarly, since $\deg(f') = n - 1$ (as noted in Section 3 of Chapter 2, because $\mathrm{char}(\mathbf{R}) = 0$), we have $p + P = n - 1$, where p (respectively, P) denotes the number of real (respectively, nonreal) roots of $f'(x)$. As $p \geq m - 1$ by the above result, we conclude that $P = n - 1 - \mathrm{p} \leq n - 1 - (m - 1) = n - m = M$, as was asserted.

We pause to mention an analytic application. Let n be a positive integer, and consider two real numbers $a < b$. Evidently, $f(x) = (x - a)^n(x - b)^n$ has but two distinct roots, namely a and b, each with multiplicity n. By reasoning as in the preceding two paragraphs, we see that $h_1(x) = f'(x)$ has roots a and b, each with multiplicity $n - 1$, as well as a simple root strictly between a and b. (Fill in the details.) Similarly, $h_2(x) = f''(x)$ has roots a and b, each with multiplicity $n - 2$, as well as two simple roots strictly between a and b. Iteration of such considerations reveals that $h_n(x) = f^{(n)}(x)$ is an n-th-degree polynomial, each of whose roots is real, simple, and contained in the open interval (a, b). Passing to the case $a = -1$, $b = 1$, and considering an associate of $h_n(x)$, we have that the n-th *Legendre polynomial*, $(1/2^n n!)(d^n((x^2 - 1)^n)/dx^n)$, is of degree n and has n simple real roots, each lying strictly between -1 and 1. As noted in Exercise 5(b), the n-th Legendre polynomial satisfies an ordinary differential equation of considerable physical importance. The above expression for the n-th Legendre polynomial is usually referred to as the *formula of Rodrigues* [6, p. 160]. For a rather different approach to the roots of Legendre polynomials, see [6, Sections 7 and 8].

To close the section, we give two further applications of the above ideas. As is the case with much of this chapter, they are drawn from [24, Chapter VI]. First, let $f(x)$ be a nonzero n-th-degree (real) polynomial with n real roots, counting multiplicities, and consider a real number $k \neq -n$. Then we claim that $G(x) = xf'(x) + kf(x)$ is also an n-th-degree polynomial with n real roots, counting multiplicities. We shall merely sketch the proof, with the following three observations: considering the sign of $G(x)/f(x)$ as x varies, note that $G(x)$ has at least one root strictly between any preassigned pair of consecutive roots of $f(x)$; since each root r of $f(x)$ of multiplicity m entails $f(x) = (x - r)^m g(x)$, where $g(r) \neq 0$, it follows that r is a root of $G(x)$ with multiplicity at least $m - 1$; and, thus, $G(x)$ has at least $n - 1$ real roots, counting multiplicities. The reader who has fathomed these cryptic hints has not only divined how to adapt the earlier proofs in this section, but is also ready to tackle the case in which not all roots need be real: see Exercise 4(b) below.

Finally, we shall consider the nature of the roots of the polynomial $f(x) = 1 + x + (x^2/2) + \ldots + (x^n/n)$, where n is a positive integer. (An alternate analysis, using Sturm's theorem, will be sketched in #4 of

Exercises 5.4.) Observe that

$$f'(x) = 1 + x + x^2 + \ldots + x^{n-1} = \frac{x^n - 1}{x - 1}.$$

In the case that n is odd, the analysis of n-th roots in Section 2 of Chapter 1 reveals that $f'(x)$ has no real roots, whence $f(x)$ has *at most* one real root. However, as $\deg(f) = n$ is odd, a result in Section 1 assures that $f(x)$ has *at least* one real root. Consequently, if n is odd, $f(x)$ has *precisely* one real root, that root is simple, and (since $f(d) > 0$ whenever $d \geq 0$) that root is negative. To treat the case of even n, notice first that $f(-1) = (\frac{1}{2} - \frac{1}{3}) + (\frac{1}{4} - \frac{1}{5}) + \ldots + \frac{1}{n} > 0$. Since $f(0) > 0$ as well, the parity lemma reveals that, counting multiplicities, $f(x)$ has an *even* number of roots in the interval $(-1, 0)$. As -1 is the only real root of $f'(x)$ (why?), $f(x)$ has at most two real roots, counting multiplicities. If $f(x)$ were blessed with real roots, those roots $r_1 < r_2$ would be distinct (since -1 is not a root of $f'(x)$) and negative (why?); moreover, by Rolle's theorem, we would have $r_1 < -1 < r_2$, thereby producing an *odd* number (namely, 1) of roots of $f(x)$ in $(-1, 0)$, contrary to what was established earlier. Consequently, if n is even, $f(x)$ has no real roots.

EXERCISES 5.2

1. For each of the following choices for $f(x)$, determine whether the number of real roots, counting multiplicities, of $f(x)$ in the interval $(-3, 2)$ is even or odd: $x^5 - 6x - 3$, $x^4 - 2x^2 + 7$, and $-x^3 + x^2 - 6x + 10$.
2. For each of the polynomials in #1, determine whether the number of positive roots, counting multiplicities, is even or odd.
3. Use Rolle's theorem to show that (ignoring multiplicities) each of the four intervals $(-\infty, -1)$, $(-1, 0)$, $(0, 1)$, and $(1, \infty)$ contains at most one root of the polynomial $f(x) = x^4 - 2x^2 + 7$. Then use the intermediate-value theorem in order to check which (if any) of these intervals actually contain a root of $f(x)$. In a similar fashion, proceed to isolate the real roots of each of the other two polynomials in #1.
4. (a) If each of the roots of the real quintic $f(x) = a_0x^5 + \ldots + a_5$ is real, prove that each of the roots of $216a_0x^5 + 125a_1x^4 + 64a_2x^3 + 27a_3x^2 + 8a_4x + a_5$ is also real. Generalize! (*Hint*: Multiply $f(x)$ by x, differentiate the product, multiply that derivative by x, differentiate the new product, ... Why don't any of these steps create nonreal roots?)
   (b) (de Gua's theorem) Suppose that, counting multiplicities, the real polynomial $f(x)$ has r positive roots and s negative roots, with 0 a root of multiplicity t. Let $g(x) = xf'(x) + kf(x)$, for some nonzero real number, k. Suppose that $g(x)$ has R positive roots, S negative roots (counting multiplicities), with 0 a root of multiplicity T. Prove that $R \geqslant r - 1$, $S \geqslant s - 1$, and $T \geqslant t$. What happens if $k = 0$?

5. (a) Construct an eighth-degree real polynomial with eight simple real roots in the interval $(-3, 2)$, by modifying the formula of Rodrigues.
   *(b) Prove that the n-th Legendre polynomial is a solution to the n-th *Legendre differential equation,* $(1-x^2)y'' - 2xy' + n(n+1)y = 0$.
6. If a ninth-degree real polynomial $f(x)$ has precisely five real roots, what are the possibilities for the number of real roots of $f'(x)$? (Count multiplicities!)

3. Variations and Descartes's Rule of Signs

We saw, in the final two results in Section 1, that information about the number of real roots of a nonzero (real) polynomial $f(x)$ is often available after comparing the signs of $f(x)$'s leading coefficient and constant term. In this section, we show how the additional consideration of the signs of $f(x)$'s intermediate coefficients leads to the notion of variation and, hence, to Descartes's rule of signs. This easily applied result will be seen to yield information about the number of positive roots and the number of negative roots of $f(x)$, information that is typically sharper than what is available using our earlier methods but also typically less precise than the dividends to be reaped in Section 4.

First, a useful reduction: one need only *divise a scheme for counting the positive roots* of an arbitrary real polynomial. Indeed, by letting $k = -1$ in method (2) of Section 1 of Chapter 2, we see that a negative number r is a root of $f(x)$ with multiplicity m if and only if the positive number $-r$ is a root of $g(x) = f(-x)$ with multiplicity m. Accordingly, *the number of negative roots* of $f(x)$ is the number of positive roots of $g(x)$, counting multiplicities in both cases. For example, the number of negative roots of $f(x) = x^6 + 8x^5 - \pi x^3 + \sqrt{2}\,x$ is the number of positive roots of $g(x) = x^6 - 8x^5 + \pi x^3 - \sqrt{2}\,x$. To complete the analysis of possible numbers of real roots, observe that *the multiplicity of* 0 *as a root* of the polynomial $f(x)$ is the largest nonnegative integer n such that $x^n \mid f(x)$. (Question: what is the multiplicity of 0 as a root of the above sextic polynomial?)

Now, the crucial definition: the (number of) *variation*(s) of a nonzero polynomial $f(x) = a_0x^n + \dots + a_n$ is obtained by first deleting the zero entries from the sequence $a_0, a_1, \dots, a_n$ and then counting the number of times that successive entries of the resulting sequence have opposite signs. For example, if $f(x) = 2x^9 - 3x^8 + 4x^6 + 7x^4 - 8x - 9$, the variation of $f(x)$ is obtained from the sequence $2, -3, 4, 7, -8, -9$, that is, from the sequence of signs $+, -, +, +, -, -$; consequently, the variation of this example is 3. Similarly, the reader should check that the numbers of variations of the polynomials $-9x^2 + 7$ and $2x^3 + 7x^2 + 6$ are 1 and 0, respectively.

It is not surprising that Descartes was intrigued by the concept of variation. Indeed, his interest in dichotomies (mind-body problem) and syntheses (analytic geometry) was noted in Section 4 of Chapter 3. What dichotomy could be more absolute than the positive-negative pairing? What mathematical method of synthesis could be more basic than counting (variations)?

Which brings us, at last, to the statement of *Descartes's rule of signs.* Let $f(x)$ be a nonzero real polynomial. If p is the number of positive roots of $f(x)$, counting multiplicities, and if v is the number of variations of $f(x)$, then $p \leq v$ and the nonnegative integer $v - p$ is even. For a proof, note first that $f(x)$'s constant term may be assumed nonzero, since factoring out a power of x from $f(x)$ leaves a quotient polynomial sharing $f(x)$'s values of v and p. Then, we are half done: namely, $v - p$ is even, by the fragment established in the third paragraph of Section 2, since v is even (respectively, odd) in case the leading coefficient and the constant term of $f(x)$ have the same (respectively, opposite) signs. Next, to prove that $p \leq v$, we proceed by mathematical induction on $n = \deg(f)$. (This is the text's second use of induction, the first having appeared in Section 6 of Chapter 3 during the proof of the fundamental theorem of algebra.) The induction-basis case, $n = 0$ (or, if you wish, $n = 1$) is a triviality. (Why?) Accordingly, we may suppose that $n > 1$ and that polynomials of degree $n - 1$ present no problem. In particular, this induction hypothesis gives $q \leq w$, where q denotes the number of positive roots of $f'(x)$, counting multiplicities, and w is the number of variations of $f'(x)$. However, $p - 1 \leq q$ by the (method of proof, using Rolle's theorem, which established the) inequality in Section 2; and $w \leq v$, directly from the formula for differentiating polynomials. Therefore, $p \leq q + 1 \leq w + 1 \leq v + 1$ and so, since $v - p$ is even, $p \leq v$, which established the induction step and completes the proof.

The intuitive appeal (not to mention the utility) of Descartes's rule of signs has prompted numerous proofs of it. Although we chose to present a proof based on Rolle's theorem, the reader may prefer the proof in [24, pp. 121–123] which is based on de Gua's theorem (#4(b) of Exercises 5.2). He/she may, in fact, prefer to read the reference, cited in Section 6 below, for the proof of the (stronger) theorem of Budan.

We next give two illustrations of Descartes's result. First, if $f(x) = 2x^9 - 3x^8 + 4x^6 + 7x^4 - 8x - 9$, then $v = 3$, and so the number of positive roots of $f(x)$, counting multiplicities, is either 1 or 3. Notice that this information is sharper than what can be gleaned from the "odd degree" result in Section 1. Moreover, $f(-x) = -2x^9 - 3x^8 + 4x^6 + 7x^4 + 8x - 9$ presents a sequence of coefficients with two variations, and so the number of negative roots of $f(x)$, counting multiplicities, is either 0 or 2. For another example, consider the polynomial $h(x) = x^4 - x - 2$ which was mentioned at the close of Section 1. Since $h(x)$ and $h(-x)$ each present exactly one variation (check), it follows that, counting multiplicities, $h(x)$ has exactly one positive root and exactly one negative root.

By way of application, we next indicate how the chapter's material to this point leads to a *sufficient* condition for the existence of a (at least two) nonreal root(s) of suitable real polynomials. (For a *necessary and sufficient* condition, see Section 4.) For motivation, return to the hospitable world of quadratics: recall that a real quadratic $ax^2 + bx + c$ has a nonreal root *if and only if* $b^2 < 4ac$. In particular, it will have a nonreal root *if* its coefficients satisfy the *strong logarithmic convexity condition,* $b^2 < ac$. We next prove the extension to arbitrary degree. Specifically, if the nonzero real polynomial $f(x) = a_0x^n + \ldots + a_n$ has the property that $(a_k)^2 < a_{k-1}a_{k+1}$ for some index k (where $1 \leq k \leq n-1$), then $f(x)$ has a (at least two) nonreal root(s).

To prove the above criterion, suppose first that $(a_{n-1})^2 < 2a_{n-2}a_n$. Let $r_1, \ldots, r_n$ denote the roots of $f(x)$, and let $g(x)$ be the polynomial constructed in Section 1 of Chapter 2 so as to have the roots $r_1^2, \ldots, r_n^2$. Then (check this, using the fact that $f(y)f(-y) = g(y^2)$), we see that $g(x) = (-1)^n(a_0)^2x^n + \ldots + (2a_{n-2}a_n - a_{n-1}{}^2)x + (a_n)^2$, which presents at most $n-1$ variations, because $2a_{n-2}a_n - a_{n-1}^2 > 0$ and $(a_n)^2 > 0$. Consequently, by Descartes's rule of signs, at most $n-1$ of the r_j^2 are positive. Moreover, none of the r_j^2 is 0 (why?), and so some r_i^2 fails to be a nonnegative real number. The the corresponding r_i is nonreal, as desired. In the general case, suppose that $(a_k)^2 < a_{k-1}a_{k+1}$, where $1 \leq k \leq n-1$. If we were to assume that each root of $f(x)$ is real, then the inequality we have established in Section 2 would yield that each root of $f^{(n-k-1)}(x) = n(n-1)\ldots(k+2)\cdot a_0x^{k+1} + \ldots + ((n-k+1)!/2)a_{k-1}x^2 + (n-k)!a_kx + (n-k-1)!a_{k+1}$ is real as well. However, the case treated earlier guarantees that $f^{(n-k-1)}(x)$ has a nonreal root, since $((n-k)!a_k)^2 < 2((n-k+1)!/2)a_{k-1})((n-k-1)!a_{k+1})$. (Check!) Accordingly, some root of $f(x)$ is indeed nonreal, as asserted. The reader may wish to read the article [26] for a different proof of this criterion.

To close the section, we offer four examples which illustrate the preceding result. First, genuinely new information: $3x^5 - 6x^4 + 5x^3 - 8x^2 + 6x - 4$ has nonreal roots because $5^2 < (-6)(-8)$ (although $(-6)^2$ is *not* less than $3 \cdot 5$, $(-8)^2$ is *not* less than $5 \cdot 6$, and 6^2 is *not* less than $(-8)(-4)$). Secondly, a caution: the strong logarithmic convexity condition must be checked for consecutive, possibly zero, coefficients. For example, the polynomial $720x^3 - 80x + 9$ fails to have nonreal roots, despite the fact that $(-80)^2 < (720)9$. Thirdly, note that the weaker hypothesis $(a_k)^2 \leq a_{k-1}a_{k+1}$ does not guarantee the presence of nonreal roots: one need only consider a quadratic example such as x^2. Lastly, although *sufficient*, the strong logarithmic convexity condition is not *necessary* for the presence of nonreal roots: consider an example such as $x^4 + 2x^3 + 3x^2 + 2x + 1$, which is (of course!) just $(x^2 + x + 1)^2$.

EXERCISES 5.3

1. What information is available from Descartes's rule of signs concerning the possible numbers of positive roots (counting multiplicities) of each of the following polynomials: $x^4 - x^2 + x - 2$, $x^9 - x^5 + x^2 + 2$, $x^5 + 2x^3 - x^2 + x - 1$, and $x^9 + x^7 + 1$? What can you say about the negative roots? What about the possibility of 0 being a root?
2. Repeat #1 for the polynomial $f(x) = x^5 + 2x^4 - x^3 - x^2 - 4$. Hence sharpen the assertions made in #1 for the polynomial $g(x) = x^4 - x^2 + x - 2$. (*Hint*: divide $f(x)$ by $g(x)$.)
3. Let $H(x)$ be the monic real polynomial whose roots are the squares of the roots of the polynomial $h(x) = x^5 + 2x^3 - x^2 + x - 1$. (Obtain $H(x)$ explicitly from $h(x)$ by means of the method introduced in Section 1 of Chapter 2.) Use Descartes's rule of signs to study the nonnegative real roots of $H(x)$. Hence conclude that $h(x)$ has four nonreal roots. Does this sharpen your findings from #1?
4. Which (if any) of the polynomials in #1 is assured of having a (at least two) complex root(s) by virtue of the "$b^2 < ac$ criterion"? Does this sharpen your findings from #1?

4. Sturm Sequences and Separation of Roots by Sturm's Theorem

Sometimes, very decent mathematical techniques may fail to answer a pressing problem. Consider, for instance, the problem of identifying the rational roots of a given rational polynomial. Swift techniques such as overhead barring (also known as reduction modulo p: see #8 of Exercises 2.1) and the weak version of Eisenstein's criterion (Section 1 of Chapter 4) often fail to discover all the rational roots. Indeed, one might argue that the exhilaratingly special nature of such methods can best be appreciated when they are contrasted with the *slow, algorithmic* rational root test (in Section 1 of Chapter 4) which *always* finds *all* the required rational roots. For the problem of separating the real roots of a real polynomial, there is a similar contrast between the available techniques. Descartes's rule of signs (Section 3) is typical of the swift, special methods, whose information payoff is frequently incomplete. It is (don't deny it!) a relief to know that there is a (necessarily slow) algorithmic approach to the root-separation problem. One such algorithm is based on a theorem of Sturm (1829). This section is devoted to stating Sturm's theorem and describing the algorithm issuing from it. The proof of Sturm's theorem is deferred to Section 5.

Consider the problem of determining the number of (real) roots of a nonzero real polynomial $f(x)$ in a given closed interval $[a, b]$. Notice, thanks to the blockbuster in Section 1 of Chapter 2, that a solution to the stated problem in case $a = 0$ and b is sufficiently large would count the number of positive roots of $f(x)$. Moreover, an ability to solve this problem for arbitrary $[a, b]$ would clearly lead to the desired separation of $f(x)$'s real roots. (Why?) Accordingly, we shall first tend to the stated problem, and then, after a description of its solution, illustrate its applicability to root separation.

Two reductions will be of enormous benefit. First, by diverting attention from $f(x)$ to each of the polynomials $g_1, g_2, g_3, \ldots$ generated therefrom in Section 3 of Chapter 2, one loses no generality in supposing that $f(x)$ has no multiple roots. Accordingly, we shall henceforth assume that $(f(x), f'(x)) = 1$. Secondly, we shall also suppose that neither a nor b is a root of $f(x)$. One loses no generality in making this additional assumption since, in case $f(a) = 0$, the factor theorem gives $f(x) = q(x) \cdot (x - a)$ and attention is then paid to the roots of the quotient polynomial, $q(x)$. Such diversion to the quotient is benign because the multiplicity of any root is finite, bounded above by $\deg(f)$.

Before stating Sturm's theorem, we need to make a definition. A *Sturm sequence for* $f(x)$ *on the interval* (a, b) consists of a finite sequence of polynomials $V_0(x) = f(x), V_1(x), V_2(x), \ldots, V_s(x)$ satisfying the following five properties:

(1) If r is any root of $f(x)$ in (a, b), then $V_0(d)/V_1(d)$ is negative (respectively, positive) for all values of d which are sufficiently close to r and less than (respectively, greater than) r.

(2) No two consecutive V's have a common root in (a, b). In other words, if $r \in (a, b)$ and $0 \leq i \leq s - 1$, then either $V_i(r) \neq 0$ or $V_{i+1}(r) \neq 0$.

(3) If $V_i(r) = 0$ for some $r \in (a, b)$ and some $1 \leq i \leq s - 1$, then $V_{i-1}(r)$ and $V_{i+1}(r)$ are of opposite sign.

(4) V_s (is of constant sign on (a, b) and) has no roots in (a, b).

(5) No $V_i(x)$ is the zero polynomial.

It is not immediately clear that Sturm sequences exist, since properties (1)–(5) are rather baroque at first sight. Indeed, the only way to appreciate (1)–(5) is to observe their role in Section 5's proof of Sturm's theorem. For the present, we shall be content to establish the existence of infinitely many Sturm sequences (relative to a, b, and $f(x)$, given as above). We shall call the *canonical Sturm sequence of* $f(x)$ the sequence $W_0, W_1, W_2, \ldots$ obtained as follows. Let $W_0(x) = f(x)$ and $W_1(x) = f'(x)$. Next, with an eye to Euclid's g.c.d. algorithm in Section 3 of Chapter 1, change notation in the computation of (f, f'), so as to obtain:

$$f(x) = q_1(x)f'(x) - W_2(x), \quad \text{where } \deg(W_2) < \deg(f');$$

$$f'(x) = q_2(x)W_2(x) - W_3(x), \quad \text{where } \deg(W_3) < \deg(W_2);$$

$$W_2(x) = q_3(x)W_3(x) - W_4(x), \quad \text{where } \deg(W_4) < \deg(W_3);$$

$$\vdots$$

$$W_{s-1}(x) = q_s(x)W_s(x), \quad \text{where } W_s(x) \text{ is a (nonzero) constant.}$$

Notice that W_2, $W_3, \ldots$ are the *negatives* of the successive remainders obtained when Euclid's algorithm is employed to compute (f, f'). Moreover, the (last nonzero) remainder $W_s(x)$ is a constant, as asserted, since $W_s(x)$ is associated to $(f, f') = 1$. The point, as you may have guessed, is this: the canonical Sturm sequence of $f(x)$ is a Sturm sequence for $f(x)$ on *any* interval (a, b), that is, a Sturm sequence for $f(x)$ on $(-\infty, \infty)$. Moreover, the same conclusion holds for any of the infinitely many sequences $k_0 W_0$, $k_1 W_1$, $k_2 W_2, \ldots, k_s W_s$ obtained from the canonical Sturm sequence and arbitrary positive reals k_0, $k_1, \ldots, k_s$. The verification of these assertions is straightforward. Indeed, (1) follows from the lemma which led to Rolle's theorem in Section 2; (2) follows since any putative common root r of two consecutive W's would entail $x - r$ dividing $(W_0, W_1) = 1$, an absurdity; (3) is evident from the explicit equations which describe the algorithm; (4) has been noted above; and (5) holds for the canonical sequence because (4) does!

Finally, we may state *Sturm's theorem*. Let V_0, $V_1, \ldots, V_s$ be a Sturm sequence for a nonzero real polynomial $f(x)$ on an open interval (a, b). Suppose (as above) that $f(x)$ has no multiple roots and that neither a nor b is a root of $f(x)$. Then the number of roots of $f(x)$ in (a, b) is $v(a) - v(b)$ where, in general, $v(d)$ denotes the number of variations in the sequence $V_0(d)$, $V_1(d), \ldots, V_s(d)$.

Before applying Sturm's theorem, we pause for two remarks. First, recall from Section 1 that if $g(x)$ is any real polynomial, then the symbols $g(-\infty)$ and $g(\infty)$ are well-defined, taking the possible values $+$ and $-$. Specifically, $g(\infty)$ was taken as the sign of $g(d)$ for all sufficiently large d; $g(-\infty)$ was defined analogously, in terms of small d. Consequently, if V_0, $V_1, \ldots, V_s$ is a Sturm sequence for $f(x)$ on $(-\infty, \infty)$, it makes sense to refer to the corresponding $v(-\infty)$ and $v(\infty)$. For instance, $v(\infty)$ counts the variations in the sequence $V_0(\infty)$, $V_1(\infty), \ldots, V_s(\infty)$. Evidently, $v(\infty) = v(b)$ for all sufficiently large b, since the sign of each $V_i(b)$ is then just the corresponding $V_i(\infty)$. As noted in this section's second paragraph, the real roots of any preassigned real polynomial $f(x) \neq 0$ may be subsumed by an open interval (a, b), for sufficiently large b and sufficiently small a. Combining these observations with Sturm's theorem shows that $v(-\infty) - v(\infty)$ $(= v(a) - v(b)$, for b sufficiently large and a sufficiently small) *counts the number of real roots of the real polynomial $f(x)$ with no multiple roots*. Exercise 1 below will give the reader practice in using this formula to recover and sharpen the import of Descartes's rule of signs.

Secondly, we note that Sturm's other theorems, typically in the area of differential equations, also concerned variations of a sort. Many readers will be familiar with an oscillation theorem of Sturm which finds use in the study of roots of the Bessel functions, as in [6, pp. 169–170]. Sturm's name is also attached, along with that of his collaborator, Liouville (who was met in a different context in Section 3 of Chapter 3), to an important class of homogeneous boundary-value problems (or systems).

Our first illustration of Sturm's theorem will be an analysis of the real roots of the quintic $f(x) = x^5 - 4x^3 + 2$. Now, to begin with, we are never simply given a Sturm sequence. (What, never? Well, hardly ever! See the next example worked out below.) Accordingly, we begin to compute the canonical Sturm sequence for $f(x)$. Routinely, $W_0 = x^5 - 4x^3 + 2$ and $W_1 = 5x^4 - 12x^2$. In order to avoid (or, at least, delay) worrying about fractions, we *multiply the already-acquired Sturm polynomials* W_0 and W_1 *by suitable positive constants* (in this case, 5 and 1, respectively). It is clear from the algorithm's equations that *subsequent Sturm polynomials* (in our case, $W_2, W_3, \ldots$) *are not affected by such simplifications*, and the requisite applications of the division algorithm become much less arduous. (Feel free to use this simplification device in the exercises, and elsewhere.) Upon dividing (5 times the old) W_0 by W_1, we find (don't we?) the remainder to be $-8x^3 + 10$, and so we would ordinarily take W_2 to be its negative, $8x^3 - 10$. However, to facilitate the arithmetic in computing variations, we multiply the last-cited polynomial by the positive factor $\frac{1}{2}$, getting $W_2 = 4x^3 - 5$. (Purists would insist on calling this positively-altered polynomial "V_2," reserving the name "W_2" for $8x^3 - 10$. No matter: signs and variations aren't affected, so denote $4x^3 - 5$ either way.) The reader is urged to check that, after modification by appropriate positive factors, the rest of the canonical Sturm sequence is given by $W_3(x) = -25x + 48x^2$, $W_4(x) = (-(25)^2/48)x + 60$, and $W_5(x) =$ some negative constant. Consider the following (hopefully, self-explanatory) table:

	W_0	W_1	W_2	W_3	W_4	W_5	ν = number of variations in row
$-\infty$	−	+	−	+	+	−	4
0	+	0	−	0	+	−	3
∞	+	+	+	+	−	−	1

The entries on the $-\infty$ row and the ∞ row were obtained by means of the results in Section 1, while direct calculation of $W_0(0), W_1(0), \ldots, W_5(0)$ led to the entries in the 0 row. By the first remark following Sturm's theorem, the number of real roots of $f(x)$ is $v(-\infty) - v(\infty) = 4 - 1 = 3$. (Incidentally, one typically sees *after* generating the canonical Sturm sequence whether Sturm's theorem applies. In our case it does, that is, $(f, f') = 1$, since $W_5 = W_s$ is a (nonzero) *constant.*) Since 0 is not a root of $f(x)$, the number of negative roots of $f(x)$ is $v(-\infty) - v(0) = 4 - 3 = 1$ (which coincides with the information to be gleaned from an application of Descartes's rule of signs). Similarly, the number of positive roots of $f(x)$ is $v(0) - v(\infty) = 3 - 1 = 2$ (*which is sharper than what Descartes's result reveals*). Consequently, there are two (conjugate) nonreal complex roots of $f(x)$. Finally, to isolate the real roots of $f(x)$, create a 1 row in the above table, hence observe that $v(1) = 2$, and conclude that the three real roots of $f(x)$ are isolated by the intervals $(-\infty, 0)$, $(0, 1)$, and $(1, \infty)$. Of course, smaller intervals could be found by creating more rows in the above table, but Sturm's theorem should be retired with gratitude at this stage of the analysis, since the intermediate-value theorem becomes a faster tool once isolation has been achieved. (Why?)

As promised, the next application of Sturm's theorem features a non-canonical Sturm sequence. Let n be a positive integer, and consider the polynomial $f(x) = 1 + x + (x^2/2!) + \ldots + (x^n/n!)$. (The reader will recall from calculus that $f(x)$ is a partial sum of the Maclaurin series for e^x.) We claim that, for each positive real number ϵ, the three polynomials $V_0(x) = f(x)$, $V_1(x) = f'(x)$, and $V_2(x) = -x^n$ constitute a Sturm sequence for $f(x)$ with respect to each of the intervals $(-\infty, -\epsilon)$ and (ϵ, ∞). (Why can't it be a Sturm sequence for $f(x)$ on $(-\infty, \infty)$?) Given the claim, we shall readily obtain the number of real roots of $f(x)$ since, as ϵ varies, $(-\infty, \epsilon)$ and (ϵ, ∞) subsume all real numbers except the nonroot, 0. (Of course, no *one* ϵ does this!) Verification of the claim, that is, that properties (1)–(5) hold on both $(-\infty, -\epsilon)$ and (ϵ, ∞), is straightforward, given the observation that $f(x) = f'(x) + (x^n/n!)$. (Provide the details.) To prepare to apply Sturm's theorem, which is applicable since $(f, f') = 1$ (why?), construct the table

	V_0	V_1	V_2	v
$-\infty$	$(-1)^n$	$(-1)^{n-1}$	$(-1)^{n+1}$	1
$-\varepsilon$	+	+	$(-1)^{n+1}$	1 if n is even; 0 if n is odd
ε	+	+	$-$	1
∞	+	+	$-$	1

whose entries have been computed for sufficiently small ϵ by means of the results in Section 1. It is now apparent that the answer to our problem depends on the parity of n. Specifically, if n is even, then $f(x)$ has no real roots, since $(v(-\infty) - v(-\epsilon)) + (v(\epsilon) - v(\infty)) = (1-1) + (1-1) = 0$ in that case. Moreover, if n is odd, then $f(x)$ has exactly one real root, and that root is negative, since $v(-\infty) - v(-\epsilon) = 1 - 0 = 1$ and $v(\epsilon) - v(\infty) = 1 - 1 = 0$.

Consider a nonzero real polynomial $f(x)$ of degree $n \geq 0$. We claim that a necessary and sufficient *condition that the n* (complex) *roots* of $f(x)$ *be* simple and all *real* is that the canonical Sturm sequence for $f(x)$ consist of $n+1$ polynomials all of whose leading coefficients have the same sign. (Supply a proof, noting that the canonical Sturm sequence consists of $s+1$ $(\leq n+1)$ polynomials. As $v(-\infty) \leq s$ and $v(\infty) \geq 0$, the requirement that $f(x)$ have n distinct real roots amounts to prescribing $v(-\infty) = n$ and $v(\infty) = 0$. Exploit the material in Section 1 to complete the proof.)

To indicate the power of Sturm's theorem, we close this section by using the criterion in the preceding paragraph to recover the main result in Section 2 of Chapter 3. Specifically, we are given a real cubic $h(x) = x^3 + px + q$ and our task is to employ $\triangle = 4p^3 + 27q^2$ to study the nature of the roots of $h(x)$. As in the earlier discussion, the cases in which $p = 0$ or $q = 0$ (or both) are treated separately. The earlier discussion showed, by means of the g.c.d. algorithm, that $\triangle = 0$ characterizes the case in which $h(x)$ has a multiple root. Those same calculations generate the Sturm polynomials $W_0(x) = x^3 + px + q$, $W_1(x) = 3x^2 + p$, $W_2(x) = -2px - 3q$ (apart from a positive factor), and (in case it's nonzero)

$$W_3(x) = \frac{\triangle}{-2p^2}.$$

It therefore follows immediately from the criterion given in the preceding paragraph that $\triangle < 0$ characterizes the case in which $h(x)$ has three distinct real roots. By the process of elimination, $\triangle > 0$ must characterize the remaining case, in which $h(x)$ has precisely one real root. The reader is invited to check that much of the analysis in Section 2 of Chapter 3 has now been obviated by the preceding discussion. Here are some additional reasons to be grateful for Sturm's theorem.

EXERCISES 5.4

1. Sharpen the consequences of Descartes's rule of signs given earlier by using Sturm's theorem to count and isolate the real roots of each of the polynomials in #1 of Exercises 5.3.
2. Repeat #1 for the polynomials in ##4(c), 5, 6 (without the third-listed polynomial!), and 7 of Exercises 2.3.
3. Repeat #1 for real polynomials of your own creation.
4. Let n be a positive integer. Recover the information in Section 2 concerning the roots of $f(x) = 1 + x + (x^2/2) + \ldots + (x^n/n)$, by use of the sequence $W_0 = f(x)$, $W_1 = f'(x)$. (*Hint*: Observe that $W_1 = (x^n - 1)/(x-1)$, a point

already put to use in Section 2 of Chapter 4. Hence, if n is odd, the sequence W_0, W_1 is a Sturm sequence for $f(x)$ on $(-\infty, \infty)$. Exploit it. If n is even, check that W_0, W_1 is a Sturm sequence for $f(x)$ on both $(-\infty, -1-\varepsilon)$ and $(-1+\varepsilon, \infty)$, provided that ε is a sufficiently small positive real number. Work out the appropriate tables, recalling that $f(-1)>0$.)

5. (a) Use the text's criterion (and not the quadratic formula) to prove that the roots of a real quadratic polynomial ax^2+bx+c are real and distinct if and only if $b^2-4ac>0$.
(b) Find necessary and sufficient conditions (on the real numbers p and q) characterizing the case in which $x^5+px^3+(p^2x/5)+q$ has five simple real roots.

6. (An example in which $f(x)$ has no multiple roots and $s<n$) Count and isolate the real roots of $f(x)=x^5+3x^2+3$.

*5. Proof of Sturm's Theorem

This section is devoted to proving the theorem of Sturm which was introduced in Section 4. Recall that $V_0, V_1, V_2, \ldots, V_s$ denotes a Sturm sequence on the interval (a,b) for a real polynomial $f(x)=V_0(x)$ which has no multiple roots. Our task is to show, in case neither a nor b is a root of $f(x)$, that the number of roots of $f(x)$ in (a,b) is $v(a)-v(b)$, where, in general, $v(d)$ counts the variations in the sequence $V_0(d)$, $V_1(d)$, $V_2(d), \ldots, V_s(d)$. The proof will consist of two cases.

Case 1: Neither a nor b is a root of any of the polynomials V_0, V_1, $V_2, \ldots, V_s$. Let $r_1, r_2, \ldots, r_m$ be the list, in increasing order, of all such roots in (a,b). (Note that m is finite, by virtue of the blockbuster in Section 1 of Chapter 2.) Next, select real numbers $c_1, c_2, \ldots, c_{m-1}$ satisfying $r_j<c_j<r_{j+1}$ for $j=1, 2, \ldots, m-1$. Setting $c_0=a$ and $c_m=b$, we evidently have the "telescoping" sum

$$\nu(a)-\nu(b)=\sum_{j=1}^{m}(\nu(c_{j-1})-\nu(c_j)).$$

We proceed to examine each of the summands on the right-hand side. Fix j, $1\le j\le m$.

Subcase (i): $V_0(r_j)\neq 0$. We shall show in this subcase (under whose assumptions, $f(x)$ has no roots in the interval $[c_{j-1}, c_j]$) that $\nu(c_{j-1})=\nu(c_j)$. Now, by hypothesis, r_j is a root of at least one of the polynomials in the given Sturm sequence. Let $V_\alpha, V_\beta, V_\gamma, \ldots, V_\lambda$ be those polynomials in the given Sturm sequence having r_j as a root, the labelling arranged so that $1\le\alpha<\beta<\gamma<\ldots<\lambda$. By property (2) of Sturm sequences, $\alpha+2\le\beta$, $\beta+2\le\gamma$, etc., and so our attempt to prove $\nu(c_{j-1})=\nu(c_j)$ may attack the following table:

	V_0	V_1	$\cdots$	$V_{\alpha-1}$	V_α	$V_{\alpha+1}$	$\cdots$	$V_{\beta-1}$	V_β	$V_{\beta+1}$	$\cdots$	V_s
c_{j-1}												
c_j												

Since no r_k is a root of any of the polynomials $V_0, \ldots, V_{\alpha-1}$, the intermediate-value theorem implies that $V_\mu(c_{j-1})$ and $V_\mu(c_j)$ are nonzero and have the same sign, for each index μ, $0 \leq \mu \leq \alpha - 1$. Accordingly, the contributions to $v(c_{j-1})$ and $v(c_j)$ from the columns indexed by μ's between 0 and $\alpha - 1$ are identical.

Next, we attend to the columns indexed by $\alpha - 1$, α, and $\alpha + 1$. We have seen, as a consequence of the intermediate-value theorem, that $V_{\alpha-1}(c_{j-1})$ and $V_{\alpha-1}(c_j)$ are nonzero and have the same sign. Similarly, since $\alpha + 1 < \beta$, we see that $V_{\alpha+1}(c_{j-1})$ and $V_{\alpha+1}(c_j)$ are nonzero and have the same sign. Thus, in order for us to show that the contributions to $v(c_{j-1})$ and $v(c_j)$ from the columns indexed by $\alpha - 1$, α, $\alpha + 1$ are each 1 (and, hence, identical), it suffices to prove that $V_{\alpha-1}(c_j)$ and $V_{\alpha+1}(c_j)$ have opposite signs. To that end, apply property (3) of Sturm sequences, after noting $V_\alpha(r_j) = 0$, to obtain that $V_{\alpha-1}(r_j)$ and $V_{\alpha+1}(r_j)$ have opposite signs. The question then amounts to demonstrating that $V_{\alpha-1}(c_j)$ shares the sign of $V_{\alpha-1}(r_j)$ and that $V_{\alpha+1}(c_j)$ shares the sign of $V_{\alpha+1}(r_j)$, and these two assertions are immediate consequences of (what else?) the intermediate-value theorem.

The contributions to $v(c_{j-1})$ and $v(c_j)$ from the columns headed by $V_{\alpha+1}, \ldots, V_{\beta-1}$ are identical: adapt the earlier analysis for the columns headed by $V_0, \ldots, V_{\alpha-1}$. Moreover, the contributions to $v(c_{j-1})$ and $v(c_j)$ from the columns indexed by $\beta - 1$, β, and $\beta + 1$ are each 1: adapt the earlier analysis for the indexes $\alpha - 1$, α, and $\alpha + 1$. Repeated adaptions of these two analyses suffice to tally the contributions to $v(c_{j-1})$ and $v(c_j)$ from the "later" indexes $\beta + 1, \ldots, s$. Corresponding contributions coincide at each (block of) step(s), and so $v(c_{j-1}) = v(c_j)$ in subcase (i), as asserted.

Subcase (ii): $V_0(r_j) = 0$. We shall show in this subcase (under whose assumptions, $f(x)$ has precisely one root in the interval $[c_{j-1}, c_j]$) that $v(c_{j-1}) = v(c_j) + 1$. To this end, note first that $V_0(c_{j-1})$ and $V_0(c_j)$ are nonzero and have opposite signs, thanks to the parity assertion of Rolle's theorem in Section 2. Moreover, $V_0(c_{j-1})$ and $V_1(c_{j-1})$ are readily seen to be nonzero and of opposite sign: by property (1) of Sturm sequences, $V_0(d)$ and $V_1(d)$ have opposite signs for d slightly to the left of r_j, and this situation persists as x moves leftward from d to c_{j-1} since, by the intermediate-value theorem, neither V_0 nor V_1 changes sign on the interval $[c_{j-1}, d]$. Consequently, the columns headed by V_0 and V_1 contribute a count of 1 to $v(c_{j-1})$. By the same token, property (1) of Sturm sequences combines with the intermediate-value theorem to show that $V_0(c_j)$ and

$V_1(c_j)$ are nonzero and of the *same* sign, whence the contribution to $v(c_j)$ from the columns headed by V_0 and V_1 is 0. Observe that, since $V_1(c_{j-1})$ and $V_1(c_j)$ have the same sign and (by (2)) no root of V_1 lies in $[c_{j-1}, c_j]$, the analysis in subcase (i) yields that the columns headed by $V_1, V_2, \ldots, V_s$ provide identical contributions, say the number n, to $v(c_{j-1})$ and $v(c_j)$. (We are *not* asserting that the sequence $V_1, V_2, \ldots, V_s$ is a Sturm sequence. Indeed, this shorter sequence may fail to have the analogue of property (1), although it *does* satisfy (2), (3), (4), and (5). The reason that the analysis in (i) adapts to the shorter sequence is that that earlier analysis made no use of (1).) Therefore, $v(c_{j-1}) = 1 + n$ and $v(c_j) = 0 + n$, whence $v(c_{j-1}) = v(c_j) + 1$ in subcase (ii), as asserted.

To dispatch case (1): by subcases (i) and (ii), $v(c_{j-1}) - v(c_j)$ is 1 or 0 according as to whether r_j is or is not a root of $V_0(x)$. Summation over j gives that $v(a) - v(b)$ is the number of roots r_j of $f(x)$ $(= V_0(x))$ in (a, b). The remaining case is:

Case 2: At least one of a, b is a root of at least one of the polynomials $V_1, V_2, \ldots, V_s$. Select sufficiently small positive numbers ϵ and δ so that all the roots of (at least one of) $V_0, V_1, V_2, \ldots, V_s$ in (a, b) are contained in $(a + \epsilon, b - \delta)$. Observe that $V_0, V_1, V_2, \ldots, V_s$ is a Sturm sequence on $(a + \epsilon, b - \delta)$ and so, by case 1, the number of roots of $f(x)$ in (a, b) is $v(a + \epsilon) - v(b - \delta)$. In order to dispatch case 2 and thus complete the proof, it is enough to show that $v(a + \epsilon) = v(a)$ and $v(b - \delta) = v(b)$. We shall treat the former, leaving the latter (which is similar) for the exercises. Consider the following table:

	V_0	...	$V_{\alpha-1}$	V_α	$V_{\alpha+1}$	...	$V_{\beta-1}$	V_β	$V_{\beta+1}$	...	V_s
a											
$a + \epsilon$											

where $V_\alpha, V_\beta, \ldots$ are those polynomials in the given Sturm sequence which have a as a root. Note that $1 \leq \alpha < \beta < \ldots < s$. As in the analysis of subcase (i) of case 1, there are two paradigm computations. First, thanks to the intermediate-value theorem, the contributions to $v(a)$ and $v(a + \epsilon)$ from the columns indexed by $0, \ldots, \alpha - 1$ are equal (details?); the same reasoning adapts to blocks of columns such as the one with indexes $\alpha + 1, \ldots, \beta - 1$. Secondly, by using (3) and the intermediate-value theorem (compare with the argument in the second paragraph treating (i)) we see that the columns indexed by $\alpha - 1$, α, $\alpha + 1$ contribute 1 to each of $v(a)$ and $v(a + \epsilon)$; similar comments hold for blocks of indexes such as $\beta - 1$, β, $\beta + 1$. Relentless applications of these paradigms show that $v(a) = v(a + \epsilon)$, as required. (Tallyho! A bad pun; let's atone with this book's only "Q.E.D.")

EXERCISES 5.5

1. Check that the text's analysis of (i) applies in degenerate cases such as $0 = \alpha - 1, \alpha + 1 = \beta - 1$, and $\beta + 1 = s$.
2. Prove that $\nu(b - \delta) = \nu(b)$ in the proof of case 2.
3. Properties (4) and (5) of Sturm sequences weren't explicitly appealed to in our proof. Were they used?
4. In the analysis of case 2, why do satisfactory ϵ and δ exist? (*Hint*: the blockbuster.)
5. Fashion a proof of case 1 for the case $m = 0$. (*Hint*: see the remarks following the first table.)

*6. Other Methods of Separating Roots: the Theorems of Vincent and Budan

This section serves as an introduction to Budan's theorem, which is a generalization of Descartes's rule of signs (Section 3); and to a result of Vincent, which permits separation of roots by techniques that are quite different from those met in applying Sturm's theorem (Section 4).

Let $f(x)$ be a nonzero (real) polynomial of degree n, and let $a < b$ be two real numbers, neither of which is a root of $f(x)$. For each real value d of x, let $w(d)$ denote the number of variations in the sequence $f(d), f'(d), f''(d), \ldots, f^{(n)}(d)$. Finally, let N denote the number of roots, counting multiplicities, of $f(x)$ in the interval (a, b). ***Budan's theorem*** asserts that $N \leq w(a) - w(b)$ and that, moreover, the nonnegative integer $w(a) - w(b) - N$ is even. A particularly clear proof of Budan's theorem is given in [15, Theorem 35, pp. 59–60].

We shall next explain how to recover Descartes's rule of signs from Budan's theorem. As above, let $f(x) = a_0x^n + a_1x^{n-1} + \ldots + a_n$ be a nonzero real polynomial. Our concern at present consists of estimating the number, M, of positive roots of $f(x)$. As explained in Section 3, we may suppose, for this purpose, that $a_n \neq 0$ (by factoring out a sufficiently high power of x from $f(x)$ and then ministering to the other factor). Next, consider any positive number b which exceeds each of the real roots of $f(x)$. (Why must such b exist? By the blockbuster, of course!) Set $a = 0$. Evidently, $M = N$, where N is as in the preceding paragraph. Accordingly, by Budan's theorem, $M \leq w(0) - w(b)$ and $w(0) - w(b) - M$ is even. To recover Descartes's result, we shall show that $w(0)$ coincides with the number of variations in the sequence $a_0, a_1, \ldots, a_n$ and that b may be (further) chosen to satisfy $w(b) = 0$.

Recall the formulas of Maclaurin from #3(b) of Exercises 2.3: $f(0) = a_n$, $f'(0) = a_{n-1}$, $f''(0) = 2a_{n-2}, \ldots, f^{(k)}(0) = k!a_{n-k}, \ldots, f^{(n)}(0) = n!a_0$. Thus, for each index k, $0 \le k \le n$, $f^{(k)}(0)$ and a_{n-k} are either both 0 or both nonzero and, in any event, $f^{(k)}(0)$ and a_{n-k} have the same sign. Consequently, $w(0)$ counts the variations in the sequence $a_n, a_{n-1}, \ldots, a_0$ or (with a nod to southpaws) what is evidently the same, the variations in the sequence $a_0, a_1, \ldots, a_n$, as asserted. Finally, to establish the assertion concerning $w(b)$, recall from Section 1 that, for all sufficiently large x, the polynomials $f(x), f'(x), \ldots, f^{(n)}(x)$ take on the sign of their leading coefficients $a_0, na_0, \ldots, n!a_0$, respectively. Suppose, in addition to the restriction made earlier, b is sufficiently large, in the sense of the preceding sentence. Then, since the positive factors $1, n, \ldots, n!$ have no effect on signs, $w(b)$ counts the variations in the sequence $a_0, a_0, \ldots, a_0$; that is, $w(b) = 0$, as asserted.

Since applications of Budan's theorem are more time consuming than corresponding uses of Descartes's result, it seems appropriate to stress two benefits of the former. First, Budan's theorem applies to an arbitrary finite interval, while Descartes's result works only for intervals such as $(0, \infty)$ and $(-\infty, 0)$. Accordingly, Budan's theorem can serve as a partial check for any application of Sturm's theorem. Secondly, since it refers to Maclaurin coefficients, one might suppose that the conclusion of Budan's theorem applies to certain analytic, nonpolynomial functions (and the roots, or *zeros*, thereof). In fact, precisely such an extension is available—it is known as the *Budan–Fourier theorem*—but limitations of space prevent a discussion of this generalization.

As was the case for Sturm's theorem, the theorem of Vincent pertains to a real polynomial without multiple roots. *Vincent's theorem* asserts, for each such polynomial $f(x)$, that there exists a (finite) sequence of substitutions $x = n_1 + (1/x_2)$, $x_2 = n_2 + (1/x_3), \ldots, x_{k-1} = n_{k-1} + (1/x_k)$, with k depending on $f(x)$ but not on the integers $n_1, \ldots, n_{k-1}$, and that, when $f(x)$ is expressed as a polynomial in x_k, the coefficients of the transformed polynomial exhibit at most one variation. A proof of Vincent's theorem is given in [24, pp. 298–304]. To indicate how Vincent's theorem is used to separate (real) roots, we confine attention to positive roots, no loss of generality thanks to the "$f(-x)$ trick." Since 1 may be assumed not to be a root of $f(x)$—otherwise, factor $x - 1$ out of $f(x)$ and attend to the other factor—positive roots either exceed 1 or are in the interval (0, 1). The former are discerned by studying positive roots of the transformed polynomial following the substitution $x = 1 + y$. By iterating such substitutions, ultimately one need attend only to roots in (0, 1). (Why? *Hint*: the greatest integer function describes the requisite number of such substitutions needed to analyze a root exceeding 1.) However, a root in (0, 1) may be pursued via the substitution $x = 1/(1 + y)$ or, what is the same, via the successive substitutions $x = 0 + (1/w)$ and $w = 1 + y$; the pursuit continues, as above, by asking whether the (necessarily positive) y-value of the target root exceeds 1. The reader is encouraged to read [24, pp. 129–136], wherein four detailed examples illustrate the above explanation of the way in which Vincent's theorem leads to separation of roots. Also recommended is the recent article [19], which explores how the above separation technique also leads to continued fraction approximations to the pursued roots.

We close with two motivating comments. First, the recommended reading in [24] will be facilitated by a familiarity with synthetic division: to gain that familiarity, read Section 1 of Chapter 6. Which leads to our second comment: synthetic division also facilitates the root-approximation method of Horner (Section 5 of Chapter 6). While separation using Vincent's theorem develops continued fraction approximations of roots, Horner's method will be seen to grind out the roots' decimal expansions.

EXERCISES 5.6

1. Apply Budan's theorem to the problems met in #1 of Exercises 5.3.
2. Prove the "evident" assertion in this section's fourth paragraph.
3. Browse through a reference such as [21, Chapter 7] in order to gain a familiarity with continued fractions, and then do the recommended reading in [19] and [24].

6

Numerical Approximation of Roots: Some Techniques and their Implementation with Hand-held Calculators

Up to this point, we have been concerned with the nature and location of solutions for real polynomial equations. In certain cases—degree less than five, for instance—it has been possible to *find* these solutions exactly. In other cases, we have seen that no general method exists for finding *exact* solutions. For example, we have the means to show that $x^6 - 2x^2 + 3x - 4 = 0$ has two real solutions and that one of them lies between -2 and -1 and the other between 1 and 2; however, unless these roots are rational (and these are not), we have seen no way of finding them exactly. In addition, we have ignored nonpolynomial equations such as $2x - \tan x = 3$ altogether. Here again, because the x-expressions refuse to combine to a form in which we can solve for x, an exact solution is not possible.

What we can do in the situations above is approximate the solutions. Because of the discussions in the first paragraph of Section 1 of Chapter 5, the present chapter's emphasis on real polynomials also treats, in principle, the task of approximating roots of complex polynomials. Many, many approximation methods exist (some of which are very ingenious), and the degree of accuracy one can attain is usually limited only by the patience of the person computing or the mechanical limitations of the instrument (computer, hand-held calculator, slide rule, etc.) being used. In this chapter, we consider just a few of the more basic approximating methods and gear the presentation to the reader with a hand-held calculator capable of arithmetic and the evaluation of trigonometric, inverse trigonometric, logarithmic, and exponential functions. Memory and programming capabilities are convenient but not necessary. We assume the reader is familiar with the use of such a calculator. Finally, for simplicity, we generally ignore questions of possible error accumulation and assume that the calculator is accurate to the limit of its display capacity. The reader who is interested in this often delicate matter might refer to a numerical analysis text such as [3].

This chapter begins with a look at synthetic division, which turns out to be an algebraic tool and a computational convenience. Sections 2 through 4 deal with three approximation methods suited to both polynomial and non-polynomial equations. Sections 5 through 7 present three methods which are useful only for solving polynomial equations and also show some techniques for algebraically manipulating the roots of such equations. The chapter concludes with a discussion of imaginary roots, speed of convergence, and error estimates.

1. Synthetic Division

Synthetic division may be thought of as a shorthand notation for computing the quotient and remainder terms when a polynomial is divided by $x - c$. For example, suppose we divide $3x^4 - 12x^2 - 4x + 5$ by $x - 2$.

$$
\begin{array}{r|l}
 & 3x^3 + 6x^2 + 0x - 4 \\
\hline
x - 2 & 3x^4 + 0x^3 - 12x^2 - 4x + 5 \\
 & \underline{3x^4 - 6x^3} \\
 & \qquad 6x^3 \\
 & \qquad \underline{6x^3 - 12x^2} \\
 & \qquad\qquad 0x^2 \\
 & \qquad\qquad \underline{0x^2 - 0x} \\
 & \qquad\qquad\qquad -4x \\
 & \qquad\qquad\qquad \underline{-4x + 8} \\
 & \qquad\qquad\qquad\qquad -3
\end{array}
$$

Now let us rewrite this scheme, dropping the divisor $x - 2$ and the quotient $3x^3 + 6x^2 - 4$. Retain only the coefficients of the other terms and slide them up to fill in the gaps.

$$\begin{array}{rrrrr} 3 & 0 & -12 & -4 & 5 \\ 3 & -6 & -12 & 0 & 8 \\ & 6 & 0 & -4 & -3 \\ & 6 & 0 & -4 & \end{array}$$

Notice that the fourth row is unnecessary since each number is only used to cancel the number above. In addition to dropping this row, move the bottom entry "3" of the first column down one place.

$$\begin{array}{rrrrr} 3 & 0 & -12 & -4 & 5 \\ & -6 & -12 & 0 & 8 \\ 3 & 6 & 0 & -4 & -3 \end{array}$$

The third row is now our answer: the coefficients of the quotient $3x^3 + 6x^2 - 4$ and the remainder -3. What arithmetic are we doing to get this answer? Well, each entry in the second row is the product of the entry just below and to the left of it with -2 (from the divisor $x - 2$)—so $8 = (-4)(-2)$. Each third-row entry is obtained by subtracting the numbers above it—so $-3 = 5 - 8$. The procedure is to write down the first row of coefficients, bring the first entry down and move left to right multiplying by -2 and subtracting. But there is one last simplification. Instead of multiplying by -2 and subtracting, we choose equivalently to multiply by $+2$ and add. The problem now looks like this

$$\begin{array}{rrrrrl} 3 & 0 & -12 & -4 & 5 & \quad \lfloor 2 \\ & 6 & 12 & 0 & -8 & \\ 3 & 6 & 0 & -4 & -3 & \end{array}$$

where 2 has been written at the right to represent the divisor $x - 2$ and remind us of the multiplication. If the multiplication and addition is done mentally or on a hand-held calculator, the result would appear like this

$$\begin{array}{rrrrrl} 3 & 0 & -12 & -4 & 5 & \quad \lfloor 2 \\ 3 & 6 & 0 & -4 & -3 & \end{array}$$

Another example should firmly fix the procedure.

Example 1:

Divide $P(x) = 2x^4 + 3x^3 - 5x + 7$ *by* $x + 3$ *to find the quotient* $Q(x)$ *and remainder* R *in the equation* $P(x) = (x + 3)Q(x) + R$.

Solution:

Use synthetic division (noting $x + 3 = x - (-3)$). Write the coefficients of $P(x)$ in a row, bring down the first entry "2." Now move left to right multiplying by -3 and adding.

$$\begin{array}{rrrrr|r} 2 & 3 & 0 & -5 & 7 & \underline{-3} \\ & & & & & \\ 2 & -3 & 9 & -32 & 103 & \end{array}$$

Thus we have $Q(x) = 2x^3 - 3x^2 + 9x - 32$ and $R = 103$. ■

In Section 5 repeated synthetic division will be used in Horner's approximation method for polynomial roots. At present, however, we recall the remainder theorem, one of whose consequences is to make synthetic division a convenient means of evaluating polynomials. If $P(x) = (x - c)Q(x) + R$, then

$$\begin{aligned} P(c) &= (c - c)Q(c) + R \\ &= R. \end{aligned}$$

Hence, in Example 1 $P(-3) = 103$. Because strings of multiplications and additions are easily performed on hand-held calculators, the synthetic division technique is an efficient means of evaluating a polynomial at a given value. Since only the remainder is of interest, the coefficients of the quotient need not be noted separately. For instance, in Example 1 in order to find $P(c)$ for a given value c, simply perform the arithmetic sequence from synthetic division

$$2 \times c + 3 \times c + 0 \times c + (-5) \times c + 7$$

on the calculator. Because of the implied parentheses in calculators, this is equivalent to

$$(((2c + 3)c + 0)c - 5)c + 7,$$

which is just an expanded form of $P(c)$.

Example 2:

If $P(x) = 3x^5 - 2x^2 + x$, *use the synthetic division technique to find* $P(5)$.

Solution:

The arithmetic sequence is

$$3 \times 5 + 0 \times 5 + 0 \times 5 + (-2) \times 5 + 1 \times 5 + 0.$$

In terms of keystrokes on many scientific calculators (Texas Instruments, Sharp, etc.), this would be

[3] [×] [5] [×] [5] [×] [5] [−] [2] [×] [5] [+] [1] [×] [5] [=]

and on an HP 25 it would be

[3] [ENTER] [5] [×] [5] [×] [5] [×] [2] [−] [5] [×] [1] [+] [5] [×]

In either case, the result is 9330. One could also find $P(5)$ directly. For instance, with a TI or Sharp calculator, enter

[3] [×] [5] [y^x] [5] [−] [2] [×] [5] [x^2] [+] [5] [=] ,

to obtain 9330 once again. ∎

EXERCISES 6.1

1. Use synthetic division to find $Q(x)$ in each of the following equations:
   (a) $x^5 + 32 = (x + 2)Q(x)$.
   (b) $2x^5 - 12x^4 + 18x^3 + 8x^2 - 25x + 3 = (x - 3)Q(x)$.
   (c) $3x^3 - 4x^2 - 27x + 36 = (3x - 4)Q(x)$.
2. Let $P(x) = x^4 + 4x^3 + 3x^2 - 4x - 5$. Use the synthetic division technique with a calculator to evaluate $P(-2.4)$, $P(-2.5)$, and $P(-2.6)$. From these results give a value which will be within 0.05 units of a solution for the equation $P(x) = 0$.
3. Let $P(x) = x^3 - x + 8$. Verify that $P(-2.1) \cdot P(-2.2) < 0$. By evaluating $P(x)$ at some or all of the values $-2.11, -2.12, \ldots, -2.19$, determine a value β with two decimal digits which is within 0.01 units of a solution for $P(x) = 0$. Now find $Q(x)$ and R such that $P(x) = (x - \beta)Q(x) + R$. Use the quadratic formula to find the two complex roots of $Q(x)$ to two decimal digits. (Because $x - \beta$ is almost a factor of $P(x)$, the roots of $Q(x)$ are approximations to the complex roots of $P(x)$.)
4. Let $P(x) = 3x^3 - 2x^2 - 5$. Find $Q_1(x)$ and R_1 so that $P(x) = (x + 1.1)Q_1(x) + R_1$. Now find $Q_2(x)$ and R_2 so that $Q_1(x) = (x + 1.1)Q_2(x) + R_2$. Finally, find constants Q_3 and R_3 so that $Q_2(x) = (x + 1.1)Q_3 + R_3$. Show by "back substituting" that $P(x) = Q_3(x + 1.1)^3 + R_3(x + 1.1)^2 + R_2(x + 1.1) + R_1$. Compare this to Taylor's expansion (see #3(c) of Exercises 2.3), $P(x) = \sum_{n=0}^{\infty} (P^{(n)}(c)/n!)(x - c)^n$ to find the derivative values $P^{(k)}(-1.1)$ for $k = 0, 1, 2, 3$.

2. The Method of Bisection

Suppose that $F(x)$ is a function which is continuous on an interval containing the two values a and b with $a < b$ and that the function values $F(a)$ and $F(b)$ have opposite signs (so that $F(a) \cdot F(b) < 0$). As noted earlier, the equation $F(x) = 0$ must have at least one solution between a and b. Now suppose we also know that there is *exactly* one solution between a and b. (This may follow from Sturm's theorem or Descartes's rule or possibly, if F is differentiable, by noting that the derivative $F'(x)$ is always positive or always negative between a and b.) Call this unique solution α.

Perhaps the simplest (but not necessarily the fastest) method of approximating α is bisection. We first decide in which half of the interval $[a, b]$ α lies by computing the function value at the midpoint, that is, $F((a + b)/2)$. If $F((a + b)/2) \cdot F(a) < 0$, then α lies in $[a, (a + b)/2]$ since $F(x)$ has a change of sign over this interval; similarly, if $F((a + b)/2) \cdot F(b) < 0$, then α lies in $[(a + b)/2, b]$. Of course, if $F((a + b)/2) = 0$, the work is done and we thankfully write $\alpha = (a + b)/2$. Suppose the first outcome occurs—that α lies in $[a_1, b_1]$ where $a_1 = a$ and $b_1 = (a + b)/2$. The bisection step is now repeated for the new interval $[a_1, b_1]$ by evaluating $F(x)$ at the new midpoint $(a_1 + b_1)/2$ to decide in which half of $[a_1, b_1]$ (therefore, in which quarter of the original $[a, b]$) α lies. Call this new interval $[a_2, b_2]$. This process may be continued, each repetition yielding a new, smaller interval containing α.

If the midpoint test is performed $n - 1$ times, the result is an interval $[a_{n-1}, b_{n-1}]$ of length $(b - a)/2^{n-1}$ containing α. For a final approximation take $\beta = (a_{n-1} + b_{n-1})/2$, the midpoint of $[a_{n-1}, b_{n-1}]$. We may now be sure that $|\beta - \alpha| < (b - a)/2^n$. (Why?) Of course, the value for n (the number of approximating steps) is determined by the person (or machine) doing the computing. The value depends upon (1) the accuracy desired and (2) the length $b - a$ of the original interval. Suppose we want our approximation β to be within E units of the true solution α. Since the error after n steps is no more than $(b - a)/2^n$, we set this expression less than or equal to E and solve for n to find

$$n \geq \log_2\left(\frac{b-a}{E}\right) = \frac{\log_{10}\left(\dfrac{b-a}{E}\right)}{\log_{10} 2}.$$

Then the number of approximating steps is taken to be the least *integer* satisfying this inequality.

Example 1:

Use the method of bisection to approximate all real solutions of $F(x) = x^4 - 4x^3 + 12 = 0$ *to an accuracy of* 10^{-2}.

Solution:

First we get an idea of the number and location of roots. Descartes's rule of signs reveals that there are no negative roots and at most two positive roots. A little trial and error (aided by consideration of $F'(x)$ if necessary) shows that $F(1) = 9$, $F(2) = -4$, and $F(4) = 12$. Thus, $F(x)$ has one root α_1 between 1 and 2 and another root α_2 between 2 and 4. (To be sure, the same information follows, *more slowly,* using the root-separation schemes in Sections 4 and 6 of Chapter 5.)

Below is the work for computing the approximation β_1 to α_1. First note the values $a = 1$, $b = 2$, and $E = 10^{-2}$, so that $\log_{10}((b - a)/E)/\log_{10} 2 \doteq 6.6$. We therefore let $n = 7$. Testing successive midpoints gives

$$F\left(\frac{3}{2}\right) = 3.5\ldots > 0, \qquad \alpha_1 \in \left[\frac{3}{2}, 2\right]$$

$$F\left(\frac{7}{4}\right) = -0.058\ldots < 0, \qquad \alpha_1 \in \left[\frac{3}{2}, \frac{7}{4}\right]$$

$$F\left(\frac{13}{8}\right) = 1.8\ldots > 0, \qquad \alpha_1 \in \left[\frac{13}{8}, \frac{7}{4}\right]$$

$$F\left(\frac{27}{16}\right) = 0.88\ldots > 0, \qquad \alpha_1 \in \left[\frac{27}{16}, \frac{7}{4}\right]$$

$$F\left(\frac{55}{32}\right) = 0.41\ldots > 0, \qquad \alpha_1 \in \left[\frac{55}{32}, \frac{7}{4}\right]$$

$$F\left(\frac{111}{64}\right) = 0.18\ldots > 0, \qquad \alpha_1 \in \left[\frac{111}{64}, \frac{7}{4}\right].$$

For the seventh and final step we take the midpoint of the last interval $\beta_1 = 223/128 = 1.742\ldots$, and this approximation is correct to within 10^{-2}.

In approximating α_2 we begin with $a = 2$ and $b = 4$; evaluating the logarithmic expression shows that n should be 8 in this case. The computation of β_2 is left as an exercise at the end of the section. ■

Note that there is no particular reason why the bisection method should be restricted to polynomial equations like Example 1. Any continuous function which we are able to evaluate is allowable.

Example 2:

A man wishes to fence a circular garden plot of radius 10 *feet, but he is* 0.8 *feet short of having enough fence. He decides to make as much of the boundary circular as he can and then use a straight line segment of fence to close the region as shown in Figure* 5. *What angle* θ *(in radians) should he allow for the straight segment?*

FIGURE 5

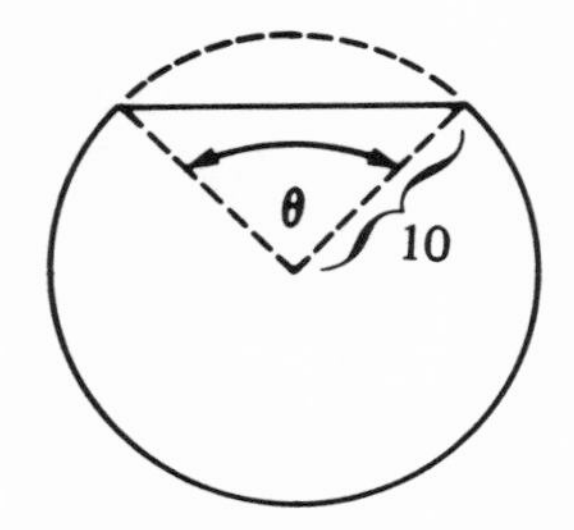

Solution:
As it would take $(2\pi)(10) = 20\pi$ feet of fence to enclose the full circle, we see that the man only has $20\pi - 0.8$ feet of fence. The circular part of the boundary is $(2\pi - \theta)(10)$ feet in length. To find the length of the straight line segment first bisect θ to form two right triangles. Then the length of the boundary segment is $20\sin(\theta/2)$ (Why?) Adding the two boundary lengths gives

$$(2\pi - \theta)(10) + 20\,sin\,\frac{\theta}{2} = 20\pi - 0.8$$

or, simplifying,

$$10\sin\frac{\theta}{2} - 5\theta + 0.4 = 0.$$

A sketch of $10\sin(\theta/2)$ and $5\theta - 0.4$ (Figure 6) indicates one solution α to this equation lying between 0 and π. With $a = 0$, $b = \pi$, and $E = 10^{-2}$, we compute

$$\log_{10}\left(\frac{b-a}{E}\right)/\log_{10}2 \doteq 8.3,$$

so that $n = 9$. The approximation with $F(\theta) = 10\sin(\theta/2) - 5\theta + 0.4$ follows.

$$F\left(\frac{\pi}{2}\right) = -0.38\ldots < 0, \qquad \alpha \in \left[0, \frac{\pi}{2}\right]$$

$$F\left(\frac{\pi}{4}\right) = 0.29\ldots > 0, \qquad \alpha \in \left[\frac{\pi}{4}, \frac{\pi}{2}\right]$$

$$F\left(\frac{3\pi}{8}\right) = 0.065\ldots > 0, \qquad \alpha \in \left[\frac{3\pi}{8}, \frac{\pi}{2}\right]$$

$$F\left(\frac{7\pi}{16}\right) = -0.12\ldots < 0, \qquad \alpha \in \left[\frac{3\pi}{8}, \frac{7\pi}{16}\right]$$

$$F\left(\frac{13\pi}{32}\right) = -0.024\ldots < 0, \qquad \alpha \in \left[\frac{3\pi}{8}, \frac{13\pi}{32}\right]$$

$$F\left(\frac{25\pi}{64}\right) = 0.022\ldots > 0, \qquad \alpha \in \left[\frac{25\pi}{64}, \frac{13\pi}{32}\right]$$

FIGURE 6

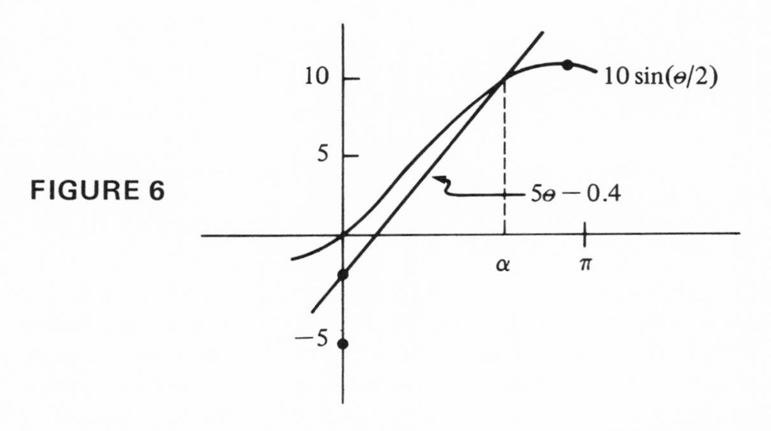

$$F\left(\frac{51\pi}{128}\right) = -0.00066\ldots < 0, \qquad \alpha \in \left[\frac{25\pi}{64}, \frac{51\pi}{128}\right]$$

$$F\left(\frac{101\pi}{256}\right) = 0.01\ldots > 0, \qquad \alpha \in \left[\frac{101\pi}{256}, \frac{51\pi}{128}\right].$$

Finally, $\beta = 203\pi/512 = 1.245\ldots$ radians (about 71.4°). ■

Before leaving the method of bisection, three observations are in order. First, the computation of n with a logarithmic expression would be useful if one wanted to know the number of steps required *without* actually making the approximation. However, in practice, there is no need to predetermine n. We need only compute until α is restricted to an interval of length less than $2E$, then approximate α by the midpoint of this interval. Secondly, one may avoid working with a lot of unnecessary digits by rounding the midpoint values to one more digit than the accuracy required. (For instance, using 1.734 instead to 111/64 in Example 1.) Finally, bisection is an easy method to remember. However, if we want a reasonably good degree of accuracy, say 10^{-5}, for a root between two consecutive integers, we need to evaluate $F(x)$ at sixteen midpoints! In the next section, we try to improve upon this rate of convergence. With bisection we compute $F(a)$ and $F(b)$ at each step and then disregard everything but the signs. Why not try to use the sizes of $F(a)$ and $F(b)$ as well?

EXERCISES 6.2

1. Finish the computation of β_2 in Example 1.
2. Use the method of bisection to approximate all real roots of $F(x) = 3x^4 - 4x^3 - 12x^2 - 5$ to within 10^{-2}. (*Hint*: Use Descartes's rule of signs.)
3. The equation $1/(\sqrt{x} + 1) = \sqrt{x - 1}$ has one real solution. Approximate it to within 10^{-1} using the method of bisection.

4. Use the method of bisection to approximate all real roots of $F(x) = x^5 - 5x + 7$ to within 10^{-1}. (*Hint*: Investigate $F'(x)$ if necessary.)
5. Suppose a circle of radius 4 has a diameter AB drawn across it. Now suppose that a point P starts at B and moves around the circle as shown in Figure 7. At the center C the angle formed by CP and CB is θ. Approximate to within 10^{-2} radians the positive angle θ for which the distance $\overline{PQ}$ from P to the diameter is half the arclength $\widehat{PB}$.
6. Use the method of bisection to approximate all real roots of $F(x) = 4x^3 - 3x^2 - 3x - 7$ to within 10^{-2}.
7. Suppose $F(x)$ is a continuous function and c is a real number. Now assume that $F(c) \cdot F(c + 4) < 0$ and that $F(x) = 0$ has exactly one solution between c and $c + 4$. Let $E = 10^{-k}$ be the accuracy required in finding this solution by the method of bisection. Find constants a and b so that the next integer larger than $a + bk$ is the number of steps required.
8. In this section it was assumed in solving $F(x) = 0$ that $F(x)$ was a continuous function with exactly one root between a and b and that $F(a) \cdot F(b) < 0$. Give examples of polynomial functions with the following behavior:
   (a) exactly one root between 0 and 1, but $F(0) \cdot F(1) > 0$,
   (b) exactly two roots between 0 and 1 with $F(0) \cdot F(1) > 0$,
   (c) exactly two roots between 0 and 1 with $F(0) \cdot F(1) < 0$.

3. Linear Interpolation or the Method of Regula Falsi

Suppose we have a continuous function $F(x)$ whose graph passes through the points $(1, -0.4)$ and $(2, 1.6)$. Since these points' y-coordinates have opposite signs, the graph crosses the x-axis somewhere between 1 and 2. If we now need to make a guess at where this solution for $F(x) = 0$ occurs without knowing anything more about $F(x)$, what should our guess be? If the midpoint 1.5 of the x-coordinates is taken, this guess cannot be off by more than 0.5. (This, of course, is the basis of Section 2's bisection method.) However, since the graph of $F(x)$ is closer to the x-axis when $x = 1$ than when $x = 2$, it seems *likely* (although not necessary) that the

FIGURE 7

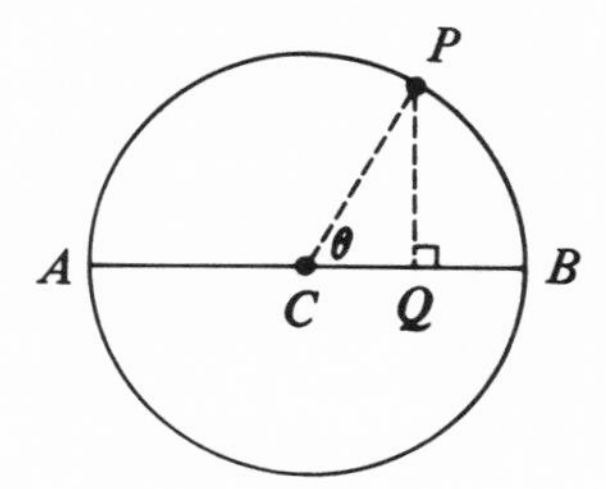

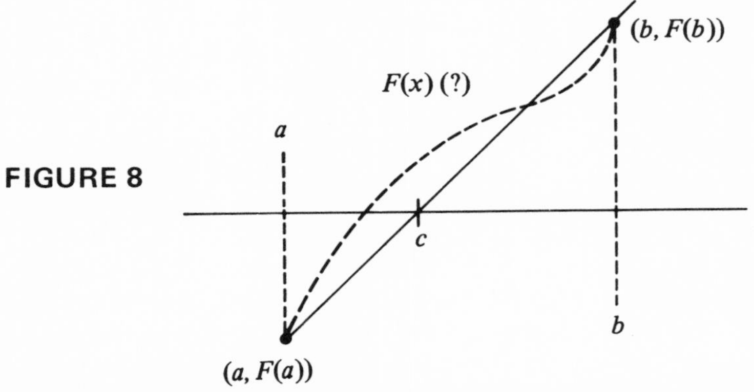

FIGURE 8

x-intercept is closer to 1 than it is to 2. One method of weighting our guess in this way is to connect the two known graph points with a straight line and then take the x-intercept of the line as our guess. This method of interpolating with a straight line between points is illustrated in Figure 8.

By similar triangles we find

$$\frac{c-a}{|F(a)|} = \frac{b-c}{|F(b)|}, \tag{3.1}$$

and solving for c gives

$$\begin{aligned} c &= \frac{a|F(b)| - b|F(a)|}{|F(b)| + |F(a)|} \\ &= a + \frac{|F(a)|}{|F(a)| + |F(b)|}(b-a). \end{aligned} \tag{3.2}$$

In the above example, with $(a, F(a)) = (1, -0.4)$ and $(b, F(b)) = (2, 1.6)$, our approximation for the x-intercept is then

$$\begin{aligned} c &= 1 + \frac{|-0.4|}{|-0.4| + |1.6|}(2-1) \\ &= 1 + \frac{0.4}{2} = 1.2. \end{aligned}$$

It should be noted that this is only a guess and in some cases not a particularly good guess. (See Example 3 of this section, for instance.) However, for most "reasonable" functions, it provides a better estimate than the bisection method. This linear interpolation method is also known as the method of *regula falsi* (meaning "false position"), since the straight line is not truly the graph of $F(x)$.

Once the approximation c for the solution α of $F(x) = 0$ has been made, one returns to the trick of the bisection method—namely, checking the sign of $F(c)$ to see whether α belongs to $[a, c]$ or $[c, b]$ according to

FIGURE 9

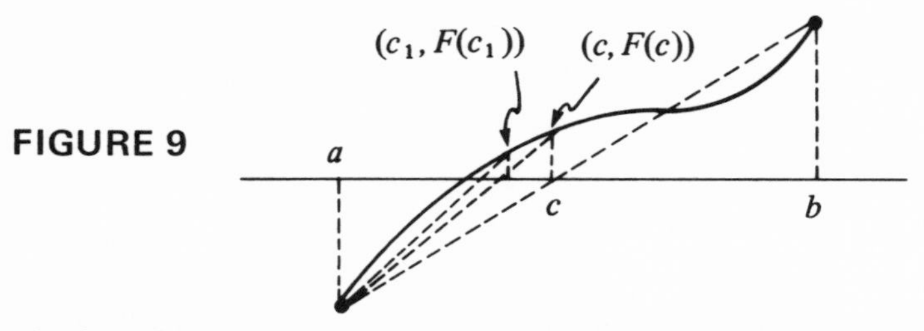

whether $F(a) \cdot F(c) < 0$ or $F(c) \cdot F(b) < 0$. Of course, whichever interval contains α is called $[a_1, b_1]$ and a new approximation c_1 is found using a_1, b_1 and (3.2). The process is repeated as illustrated in Figure 9 until the desired accuracy is obtained. One advantage of the bisection method which is now lost is that the accuracy of the approximation at each step is uncertain. Therefore, the following plan will be used in an attempt to reduce button pushing and have some idea of the accuracy attained at each step. Suppose the solution is desired to four decimal places. Each approximation c_i will be written with four decimal digits, but in computing the next approximation c_{i+1} only as many digits of c_i will be used as would seem to be accurate by comparing with previous approximations. (Ordinarily, we will try to increase the accuracy by one decimal digit with each step.) Also, in evaluating $|F(a)|/(|F(a)| + |F(b)|)$, the function values will be rounded off to the first two significant digits. When it seems that the desired accuracy has been reached, we check this by verifying a change of sign near the latest approximation. One nice feature of linear interpolation is that a computational error or sloppiness in rounding off can slow down the process, but usually not affect the final answer. As long as each c_i is between a_i and b_i, any errors are usually self-correcting.

Example 1:

If $F(x) = x^4 - 4x^3 + 12$, *approximate the solution of* $F(x) = 0$ *which lies between* 1 *and* 2 *to an accuracy of* 10^{-3} *using linear interpolation.*

Solution:

Beginning with $a = 1$ and $b = 2$, we record the following computation.

$$F(1) = 9 \qquad F(2) = -4$$

$$c = 1 + \frac{9}{9+4}(2-1) \doteq 1.692.$$

Attempting to add one decimal digit of accuracy, round c to 1.7. $F(1.7) \doteq 0.70$, so that $[a_1, b_1] = [1.7, 2]$.

$$F(1.7) \doteq 0.70 \qquad F(2) = -4$$

$$c_1 = 1.7 + \frac{0.70}{0.70 + 4}(0.3) \doteq 1.745.$$

Noting $|c_1 - c| \doteq -0.05$, we expect accuracy to about two decimal places. $F(1.75) \doteq -0.059$, so that $[a_2, b_2] = [1.7, 1.75]$.

$$F(1.7) \doteq 0.70 \qquad F(1.75) \doteq -0.059$$

$$c_2 = 1.7 + \frac{0.70}{0.70 + 0.059}(0.05) \doteq 1.746.$$

Since $|c_2 - c_1| \doteq 0.001$, accuracy is expected to about three decimal places and this is verified by checking for a change of sign near c_2 as follows.

$$F(1.746) \doteq 0.0026 \qquad F(1.747) \doteq -0.013.$$

Hence $\alpha = 1.746\ldots$, and for our approximation we take $\beta = 1.746$ since $|F(1.746)| < |F(1.747)|$. (Actual value: $\alpha = 1.74617\ldots$.) Note that this same solution to one less digit of accuracy was found with more work in Example 1 of Section 2. ■

We next present a nonpolynomial problem.

Example 2:

If $F(x) = \cos^2 x - x$, *approximate all solutions of* $F(x) = 0$ *to an accuracy of* 10^{-4} *using linear interpolation.*

Solution:

Again, the number and rough location of the roots must be determined first. This might be accomplished in either of the following two ways.

(1) Since $0 \le \cos^2 x \le 1$, we only expect roots between 0 and 1. But $dF/dx = -2\cos x \sin x - 1 = -\sin 2x - 1$, so that $F(x)$ is a decreasing function on $0 \le x \le 1$. Checking $F(0) > 0$ and $F(1) < 0$ shows that there is exactly one root of $F(x)$ between 0 and 1.

(2) Consider Figure 10. We start with $a = 0$ and $b = 1$.

$$F(0) = 1 \qquad F(1) \doteq -0.71$$

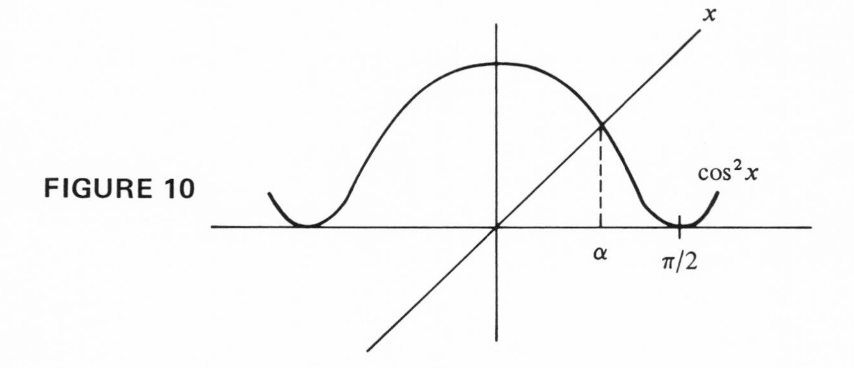

FIGURE 10

$$c = 0 + \frac{1}{1 + 0.71}(1) \doteq 0.5848.$$

$F(0.6) \doteq 0.081$, so that $[a_1, b_1] = [0.6, 1]$.

$$F(0.6) \doteq 0.081 \qquad F(1) \doteq -0.71$$

$$c_1 = 0.6 + \frac{0.081}{0.081 + 0.71}(0.4) \doteq 0.6410.$$

$F(0.64) \doteq 0.0034$, so that $[a_2, b_2] = [0.64, 1]$.

$$F(0.64) \doteq 0.0034 \qquad F(1) \doteq -0.71$$

$$c_2 = 0.64 + \frac{0.0034}{0.0034 + 0.71}(0.36) \doteq 0.6417.$$

Since $|c_2 - c_1| \doteq 0.0007$, accuracy in c_2 might be expected to about four decimal places. Accordingly, we check for a change of sign near c_2.

$$F(0.6417) \doteq 0.000028 \qquad F(0.6418) \doteq -0.00017.$$

Therefore $\alpha = 0.6417\ldots$, and β is taken as $\beta = 0.6417$ since $|F(0.6417)| < |F(0.6418)|$. (Actual value: $0.641714\ldots$.) ■

Finally, are there situations where the linear interpolation method might behave badly? Well, since linear interpolation means that we are assuming the graph between two known points behaves like a straight line, we might expect the method to converge slowly in a case like the one in Figure 11 where the actual graph stays away from the straight line approximation. Of course, for $F(x) = 2x^4 - 1$ the root between 0 and 1 is $(\frac{1}{2})^{1/4}$. However, to illustrate what happens to the rate of convergence we attempt to find this root by the linear interpolation method in the next example.

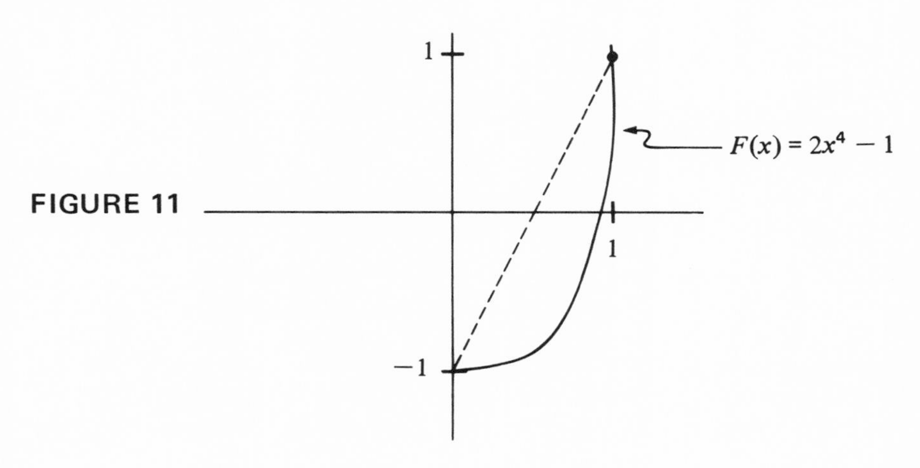

FIGURE 11

Example 3:

Approximate the solution of $2x^4 - 1 = 0$ *between* 0 *and* 1 *to an accuracy of* 10^{-3}.

Solution:

We begin with $a = 0$ and $b = 1$.

$$F(0) = -1 \qquad F(1) = 1$$

$$c = 0.5.$$

$F(0.5) \doteq -0.88$, so that $[a_1, b_1] = [0.5, 1]$.

$$F(0.5) \doteq -0.88 \qquad F(1) = 1$$

$$c_1 = 0.5 + \frac{0.88}{0.88 + 1}(0.5) \doteq 0.734.$$

Since $|c_1 - c| \doteq 0.23$, round r_1 to 0.7. $F(0.7) \doteq -0.52$, so that $[a_2, b_2] = [0.7, 1]$.

$$F(0.7) \doteq -0.52 \qquad F(1) = 1$$

$$c_2 = 0.7 + \frac{0.52}{0.52 + 1}(0.3) \doteq 0.803.$$

Since $|c_2 - c_1| \doteq 0.1$, try rounding c_2 to 0.80. $F(0.80) \doteq -0.18$, consequently $[a_3, b_3] = [0.80, 1]$.

$$F(0.80) \doteq -0.18 \qquad F(1) = 1$$

$$c_3 = 0.80 + \frac{0.18}{0.18 + 1}(0.2) \doteq 0.831.$$

Since $|c_3 - c_2| \doteq 0.03$, round c_3 to 0.83. $F(0.83) \doteq -0.051$, so that $[a_4, b_4] = [0.83, 1]$.

$$F(0.83) \doteq -0.051 \qquad F(1) = 1$$

$$c_4 = 0.83 + \frac{0.051}{0.051 + 1}(0.17) \doteq 0.838.$$

$F(0.838) \doteq -0.014$, so that $[a_5, b_5] = [0.838, 1]$.

$$F(0.838) \doteq -0.014 \qquad F(1) = 1$$

$$c_5 = 0.838 + \frac{0.014}{0.014 + 1}(0.162) \doteq 0.840.$$

Since $|c_5 - c_4| = 0.002$, we check for a change of sign near c_5.

$$F(0.840) \doteq -0.0043 \qquad F(0.841) \doteq 0.00050.$$

Hence, we take $\beta = 0.841$. (Actual value: 0.84089 . . . .) Notice that the sequence of approximations was 0.5, 0.7, 0.80, 0.83, 0.838, 0.840, 0.841 so that it is taking roughly *two* additional steps for each additional digit of accuracy. Compare this with Examples 1 and 2. There are cases where the convergence will be even slower. In such cases, a return to the method of bisection might be in order, or we might use the method of the next section. ■

EXERCISES 6.3

1. In Example 1, use linear interpolation to approximate the root of $F(x)$ which lies between 3 and 4 to an accuracy of 10^{-3}.
2. Use linear interpolation to approximate all real roots of $F(x) = x^3 + x^2 - 2x - 1$ to an accuracy of 10^{-3}.
3. Use linear interpolation to approximate both roots of $F(x) = e^x + x^2 - 3$ to an accuracy of 10^{-3}.
4. Use linear interpolation to find the solution of $\sqrt{x+1} - \sqrt{x} = 0.2$ to an accuracy of 10^{-4}. (To avoid slow convergence in this problem, be sure to begin with a and b consecutive integers.)
5. The volume of a sphere of radius R is $(4/3)\pi R^3$. Find the positive value x so that the volume of the sphere whose radius is two units more than x is four times the volume of the sphere whose radius is two units less than x. Use linear interpolation and find x to an accuracy of 10^{-4} units.
6. Two vertical towers stand 25 feet apart and one is 12 feet taller than the other. From the top of each tower a cable stretches to the bottom of the other tower. If the angle which one cable makes with the ground is 1.1 times as large as the angle which the other cable makes with the ground, how tall are the towers? Use linear interpolation and find the heights to an accuracy of 10^{-3} feet. (*Hint*: Use inverse trigonometric functions.)
7. The polynomial $P(x) = 27x^4 - 27x^3 - 18x^2 + 28x - 8$ has a root of multiplicity 3 at 2/3. Beginning with $a = 0$ and $b = 1$, use linear interpolation to find c, c_1, and c_2 to three decimal places. Note that $P'(2/3)$ must be zero and make a sketch of $P(x)$ on $[0, 1]$ to indicate why the rate of convergence is so slow.
8. The equation $F(x) = \sqrt{x} + 1 - \ln(x + 3)$ has a root between 0 and 1. Use linear interpolation to find the approximations c, c_1, c_2, and c_3 to three decimal places. Note the slow convergence. Now use bisection to finish approximating the root to an accuracy of 10^{-3} units.

4. Newton's Method (Following One's Nose)

In the linear interpolation method of approximating a solution for $F(x) = 0$, each approximation step was based upon two function values, $F(a)$ and $F(b)$. In Newton's method each step will be based upon the function value and *derivative* value at *one* point, that it, $F(a)$ and $F'(a)$. Suppose we know that $F(x) = 0$ has a solution in the vicinity of a point a_0, and we compute $F(a_0)$ and $F'(a_0)$. Now, the value $F'(a_0)$ is the slope of the tangent line to the graph of $F(x)$ at the point $(a_0, F(a_0))$, and may be thought of as giving the direction in which the graph is going at that particular point. What then is the most reasonable way of approximating the point where $F(x)$ crosses the x-axis using only the values $F(a)$ and $F'(a)$? Begin at $(a_0, F(a_0))$, look in the direction of the graph at that point and follow the straight line in that direction to the x-axis. What we are doing, of course, is finding the x-intercept (call it a_1) of the tangent line to $F(x)$ at $(a_0, F(a_0))$. The next step, naturally, is to compute $F(a_1)$ and $F'(a_1)$ and follow the tangent line to $F(x)$ at $(a_1, F(a_1))$ to a new (and, we hope, better) approximation to the actual solution α. This successive approximation scheme is know as Newton's method and is illustrated in Figure 12. What is needed next is a formula with which the values a_i given geometrically can be computed, starting from the known value a_0. Since the line through the point $(a_i, F(a_i))$ with slope $F'(a_i)$ has the equation

$$y - F(a_i) = F'(a_i)(x - a_i)$$

(point-slope form), the x-intercept of this line (that is, a_{i+1}) may be found by letting $y = 0$ and solving for x:

$$0 - F(a_i) = F'(a_i)(x - a_i)$$

$$x = a_i - \frac{F(a_i)}{F'(a_i)} .$$

In general, then, a_{i+1} is found from a_i by computing

$$a_{i+1} = a_i - \frac{F(a_i)}{F'(a_i)} .$$

Before looking at examples, we mention three things to watch for. First, unlike the methods in Sections 2 and 3, Newton's method does not depend upon checking for changes in sign since at each step we work with only one x-value. Secondly, Newton's method does not always work! That is, the sequence of approximations a_0, a_1, a_2, . . . may not converge to anything. Instances of this type of behavior are pointed out in Examples 2

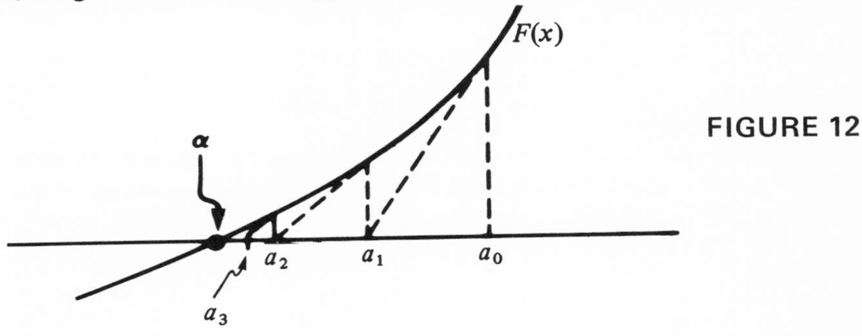

FIGURE 12

and 3 below. Basically, the problem is that the graph of $F(x)$ might not follow its tangent line very well. This possibility of failure is in contrast to the methods of bisection and linear interpolation where the solution is "trapped" by noting changes of sign and convergence is assured. This somewhat negative feature of Newton's method is not as bad as it seems, since success can ordinarily be obtained if we only start with an initial approximation a_0 which is close enough to the true solution α. (How close? See Section 9.) Finally, a very positive feature of Newton's method is that when it *does* converge it usually converges rapidly. In fact, for reasons which are explained in Section 9, the convergence is usually "quadratic," meaning that each approximation is accurate to roughly twice as many decimal places as the preceding approximation. We will keep this in mind in deciding when to stop the approximation process.

Example 1:

Find all real roots of $F(x) = 2x^3 + 15x^2 - 57x + 30$ *to an accuracy of* 10^{-8}.

Solution:

The techniques to Sections 1 and 2 of Chapter 3 reveal that $F(x)$ has three real roots, say $\alpha_3 < \alpha_2 < \alpha_1$. With a little trial and error we find that $F(-11) < 0$, $F(-10) > 0$, $F(0) > 0$, $F(1) < 0$, $F(2) < 0$, $F(3) > 0$; so that $2 < \alpha_1 < 3$, $0 < \alpha_2 < 1$, and $-11 < \alpha_3 < -10$. The next step is to write out the iteration formula for Newton's method *and simplify*.

$$G(x) = x - \frac{F(x)}{F'(x)} = x - \frac{2x^3 + 15x^2 - 57x + 30}{6x^2 + 30x - 57}$$

$$= \frac{4x^3 + 15x^2 - 30}{6x^2 + 30x - 57}.$$

Now use this formula to approximate α_1. Since $2 < \alpha_1 < 3$, let the initial approximation be $a_0 = 2.5$. Remember that we usually expect to double the number of accurate decimal digits with each step and that we round off to the expected number of accurate digits to minimize button pushing.

$$a_0 = 2.5,$$

$$a_1 = G(2.5) \doteq 2.27477478,$$

$$a_2 = G(2.3) \doteq 2.24092364,$$

$$a_3 = G(2.24) \doteq 2.23844071,$$

$$a_4 = G(2.2384) \doteq 2.23843900.$$

We expect that a_4 is accurate to about eight decimal places and check for a change of sign near a_4 in the original function $F(x)$, remembering $F(3) > 0$ and $F(2) < 0$.

$$F(2.23843900) \doteq -0.0000038 \qquad F(2.23843901) \doteq 0.00000065.$$

The approximation to α_1 is therefore $\beta_1 = 2.23843901$.

The process of approximating α_2 follows:

$$a_0 = 0.5$$

$$a_1 = G(0.5) \doteq 0.63580247,$$

$$a_2 = G(0.6) \doteq 0.64429967,$$

$$a_3 = G(0.64) \doteq 0.64532754,$$

$$a_4 = G(0.6453) \doteq 0.64534276,$$

$$F(0.64534276) \doteq 0.00000014 \qquad F(0.64534277) \doteq -0.00000022.$$

Ths approximation to α_2 is therefore $\beta_2 = 0.64534276$.

Before considering α_3, suppose we try something. Recall that beginning at $a_0 = 2.5$, the Newton sequence converged to α_1; and beginning at $a_0 = 0.5$, it converged to α_2. If it begins at a point more or less midway between α_1 and α_2, will the sequence converge to whichever root is slightly closer? Let us try, using $a_0 = 1.45 \doteq (\alpha_1 + \alpha_2)/2$.

$$a_0 = 1.45,$$

$$a_1 = G(1.45) \doteq -15.51638418 \quad (!)$$

What happened is illustrated in Figure 13 in which the indicated tangent line has x-intercept at $-15.516\ldots$. If we continue computing the Newton sequence, we will find that it converges to α_3. Hence, we may begin in the vicinity of one root and jump to another root. When this happens, of course, we merely have to start with a value of α_0 closer to the root we seek. The approximation of α_3 is left as an exercise at the end of the section. ■

As with bisection and linear interpolation, Newton's method is not restricted to polynomial equations.

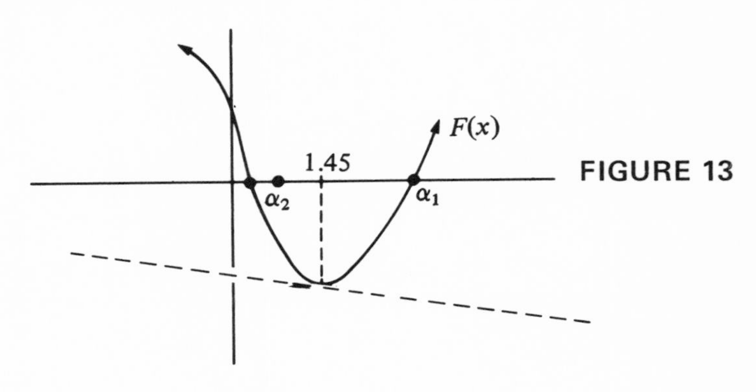

FIGURE 13

FIGURE 14

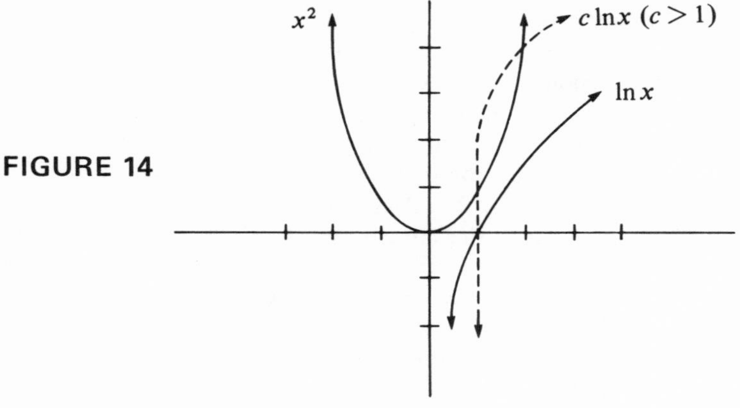

Example 2:

Find all solutions of $F(x) = 8 \ln x - x^2 = 0$ *to an accuracy of* 10^{-8}.

Solution:

To get an idea of how many solutions there might be, think of the graphs of x^2, $\ln x$, and $c \ln x$ which are shown in Figure 14. It appears that if c is large enough, then $c \ln x$ will intersect x^2 twice. Testing some values shows that $F(1) < 0$, $F(2) > 0$, and $F(3) < 0$ so that $1 < \alpha_1 < 2$ and $2 < \alpha_2 < 3$. (Alternately, consider $F'(x)$ to help sketch $F(x)$ and locate the roots.) Now find the iteration formula *and simplify*.

$$G(x) = x - \frac{F(x)}{F'(x)} = x - \frac{8 \ln x - x^2}{\frac{8}{x} - 2x}$$

$$= \frac{8x - 8x \ln x - x^3}{8 - 2x^2} .$$

To find α_1 begin with $a_0 = 1.5$ and compute the following sequence:

$$a_0 = 1.5,$$

$$a_1 = G(1.5) \doteq 1.07411963,$$

$$a_2 = G(1.1) \doteq 1.18822051,$$

$$a_3 = G(1.19) \doteq 1.19563557,$$

$$a_4 = G(1.1956) \doteq 1.19566376,$$

$$F(1.19566376) \doteq 0.00000001 \qquad F(1.19566377) \doteq -0.00000004,$$

$$\beta_1 = 1.19566376.$$

The solution α_2 may be found in a similar manner. This is left as an exercise at the end of the section.

Suppose for a moment though that we consider what happens if we begin with initial approximations of $a_0 = 2$ or $a_0 = 1.9$. If $a_0 = 2$, then in computing $a_1 = G(2)$ our calculator indicates an error. A careful look at $G(x)$ shows that the denominator is 0 when $x = 2$ so that the division is undefined. That is, we have picked the one value where $F(x)$ has a horizontal tangent (which thus has no intercept). If we try moving a little so that $a_0 = 1.9$, then in computing a_1 we find $a_1 = G(1.9) = -1.62 \ldots$. An attempt to compute $a_2 = G(a_1)$ will give an error indication since $\ln x$ is only defined for $x > 0$. Thus, the tangent line to $F(x)$ at $x = 1.9$ has an intercept outside the domain of $F(x)$. These occurrences for $a_0 = 2$ and $a_0 = 1.9$ are illustrated in Figure 15. Of course, these "failures" are really failures in the choice of a_0, not in Newton's method. The remedy is simply to start with a value of a_0 closer to the solution being sought. ■

As an instance when a Newton sequence fails to converge in a slightly different way, consider the following simple example.

Example 3:

Find the solution of $F(x) = 2x^{1/5} - 1 = 0$ *to an accuracy of* 10^{-6}.

Solution:

Of course, we may simply solve this equation to find $\alpha = \frac{1}{32}$, but suppose we try Newton's method instead. The iteration formula is

$$G(x) = x - \frac{2x^{1/5} - 1}{\frac{2}{5}x^{-4/5}} = -4x + \frac{5}{2}x^{4/5}.$$

Since $F(0) = -1$ and $F(1) = 1$, let $a_0 = 0.5$. The following sequence results. ($a_i^{4/5}$ is evaluated by using logarithms or by the y^x key if the calculator has one. Be careful about negative values.)

$$a_0 = 0.5,$$

$$a_1 \doteq -0.564127,$$

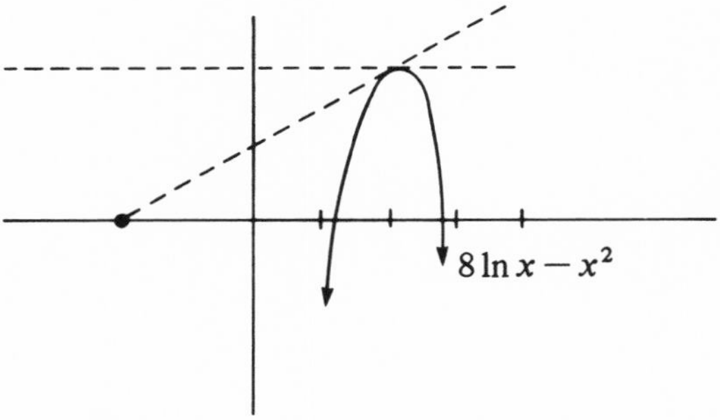

FIGURE 15

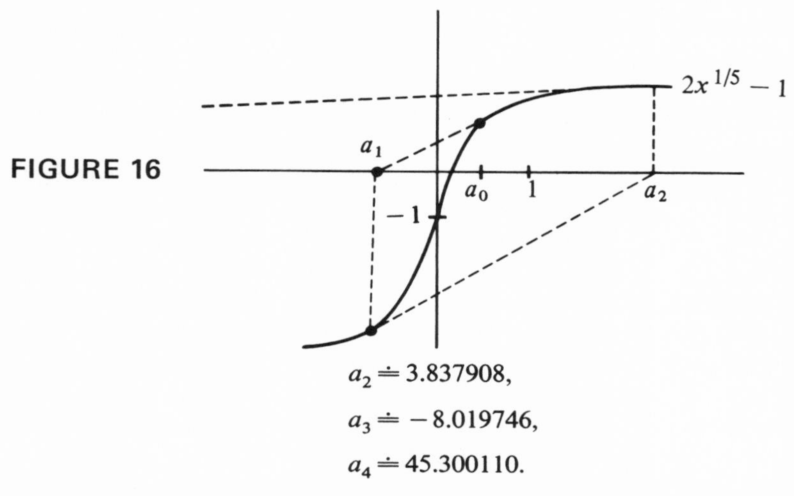

FIGURE 16

$$a_2 \doteq 3.837908,$$

$$a_3 \doteq -8.019746,$$

$$a_4 \doteq 45.300110.$$

In this case the approximations are oscillating back and forth *away from* the solution as illustrated in Figure 16. Once more, the cure for this behavior is to simply choose the value for a_0 close enough to the true solution. For example, with $a_0 = 0.1$, we have

$$a_0 = 0.1,$$

$$a_1 \doteq -0.003777,$$

$$a_2 \doteq 0.043926,$$

$$a_3 \doteq 0.029468,$$

$$a_4 \doteq 0.031208,$$

$$a_5 \doteq 0.031250,$$

$$\beta = 0.031250 \qquad \left(\alpha = \frac{1}{32} = 0.03125\right).$$

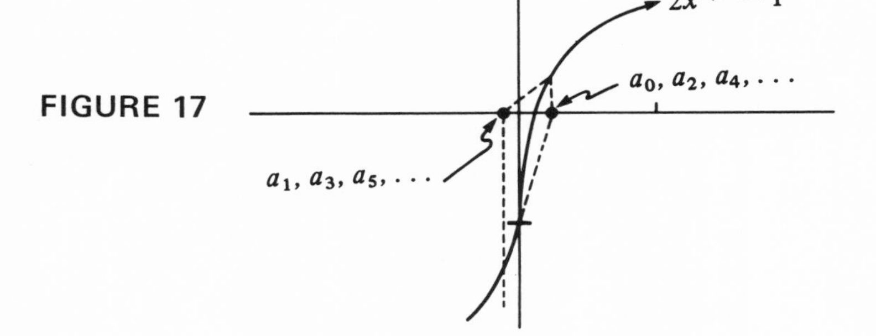

FIGURE 17

In fact, it is interesting to note that there is a value of a_0 between 0.1 and 0.5 (near 0.108) for which the approximations form a closed loop (see Figure 17).

As a final example, consider the following special case where the Newton sequence converges, but slowly.

Example 4:
Find the root of $F(x) = 27x^4 - 126x^2 + 80x + 32$ *between* 1 *and* 2 *to an accuracy of* 10^{-8}.

Solution:
If we take time to check for a change of sign to verify the existence of the root, we find that all function values seem to be nonnegative. In fact, one may use the algorithm of Section 3 in Chapter 2 to show that $F(x)$ has a multiple root (although we would not ordinarily do this). Hence between 1 and 2 the graph of $F(x)$ might appear as shown in Figure 18. For this type of root, the methods of bisection and linear interpolation fail since there is no change of sign. However, there is no reason why Newton's method should not work (think of the geometry).

We find the iteration formula and begin with $a_0 = 1.5$.

$$G(x) = x - \frac{27x^4 - 126x^2 + 80x + 32}{108x^3 - 252x + 80}$$

$$= \frac{81x^4 - 126x^2 - 32}{108x^3 - 252x + 80},$$

$$a_0 = 1.5,$$

$$a_1 = G(1.5) \doteq 1.42199248,$$

$$a_2 = G(1.42) \doteq 1.37820807.$$

It appears that a_1 was correct to about one decimal place, and we assume that a_2 is correct to about two.

$$a_3 = G(1.38) \doteq 1.35712993,$$

$$a_4 = G(1.36) \doteq 1.34682075.$$

It seems that we are having trouble even getting the *second* decimal digit. (This slow rate of convergence is typical in a function F for which

FIGURE 18

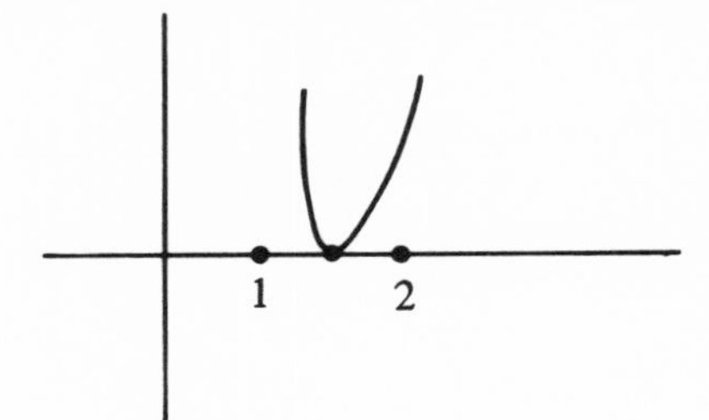

$F'(\alpha) = 0$ (or nearly 0). The reason is investigated in Section 9.) One may check that the true root is $\alpha = \frac{4}{3}$, and the sequence will slowly converge to this value. However, a look at Figure 18 indicates that a computational problem is going to arise. As the a_i approach α, $F(a_i)$ and $F'(a_i)$ each approach 0. This means that in computing $a_{i+1} = a_i - (F(a_i)/F'(a_i))$, we are dividing very small numbers. On the calculator, we then get fewer significant digits due to the initial zeros in the numerator and denominator. For example, in computing $G(1.33334)$ we obtain

$$\frac{0.00288003}{0.00216001} = 1.333341\ldots,$$

so that we have actually moved away from the true root. Since the numerator and denominator have only six significant digits, we may only expect that many in the quotient. Some choices of a_0 give convergence to 1.33333333, while others give erratic behavior near this value or even jump to the other root of $F(x)$. This is a mechanical problem, not a theoretical one. In any event one should be quite careful when the *derivative* value is small near a root of $F(x)$. ■

EXERCISES 6.4

1. Find α_3 to an accuracy of 10^{-7} in Example 1.
2. Find α_2 to an accuracy of 10^{-8} in Example 2.
3. Find all positive roots of $F(x) = x^5 - 4x^3 + 2$ to an accuracy of 10^{-8}.
4. Find the negative root of $F(x) = 2x^3 + 3x^2 - 18x + 7$ to an accuracy of 10^{-8}.
5. Find the solution of $1/(\sqrt{x} + 1) = x^2$ to an accuracy of 10^{-4}.
6. Find all solutions of $(x - 1)^2 = \sin x$ to an accuracy of 10^{-8}.
7. A rectangular box is 15 inches long, 10 inches wide, and 8 inches high. Each of the three dimensions is to be increased by the same amount x so as to increase the volume of the box by 300 cubic inches. Find x to an accuracy of 10^{-4}.
8. The value of a savings account with continuous compounding is Ae^{rt}, and the value with simple interest (before the end of the first compounding period) is $A + Art$. In each case A is the initial deposit, r is the decimal rate of interest (5% = 0.05) and t is the time since deposit measured in years. After how many days (to the nearest day) is an account with continuous compounding at 5.3% worth the same as an account with simple interest at 5.4%?
9. Find the first five terms of the Newton sequence for $F(x) = 2x^4 - 1$ starting with $a_0 = 1$. Compare this with the sequence of Example 3 in Section 3.
10. Look back at $F(x)$ in #7 of Exercises 6.3, which has a root of multiplicity 3 at 2/3. Let $a_0 = 0.66$ and find $a_1 = G(a_0)$ and $a_2 = G(a_1)$ from Newton's method. Now evaluate $G(0.6666)$ and $G(0.6665)$. Why is the sequence behaving this way?

11. Let $F(x) = 4\sin^{-1}x - 4x - 1$. Find $G(x)$ from Newton's method. If $a_0 = 0$, find $a_1 = G(a_0)$. If $a_0 = 0.5$, find $a_1 = G(a_0)$ and $a_2 = G(a_1)$. Recall $d\sin^{-1}x/dx = 1/\sqrt{1-x^2}$.

5. Horner's Method

Up to this point, we have considered approximation methods which work for both polynomial equations and nonpolynomial equations. For the methods of bisection and linear interpolation, the only restriction was that there be a solution which could be isolated by a change of sign in a continuous function. In Newton's method not even a change of sign was required, but only the ability to differentiate. We now turn to Horner's method which applies *only* to polynomial equations and requires that the solution being sought can be isolated by a change in sign. We consider this method because it is an application of an algebraic manipulation we introduced earlier and because it has the interesting property of allowing us to find solutions to an accuracy beyond the display capacity of the calculator.

The two basic algebraic procedures used in Horner's method for a polynomial $P(x)$ are procedures which were considered in Chapter 2; namely (1) decreasing the roots of $P(x)$ by c and (2) expanding the roots of $P(x)$ by a factor of 10. Finding the coefficients of the new polynomial in the first procedure is a matter of repeated polynomial division by $x - c$. (Why?) Of course, this may now be performed more conveniently using synthetic division. The following example illustrates the procedure.

Example 1:
Decrease the roots of $P(x) = 2x^4 - 9x^3 + 2x^2 + 3$ *by* 6.

Solution:
Divide $P(x)$ by $x - 6$. Then divide the resulting quotient by $x - 6$. Keep dividing successive quotients by $x - 6$ until the result is a constant and a remainder. The successive divisions are written synthetically as follows:

2	−9	2	0	3	6
2	3	20	120	(723)	
2	15	110	(780)		
2	27	(272)			
(2)	(39)				

The remainders and first constant "2" have been circled since these are the coefficients of the polynomial

$$P_1(x) = 2x^4 + 39x^3 + 272x^2 + 780x + 723$$

which is being sought. That is, the roots of $P_1(x)$ are 6 units less than the roots of $P(x)$. ■

Now, to Horner's method. Suppose that a real polynomial $P(x)$ has exactly one root α (not counting multiplicities) between the nonnegative integers d and $d+1$. (If we were seeking a negative root, we would consider $P(-x)$.) Suppose also that $P(d) \cdot P(d+1) < 0$. We know that the integer part of the root α is d. To find the first decimal digit d_1 perform the following steps.

(1) Find a new polynomial $P_1(x)$ by first decreasing the roots of $P(x)$ by d and then expanding the new roots by a factor of 10. Note that $P_1(x)$ now has a root between 0 and 10 with the same digits as those following d in α.

(2) Determine the integer value d_1 between 0 and 9 such that $P_1(x)$ changes sign between d_1 and $d_1 + 1$.

Steps (1) and (2) may now be repeated starting with $P_1(x)$ and d_1 to find $P_2(x)$ and d_2, the second decimal digit of α. Continue to the accuracy desired.

Before illustrating Horner's method with an example, we make an observation about step (2). It is not necessary to evaluate the polynomial at each integer 0 through 9 looking for a change of sign. Rather, motivated by our experience in Section 3, we may make a preliminary guess using linear interpolation. For example, in looking for d_1, the preliminary estimate is $(10)(|P(d)|/(|P(d)| + |P(d+1)|))$. (Why?)

Example 2:

Use Horner's method to find the root of $P(x) = x^3 - 3x^2 + 2x - 7$ *between* 3 *and* 4 *to an accuracy of* 10^{-11}.

Solution:

Note $P(3) = -1$ and $P(4) = 17$. Perform steps (1) and (2) to find $P_1(x)$. Synthetically,

$$\begin{array}{rrrrl} 1 & -3 & 2 & 7 & \quad \underline{|3} \\ 1 & 0 & 2 & -1 & \\ 1 & 3 & 11 & & \\ 1 & 6 & & & \end{array}$$

and expanding the roots of $x^3 + 6x^2 + 11x - 1$ by a factor of 10 gives

$$P_1(x) = x^3 + 60x^2 + 1100x - 1000.$$

$P_1(x)$ has a root between 0 and 10 with the same digits as those following "3" in α. To estimate the next digit d_1 we evaluate

$$(10)\left(\frac{|P(3)|}{|P(3)| + |P(4)|}\right) = \frac{10}{18} = \frac{5}{9}.$$

Since 0 and 1 are the digits on either side of this value, we suspect $d_1 = 0$. This is verified by checking that $P_1(0) = -1000$ and $P_1(1) = 161$ have opposite signs.

Now, repeat the steps of Horner's method for $P_1(x)$ and d_1. Decreasing the roots by 0 does nothing and expanding the roots by a factor of 10 gives

$$P_2(x) = x^3 + 600x^2 + 10000x - 1000000.$$

Because

$$(10)\left(\frac{|P_1(0)|}{|P_1(0)| + |P_1(1)|}\right) = \frac{10000}{1161} \doteq 8.6,$$

we suspect $d_2 = 8$ and verify this with $P_2(8) = -81088$ and $P_2(9) = 39329$. Repeating Horner's method for $P_2(x)$ and d_2 gives

1	600	110000	−1000000	⌊8
1	608	114864	−81088	
1	616	119792		
1	624			

and $P_3(x) = x^3 + 6240x^2 + 11979200x - 81088000$. Estimating d_3 gives

$$(10)\left(\frac{|P_2(8)|}{|P_2(8)| + |P_2(9)|}\right) = \frac{81088}{81088 + 39329} \doteq 6.7,$$

and we verify that $d_3 = 6$ by $P_3(6) = -8987944$ and $P_3(7) = 3072503$.

It is now known that α is $3.086d_4d_5d_6\ldots$. Notice that each successive polynomial $P_k(x)$ tends to have more digits in its coefficients. Another use of Horner's method to find d_4 will likely result in coefficients longer than the display capacity of the calculator. Rather than attempt some rounding-off procedure (our calculations have been exact so far), we stop using Horner's method. Instead, we resort to Newton's method for the root of $P_3(x)$. Because the third and fourth coefficients of $P_3(x)$ dominate the first two, $P_3(x)$ is almost linear for $0 < x < 10$. So we expect fast convergence (think of the picture for Newton's method). In fact, we have

$$G(x) = x - \frac{x^3 + 6240x^2 + 11979200x - 81088000}{3x^2 + 12480x + 11979200}$$

$$= \frac{2x^3 + 6240x^2 + 81088000}{3x^2 + 12480x + 11979200}$$

$a_0 = 6.5$ (since P_3 has a root between 6 and 7),

$a_1 = G(6.5) \doteq 6.74537112,$

$a_2 = G(6.7453) \doteq 6.74533988.$

We kept four decimal digits of a_1 to compute d_2 because we expected fast convergence and wanted eight accurate digits for a_2. One may check by change of sign that a_2 is accurate to eight digits. Hence, this combined Horner-Newton method supplies

$$\beta = 3.08674533988$$

with $|\beta - \alpha| < 10^{-11}$. ■

EXERCISES 6.5

1. $P(x) = 4x^3 - 11x^2 + 6x - 2$ has a root between 2 and 3. Find the polynomial $P_1(x)$ whose roots are those of $P(x)$ reduced by 2 and expanded by a factor of 10.
2. $P(x) = x^3 - 101x^2 + x - 105$ has a root between 101 and 102. Find the polynomial $P_1(x)$ whose roots are those of $P(x)$ reduced by 101. Apply Newton's method for $P_1(x)$ to find the next eight digits of the root of $P(x)$.
3. Find the real root of $P(x) = x^3 + 25x - 90$ to an accuracy of 10^{-10} using the method of Example 2 and applying Newton's method to $P_2(x)$.
4. Use the method of Example 2 to find the negative root of $P(x) = x^3 - 3x + 1$ to an accuracy of 10^{-11} using Newton's method on $P_3(x)$. (*Hint*: consider $P(-x)$.)
5. Find the positive root of $P(x) = x^4 + 3x^3 + x^2 + x - 63$ to an accuracy of 10^{-10} using the method of Example 2.
6. The two basic operations of Horner's method are decreasing the roots of a polynomial by c and expanding the roots of a polynomial by a factor of 10. Do these operations commute–that is, does the same new polynomial result regardless of the order in which the operations are performed? Why?

6. Bernoulli Iteration

In Section 4 of Chapter 2, the root-approximation method known as Bernoulli iteration was previewed in connection with the symmetric functions s_k and the k-th power functions p_k. Recall that *if* an n-th-degree real polynomial has n real nonnegative roots $r_1, \ldots, r_n$ and $r_1 > r_2 \geq r_3 \geq \ldots \geq r_n$, then

$$\lim_{k\to\infty} \frac{p_{k+1}}{p_k} = \lim_{k\to\infty} \sqrt[k]{p_k} = r_1,$$

so that the dominant root r_1 may be estimated by p_{k+1}/p_k or $\sqrt[k]{p_k}$ for some large value of k. How do the ratios p_{k+1}/p_k behave for other arrangements of the roots? Before considering a few cases, let us agree to assume that $p_k \neq 0$ for all sufficiently large k.

As a first case, suppose the real polynomial $P(x)$ has a real root r_1 of multiplicity 1 whose magnitude is larger than that of all other roots, be they real or complex; that is, $|r_1| > |r_2| \geq \ldots |r_n|$. Then, just as before,

$$\frac{p_{k+1}}{p_k} = (r_1)\frac{1 + (r_2/r_1)^{k+1} + \ldots + (r_n/r_1)^{k+1}}{1 + (r_2/r_1)^k + \ldots + (r_n/r_1)^k}$$

$$\to r_1 \quad \text{as } k \to \infty$$

since

$$\left|\left(\frac{r_j}{r_1}\right)^k\right| = \left(\frac{|r_j|}{|r_1|}\right)^k \to 0.$$

Example 1:

Approximate the largest root of the polynomial $P(x) = x^3 + 7x^2 + x + 6$ by computing p_5/p_4.

Solution:

Computing from the formulas of Section 4 in Chapter 2, we find $s_0 = 1$, $s_1 = -7$, $s_2 = 1$, $s_3 = -6$,

$$p_1 = -7,$$

$$p_2 = (-7)(-7) - (2)(1) = 47,$$

$$p_3 = (47)(-7) - (-7)(1) + (3)(-6) = -340,$$

$$p_4 = (-340)(-7) - (47)(1) + (-7)(-6) = 2375,$$

$$p_5 = (2375)(-7) - (-340)(1) + (47)(-6) = -16567.$$

The sequence of ratios p_{k+1}/p_k for $k = 1, 2, 3, 4$ is -6.7142, -7.2340,

-6.9853, -6.9756. Hence the approximation p_5/p_4 is -6.9756. (The actual value of the root of largest magnitude is $-6.97988\ldots$.) The other two roots are imaginary and, of course, complex conjugates (roughly $-0.01 \pm 0.93i$). ■

As a new possibility, consider the case in which the dominant root is real and of multiplicity 2, so that $|r_1| = |r_2| > |r_3| \geq \ldots \geq |r_n|$ and $r_1 = r_2$. In this case since $r_2/r_1 = 1$,

$$\frac{p_{k+1}}{p_k} = (r_1)\left[\frac{2 + (r_3/r_1)^{k+1} + \ldots + (r_n/r_1)^{k+1}}{2 + (r_3/r_1)^k + \ldots + (r_n/r_1)^k}\right]$$

$$\rightarrow r_1 = r_2 \quad \text{as } k \rightarrow \infty.$$

So again we have convergence of p_{k+1}/p_k to the dominant root.

As a third case, suppose there are two dominant real roots of opposite sign, so that $|r_1| = |r_2| > |r_3| \geq \ldots \geq |r_n|$ and $r_1 = -r_2$. Then we have

$$\frac{p_{k+1}}{p_k} = (r_1)\left[\frac{1 + (-1)^{k+1} + (r_3/r_1) + \ldots + (r_n/r_1)^{k+1}}{1 + (-1)^k + (r_3/r_1)^k + \ldots + (r_n/r_1)^k}\right].$$

For large *even* values of k this ratio is close to 0, while for large *odd* values of k the ratio is large in absolute value. (Why?) Note, however, that

$$\frac{p_{k+1}}{p_{k-1}} = \frac{p_{k+1}}{p_k} \cdot \frac{p_k}{p_{k-1}} \doteq (r_1)^2\left[\frac{1 + (-1)^{k+1}}{1 + (-1)^{k-1}}\right]$$

$$\doteq (r_1)^2$$

if k is a large odd value.

As a final case, suppose the dominant root of the real polynomial $P(x)$ is imaginary. Then its conjugate is also a root and has the same magnitude, so that $|r_1| = |r_2| > |r_3| \geq \ldots \geq |r_n|$ and $r_1 = r_2^*$. If we write the dominant roots in polar form $r_1 = R \operatorname{cis} \theta$ and $r_2 = R \operatorname{cis}(-\theta)$ and ignore the terms in p_{k+1}/p_k which tend to 0 as $k \rightarrow \infty$, then p_{k+1}/p_k looks like

$$(R \operatorname{cis} \theta)\left[\frac{1 + (\operatorname{cis}(-2\theta))^{k+1}}{1 + (\operatorname{cis}(-2\theta))^k}\right] = (R)\frac{\operatorname{cis}((k+1)\theta) + \operatorname{cis}(-(k+1)\theta)}{\operatorname{cis}(k\theta) + \operatorname{cis}(-k\theta)}$$

$$= (R)\frac{\cos((k+1)\theta)}{\cos(k\theta)}.$$

In general this will not converge as $k \rightarrow \infty$, but will drift about in a somewhat periodic fashion.

EXERCISES 6.6

1. Approximate the dominant root of $P(x) = x^3 - 8x^2 + 7x - 1$ by computing p_5/p_4. Check for a change of sign to determine the accuracy of this approximation.
2. Compute $p_1, p_2, \ldots, p_7$ for $P(x) = 2x^3 - x^2 - 10x + 5$. What does this behavior indicate about the dominant root(s) of $P(x)$? Can you approximate the dominant root(s)? (*Hint*: Consider p_6/p_4.)
3. Compute p_5/p_4 for $P(x) = 8x^3 + 52x^2 + 70x - 49$. From this estimate try to guess the actual value of the dominant root and then find the remaining two roots.
4. Compute $p_1, p_2, \ldots, p_6$ for $P(x) = x^2 + x + 1$. Write the sequence of terms p_{k+1}/p_k for $k = 1, \ldots, 6$.
5. Compute $p_1, \ldots, p_7$ for $P(x) = 2x^3 - 11x^2 + 14x + 10$. What do the values p_{k+1}/p_k for $k = 1, \ldots, 6$ suggest about the dominant root(s) of $P(x)$?
6. If a polynomial of degree five has roots $8, 8, -8, 1, -1$, how will the ratios p_{k+1}/p_k behave for large k?
7. Find a polynomial of degree four for which $p_k = 0$ for infinitely many values of k.

7. Root Squaring and the Method of Graeffe

Suppose a polynomial $P(x)$ is given. We saw in Section 1 of Chapter 2 that if $Q(y^2) = P(y)P(-y)$, then $Q(x)$ is a polynomial whose roots are the squares of the roots of $P(x)$. This root-squaring technique is the basis of a method for approximating the roots of real polynomials which is known as Graeffe's method. For reasons of convenience of signs, the actual manipulation will be that of finding a polynomial whose roots are the *negatives* of the squares of the roots of $P(x)$, as in the following example.

Example 1:

Let $P(x) = 2x^3 - 4x^2 + 3x - 1$ *and find a polynomial* $P_1(x)$ *whose roots are the negatives of the squares of the roots of* $P(x)$.

Solution:

We first find $Q(y^2) = P(y)P(-y)$. If r_1, r_2, r_3 are the roots of $P(x)$, then $(r_1)^2, (r_2)^2, (r_3)^2$ are the roots of $Q(x)$. To get the negatives of these squares, of course, let $P_1(x) = Q(-x)$.

$$Q(y^2) = (2y^3 - 4y^2 + 3y - 1)(-2y^3 - 4y^2 - 3y - 1)$$

$$= (2)(-2)y^6 + [(-4)(-4) + (3)(-2) + (2)(-3)]y^4$$

$$= +[(3)(-3) + (-1)(-4) + (-4)(-1)]y^2 + (-1)^2$$

$$Q(x) = -(2)^2x^3 + [(-4)^2 - 2(2)(3)]x^2 + [-(3)^2 + 2(-4)(-1)]x + (-1)^2$$

$$Q(-x) = (2)^2x^3 + [(-4)^2 - 2(2)(3)]x^2 + [(3)^2 - 2(-4)(-1)]x + (-1)^2$$

$$P_1(x) = Q(-x) = 4x^3 + 4x^2 + x + 1.$$

The form for $Q(-x)$ in the example was chosen to illustrate the following formula which may be checked by careful inspection. If $P(x) = a_0x^n + \ldots + a_{n-1}x + a_n$, then $P_1(x) = b_0x^n + \ldots + b_{n-1}x + b_n$, where

$$\begin{aligned}
b_0 &= (a_0)^2, \\
b_1 &= (a_1)^2 - 2(a_0)(a_2), \\
b_2 &= (a_2)^2 - 2(a_1)(a_3) + 2(a_0)(a_4), \\
&\vdots \\
b_{n-1} &= (a_{n-1})^2 - 2(a_{n-2})(a_n), \\
b_n &= (a_n)^2.
\end{aligned} \tag{7.1}$$

Notice how the expression for each b_i grows symmetrically about a_i until we reach the "edge" of $P(x)$. The formulas should make computing $P_1(x)$ considerably easier than the polynomial multiplication of Example 1. The whole process could be repeated starting with $P_1(x)$ to find $P_2(x)$, the polynomial whose roots are the negative squares of the roots of $P_1(x)$ and therefore the negative fourth powers of the roots of $P(x)$. In general, if $P(x)$ has roots $r_1, \ldots, r_n$, we may compute the sequence $P(x)$, $P_1(x)$, $P_2(x), \ldots$ where $P_m(x)$ has roots $-(r_1)^{2^m}, \ldots, -(r_n)^{2^m}$.

How does finding $P_m(x)$ help in approximating the roots of $P(x)$? Suppose for the moment that $P(x)$ has roots $r_1, \ldots, r_n$ such that no two have the same magnitude. That is, suppose $|r_1| > |r_2| > \ldots > |r_n|$. If $P_m(x) = b_0x^n + \ldots + b_{n-1}x + b_n$ has roots $-(r_1)^{2^m}, \ldots, -(r_n)^{2^m}$, then we know from our work with symmetric functions in Section 4 of Chapter 2 that

$$\frac{b_1}{b_0} = -s_1(-(r_1)^{2^m}, \ldots, -(r_n)^{2^m})$$

$$= (r_1)^{2^m}\left[1 + \left(\frac{r_2}{r_1}\right)^{2^m} + \ldots + \left(\frac{r_n}{r_1}\right)^{2^m}\right]$$

$$= (r_1)^{2^m}\left[1 + \sum_{i \neq 1}\left(\frac{r_i}{r_1}\right)^{2^m}\right],$$

$$\frac{b_2}{b_0} = s_2(-(r_1)^{2^m}, \ldots, -(r_n)^{2^m})$$

$$= (r_1 r_2)^{2^m}\left[1 + \sum_{\{i, j\} \neq \{1, 2\}}\left(\frac{r_i r_j}{r_1 r_2}\right)^{2^m}\right], \tag{7.2}$$

$$\frac{b_3}{b_0} = -s_3(-(r_1)^{2^m}, \ldots, -(r_n)^{2^m})$$

$$= (r_1 r_2 r_3)^{2^m}\left[1 + \sum_{\{i, j, k\} \neq \{1, 2, 3\}}\left(\frac{r_i r_j r_k}{r_1 r_2 r_3}\right)^{2^m}\right],$$

$$\vdots$$

$$\frac{b_n}{b_0} = (-1)^n s_n(-(r_1)^{2^m}, \ldots, -(r_n)^{2^m})$$

$$= (r_1 r_2 \ldots r_n)^{2^m}.$$

Looking at the first equation and remembering that r_1 is the dominant root, we find that b_1/b_0 is approximately $(r_1)^{2^m}$ if m is large. (In what sense?) Because r_1 and r_2 are the largest and next largest of the roots, b_2/b_0 is approximately $(r_1 r_2)^{2^m}$ for large m. In the same way we find that for $1 \leq j \leq n$, b_j/b_0 is approximately $(r_1 r_2 \ldots r_j)^{2^m}$ for large m. To recover the roots simply note that if $b_{j-1}/b_0 \doteq (r_1 r_2 \ldots r_{j-1})^{2^m}$ and $b_j/b_0 \doteq (r_1 r_2 \ldots r_j)^{2^m}$, then $b_j/b_0 \doteq (b_{j-1}/b_0)(r_j)^{2^m}$ or $(r_j)^{2^m} \doteq b_j/b_{j-1}$. In summary, *when* the roots of $P(x)$ satisfy $|r_1| > |r_2| > \ldots > |r_n|$, then from the coefficients $b_0, \ldots, b_n$ of $P_m(x)$ the following approximations result

$$r_1 \doteq \left(\frac{b_1}{b_0}\right)^{1/2^m},$$

$$r_2 \doteq \left(\frac{b_2}{b_1}\right)^{1/2^m},$$

$$\vdots \tag{7.3}$$

$$r_n \doteq \left(\frac{b_n}{b_{n-1}} \right)^{1/2^m}.$$

There is still a small problem, however. Which 2^m-th root should we take for each approximation? Under our assumption that $|r_1| > |r_2| > \ldots > |r_n|$ the roots must be real (why?), so let us assume that we can tell either by substitution in $P(x)$ or by one of our separation of roots techniques whether to take the positive or the negative 2^m-th root for our approximation.

Example 2:

Approximate the roots of $P(x) = 2x^3 - 7x^2 - 13x + 16$ *to three decimal places by computing* $P_3(x)$ *from Graeffe's method.*

Solution:

Noting changes of sign, one may check that $4 < r_1 < 5$, $-2 < r_2 < -1$, and $0 < r_3 < 1$. Computing $P_1(x)$ from formulas (7.1) gives

$$b_0 = (2)^2 = 4,$$

$$b_1 = (-7)^2 - 2(2)(-13) = 101,$$

$$b_2 = (-13)^2 - 2(-7)(16) = 393,$$

$$b_3 = (16)^2 = 256,$$

$$P_1(x) = 4x^3 + 101x^2 + 393x + 256.$$

Next find $P_2(x)$. We keep four significant digits since we wish to approximate three decimal places.

$$b_0 = (4)^2 = 16,$$

$$b_1 = (101)^2 - 2(4)(393) = 7057,$$

$$b_2 = (393)^2 - 2(101)(256) \doteq 1027 \times 10^2,$$

$$b_3 = (256)^2 \doteq 6554 \times 10,$$

$$P_2(x) \doteq 16x^3 + 7057x^2 + (1027 \times 10^2)x + (6554 \times 10).$$

Finally, compute $P_3(x)$.

$$b_0 = (16)^2 = 256,$$

$$b_1 \doteq (7057)^2 - (2)(16)(1027 \times 10^2) \doteq 4651 \times 10^4,$$

$$b_2 \doteq (1027 \times 10^2)^2 - 2(7057)(6554 \times 10) \doteq 9622 \times 10^6,$$

$$b_3 \doteq (6554 \times 10)^2 \doteq 4295 \times 10^6,$$

$$P_3(x) \doteq 256x^3 + (4651 \times 10^4)x^2 + (9622 \times 10^6)x + 4295 \times 10^6.$$

From the coefficients of P_3 approximate the roots using (7.3) and choose the signs based on the initial comments.

$$r_1 \doteq \left(\frac{4651 \times 10^4}{256} \right)^{1/8} \doteq 4.544,$$

$$r_2 \doteq -\left(\frac{9622 \times 10^6}{4651 \times 10^4} \right)^{1/8} \doteq -1.947,$$

$$r_3 \doteq \left(\frac{4295 \times 10^6}{9622 \times 10^6} \right)^{1/8} \doteq 0.904.$$

(The actual roots are 4.543 . . . , −1.947 . . . , 0.904 . . . .) ■

The reader might now be asking how many polynomials it is necessary to find; that is, when is m large enough? In Example 2, $P_3(x)$ was specified. There is a general indicator, however, of when m is large enough for $P_m(x)$ to result in "good" approximations to the roots. Suppose the ratio b_1/b_0 in $P_{m-1}(x)$ is approximately $(r_1)^{2^{m-1}}$ and the ratio b_1/b_0 in $P_m(x)$ is approximately $(r_1)^{2^m} = ((r_1)^{2^{m-1}})^2$. Then b_1/b_0 in $P_m(x)$ is roughly the *square* of b_1/b_0 in $P_{m-1}(x)$. But since b_0 in $P_m(x)$ *is* the square of b_0 in $P_{m-1}(x)$, b_1 in $P_m(x)$ must be approximately the square of b_1 in $P_{m-1}(x)$. The same behavior can be shown for the other coefficients as well. In other words, *as $P_m(x)$ becomes a "good" approximating tool, the coefficients of $P_m(x)$ are approximately the squares of the corresponding coefficients in $P_{m-1}(x)$.* Looking back at formulas (7.1) this means that the terms after $(a_i)^2$ in completing b_i become small relative to $(a_i)^2$. (Check Example 2 again.) Of course, all of *this applies only when no two roots have the same magnitude.*

The matter of using Graffe's method to find the imaginary roots of a real polynomial will be considered in the next section. As a final consideration in this section, however, take the case of real roots where two of the roots have the same magnitude, say $|r_1| = |r_2| > |r_3| > \ldots > r_n|$. Is there some clue in the calculation of the polynomials $P_1(x)$, $P_2(x)$, . . . which indicates that the two largest roots are of equal magnitude? Yes! Let $a_0, \ldots, a_n$ be the coefficients of $P_{m-1}(x)$ and $b_0, \ldots, b_n$ the coefficients of $P_m(x)$. Then from equations (7.1) we know that

$$b_0 = (a_0)^2 \qquad b_1 = (a_1)^2 - 2(a_0)(a_2).$$

Dividing the second equation by the first gives

$$\frac{b_1}{b_0} = \left(\frac{a_1}{a_0}\right)^2 - 2\left(\frac{a_2}{a_0}\right).$$

When $|r_1| = |r_2|$ we find from (7.2) that

$$\frac{a_1}{a_0} \doteq 2(r_1)^{2^{m-1}}$$

$$\frac{a_2}{a_0} \doteq (r_1 r_2)^{2^{m-1}} = (r_1)^{2^m}$$

for large m. Therefore, substitution gives

$$\frac{b_1}{b_0} \doteq \left(2(r_1)^{2^{m-1}}\right)^2 - 2(r_1)^{2^m}$$

$$\doteq 2(r_1)^{2^m} = \frac{1}{2}\left(2(r_1)^{2^{m-1}}\right)^2$$

$$\doteq \frac{1}{2}\left(\frac{a_1}{a_0}\right)^2$$

and because $b_0 = (a_0)^2$ we may conclude that $b_1 \doteq \frac{1}{2}(a_1)^2$. So the coefficient b_1 in $P_m(x)$ is approximately *half* the square of the corresponding coefficient in $P_{m-1}(x)$. *This behavior holds for other pairs of roots of equal magnitude.* For example, if

$$|r_1| > |r_2| > \ldots > |r_j| = |r_{j+1}| > \ldots > |r_n|,$$

then b_j in $P_m(x)$ will be approximately half the square of b_j in $P_{m-1}(x)$ for large m.

Finally, having an indicator for a pair of real roots of equal magnitude, how do we go about finding the values of these roots? Suppose again that $|r_1| = |r_2| > \ldots > |r_n|$. Then from (7.3) we see that $b_2/b_0 \doteq (r_1 r_2)^{2^m} = (r_1)^{2^{m+1}}$. Therefore, $|r_1| = |r_2| \doteq (b_2/b_0)^{1/2^{m+1}}$ and as usual we must decide on the signs for r_1 and r_2. In the more general case $|r_1| > |r_2| > \ldots > |r_j| = |r_{j+1}| > \ldots > |r_n|$ we have

$$|r_j| = |r_{j+1}| \doteq \left(\frac{b_{j+1}}{b_{j-1}}\right)^{1/2^{m+1}}, \tag{7.4}$$

$$r_k \doteq \left(\frac{b_k}{b_{k-1}}\right)^{1/2^m}, \qquad \text{for } k \neq j, j+1.$$

Example 3:

Approximate the roots of $P(x) = x^4 - 4x^3 - 13x^2 + 28x + 42$ *to three decimal places using Graeffe's method.*

Solution:

By checking for changes of sign we find $5 < r_1 < 6$, $2 < r_2 < 3$, $-3 < r_3 < -2$, and $-2 < r_4 < -1$. The following tabular form for the computation of $P_1(x)$, $P_2(x)$, . . . should be self-explanatory with the calculations from (7.1) appearing between the rows of coefficients for $P_m(x)$ and $P_{m-1}(x)$.

P	1	-4	-13	28	42
	1	16	169	784	1764
		26	224	1092	
			84		
P_1	1	42	477	1876	1764
	1	1764	$2275 \cdot 10^2$	$3519 \cdot 10^3$	$3112 \cdot 10^3$
		-954	$-1576 \cdot 10^2$	$-1683 \cdot 10^3$	
			$35 \cdot 10^2$		
P_2	1	810	$734 \cdot 10^2$	$1836 \cdot 10^3$	$3112 \cdot 10^3$
	1	$6561 \cdot 10^2$	$5388 \cdot 10^6$	$3371 \cdot 10^9$	$9685 \cdot 10^9$
		$-1468 \cdot 10^2$	$-2974 \cdot 10^6$	$-457 \cdot 10^9$	
			$6 \cdot 10^6$		
P_3	1	$5093 \cdot 10^2$	$2420 \cdot 10^{16}$	$2914 \cdot 10^9$	$9685 \cdot 10^9$
	1	$2594 \cdot 10^8$	$5856 \cdot 10^{15}$	$8491 \cdot 10^{21}$	$9380 \cdot 10^{22}$
		$-48 \cdot 10^8$	$-2986 \cdot 10^{15}$	$-47 \cdot 10^{21}$	
			$2 \cdot 10^{13}$		
P_4	1	$2546 \cdot 10^8$	$2888 \cdot 10^{15}$	$8444 \cdot 10^{21}$	$9380 \cdot 10^{22}$

Notice from the calculations between the rows for $P_3(x)$ and $P_4(x)$ that each coefficient of $P_4(x)$ is roughly the square of the corresponding coefficient of $P_3(x)$, *except* the third coefficient. This coefficient is roughly *half* the square of the third coefficient in $P_3(x)$. This indicates that r_2 and r_3 are of equal magnitude, and we make the following approximations from (7.4) with $j = 2$.

$$r_1 \doteq (2546 \times 10^8)^{1/16} \doteq 5.1625 \ldots \doteq 5.163,$$

$$|r_2| = |r_3| \doteq ((8444 \times 10^{21})/(2546 \times 10^8))^{1/32}$$

$$\doteq 2.6455 \ldots \doteq 2.646,$$

$$r_2 \doteq 2.646,$$

$$r_3 \doteq -2.646,$$

$$r_4 \doteq -((9380 \times 10^{22})/(8444 \times 10^{21}))^{1/16}$$

$$\doteq -1.1623 \ldots \doteq -1.162.$$

(The actual roots are $5.1622 \ldots$, $\pm 2.6457 \ldots$, $-1.1622 \ldots$.) ■

The cases of real roots have not, of course, been exhausted. What if three roots have equal magnitude or there are two pairs of equal roots? For these cases, one must look back to equations (7.3) and devise the correct approximation formula as we did for (7.4).

Finally, we remark that Graeffe's method has the advantages of approximating *all* roots simultaneously and giving an indication of multiple roots.

EXERCISES 6.7

1. Use Graeffe's method to approximate the roots of the following polynomials to three decimal places. Each polynomial has only real roots and at most one pair of roots of equal magnitude.
   (a) $x^3 - 4x^2 - 11x + 30$,
   (b) $2x^3 - 5x^2 - 16x + 9$,
   (c) $12x^3 - 112x^2 + 279x - 81$,
   (d) $x^3 - 6x + 3$,
   (e) $27x^3 + 171x^2 + 105x + 17$ (use P_2),
   (f) $x^4 + 4x^3 - 14x^2 - 40x + 40$.
2. Suppose $P(x)$ has real roots $|r_1| = |r_2| > |r_3| > \ldots > |r_n|$. Equations (7.4) give a formula for approximating $|r_1| = |r_2|$ using $P_m(x)$. Derive another formula by considering the expression for b_1/b_0 in equations (7.2).
3. Suppose $P(x)$ has roots $|r_1| = |r_2| = |r_3| > |r_4| > \ldots > |r_n|$. Show that for large m, $|r_1| = |r_2| = |r_3| \doteq (b_3/b_0)^{1/(3)(2^m)}$.

4. Approximate the roots of $P(x) = x^5 - 60x^3 + 195x^2 + 10x - 396$ by computing $P_4(x)$ and using Graeffe's method.
5. The polynomial $P(x) = 8x^3 + 11x^2 - 105x - 145$ has roots $(-5 \pm \sqrt{5})/2$ and 29/8. Label these roots r_1, r_2, and r_3 so that $r_1 > r_2 > r_3$. Since no two roots have the same magnitude, $b_1/b_0 \doteq (r_1)^{2^m}$ in $P_m(x)$ for large values of m. Write out the expression for b_1/b_0 in $P_m(x)$ from equations (7.2) using the roots given above. Now determine how large m must be so that $b_1/b_0 \leqslant (1.1)(r_1)^{2^m}$. Is Graeffe's method efficient when two roots (like 29/8 and $(-5 - \sqrt{5})/2$) have magnitudes which are nearly but not quite equal?

*8. In Search of our Imaginary Roots

To this point we have been concerned with approximating the real roots of real polynomials. Now consider the problem of approximating imaginary roots. If $P(x)$ has exactly one pair of complex conjugate roots, then one method of approximating these roots was given in #3 of Exercises 6.1. Namely, in that case, we approximate the real roots $r_1, \ldots, r_{n-2}$ by any convenient means, obtaining approximations $\beta_1, \ldots, \beta_{n-2}$. Next factor $P(x)$ *approximately* as $(x - \beta_1) \ldots (x - \beta_{n-2})$ $Q(x)$ with $Q(x)$ quadratic. To do this, we divide $P(x)$ successively by $x - \beta_1, \ldots, x - \beta_{n-2}$ ignoring the remainder (which is small) at each stage. $Q(x)$ is now a quadratic polynomial whose roots are approximately the imaginary roots of $P(x)$.

Example 1:

Approximate all roots of the polynomial $P(x) = 6x^4 - 7x^3 + 25x^2 + (10 + 5\sqrt{2})x - 25$. *Approximate the real roots to an accuracy of* 10^{-4}.

Solution:

Descartes's rule of signs tells us there is exactly one negative (real) root and either one or three positive (real) roots. Write $P'(x)$ as follows:

$$P'(x) = 24x^3 - 21x^2 + 50x + 10 + 5\sqrt{2}$$

$$= (24x - 21)x^2 + 50x + 10 + 5\sqrt{2}$$

$$= 24x^3 + (-21x + 50)x + 10 + 5\sqrt{2}\ .$$

These expressions show $P'(x) > 0$ for $x \geq 0$, and so $P(x)$ has only one positive root. Any of our earlier methods will show the real roots to be -1.0496 and 0.7345 correct to four decimal places. Dividing $P(x)$ by the approximate factors $x + 1.0496$ and $x - 0.7345$ gives

6	−7	25	$10+\sqrt{2}$	−25	−1.0496
6	−13.2976	38.9572	−23.8184		0.7345
6	−8.8906	32.4270			

so that $P(x) \doteq (x + 1.0496)(x - 0.7345)(6x^2 - 8.8906x + 32.4270)$. The imaginary roots of $P(x)$ are then approximately

$$\frac{8.8906 \pm \sqrt{(8.8906)^2 - (4)(6)(32.4270)}}{(2)(6)} = 0.7409 \pm 2.2035i.$$

So $\beta_1 = -1.0496$, $\beta_2 = 0.7345$, $\beta_3 = 0.7409 - 2.2035i$, and $\beta_4 = 0.7409 + 2.2035i$. We know that β_1 and β_2 are correct to four decimal places. What accuracy might we expect in β_3 and β_4? Although it is not particularly convenient, the accuracy may be estimated in the following way. If $\alpha_1, \ldots, \alpha_4$ are the true roots, then

$$P(\beta_4) = (\beta_4 - \alpha_1)(\beta_4 - \alpha_2)(\beta_4 - \alpha_3)(\beta_4 - \alpha_4)$$

or

$$|\beta_4 - \alpha_4| = \left| \frac{P(\beta_4)}{(\beta_4 - \alpha_1)(\beta_4 - \alpha_2)(\beta_4 - \alpha_3)} \right|.$$

Computing $P(\beta_4)$ using De Moivre's theorem and approximating $\beta_4 - \alpha_i$ by $\beta_4 - \beta_i$ we obtain

$$|\beta_4 - \alpha_4| \doteq \left| \frac{-0.0086 + 0.0006i}{(1.7905 + 2.2035i)(0.0057 + 2.2035i)(4.407i)} \right|$$

$$\doteq 0.0003.$$

Of course, the same estimate holds for $|\beta_3 - \alpha_3|$.■

The Graeffe method of Section 7 may also be used to approximate the imaginary roots of a real polynomial. Suppose the roots of $P(x)$ are such that r_j is imaginary with $|r_1| > \ldots > |r_j| = |r_{j+1}| > \ldots > |r_n|$ and $r_j = r_{j+1}{}^*$. Then all the roots except r_j, r_{j+1} must then be real. If $r_j = R\operatorname{cis}(\theta)$ and $r_{j+1} = R\operatorname{cis}(-\theta)$ in polar form, then from equations (7.2) we find that in $P_m(x)$ for large m

$$\frac{b_j}{b_0} \doteq (r_1 \ldots r_j)^{2^m}\left[1 + \left(\frac{r_{j+1}}{r_j}\right)^{2^m}\right]$$

$$= (r_1 \ldots r_{j-1})^{2^m}\left[(r_j)^{2^m} + (r_{j+1})^{2^m}\right]$$

$$= (r_1 \ldots r_{j-1})^{2^m}(R)^{2^m}(2\cos(2^m\theta)).$$

The presence of the cosine term will then cause b_j/b_0 to be sometimes positive and sometimes negative depending upon the value of m. *This is the indicator for the presence of imaginary roots.* If all roots are real, the coefficients of the polynomials $P_m(x)$ (for $m \geq 1$) must always be positive (from equations (7.2)). However, if even one of these coefficients is negative, the original $P(x)$ has imaginary roots.

Now, having this indicator for imaginary roots, let us determine how to approximate the roots (still for the case $|r_1| > \ldots > |r_j| = |r_{j+1}| > \ldots > |r_n|$, $r_{j+1} = r_j^*$). Exactly as in Section 7 for real roots of equal magnitude, it can be shown that b_j in $P_m(x)$ is approximately the square of b_j in $P_{m-1}(x)$ *times* $\cos(2^m\theta)/2\cos^2(2^{m-1}\theta)$; and that

$$(r_k)^{2^m} \doteq \frac{b_k}{b_{k-1}}, \quad \text{for } k \neq j, j+1; \tag{8.1}$$

$$(r_j r_{j+1})^{2^m} \doteq \frac{b_{j+1}}{b_{j-1}}.$$

Hence we approximate r_k for $k \neq j, j+1$ just as before and notice that

$$|r_j| = |r_{j+1}| \doteq \left(\frac{b_{j+1}}{b_{j-1}}\right)^{1/2^{m+1}}$$

Knowing the magnitude of r_j and r_{j+1}, our only remaining problem is to determine the real and imaginary parts of r_j and r_{j+1}. For this we recall that the sum of the roots equals $-a_1/a_0$ in the original polynomial $P(x)$. Hence if $\sum' r_k$ denotes the sum of *real* roots only,

$$\sum_{k=1}^{n} r_k = 2\,\mathrm{Re}(r_j) + \sum{}' r_k = -\frac{a_1}{a_0}$$

and so

$$\mathrm{Re}(r_j) = \frac{1}{2}\left[-\frac{a_1}{a_0} - \sum{}' r_k\right].$$

To find $\mathrm{Im}(r_j)$, simply note that

$$|r_j|^2 = (\mathrm{Re}(r_j))^2 + (\mathrm{Im}(r_j))^2$$

so

$$\mathrm{Im}(r_j) = \pm\left(|r_j|^2 - (\mathrm{Re}(r_j))^2\right)^{1/2}.$$

Example 2:

Approximate the roots of $P(x) = 4x^4 + 12x^3 - 7x^2 + 24x - 9$ *to three decimal digits.*

Solution:
The computations for Graeffe's method are tabulated below showing only the relevant coefficients.

P	4	12	−7	24	−9
P_1	16	200	−599	450	81
P_2	256	$5917 \cdot 10$	$1814 \cdot 10^2$	$2995 \cdot 10^2$	6561
P_3	$6554 \cdot 10$	$3408 \cdot 10^6$	$-2534 \cdot 10^6$	$8732 \cdot 10^7$	$4305 \cdot 10^4$
P_4	$4295 \cdot 10^6$	$1161 \cdot 10^{16}$	$-5888 \cdot 10^{17}$	$7625 \cdot 10^{18}$	$1853 \cdot 10^{12}$
P_5	$1845 \cdot 10^{16}$	$1348 \cdot 10^{35}$	$1696 \cdot 10^{38}$	$5814 \cdot 10^{40}$	$3434 \cdot 10^{27}$

As before, we stop at $P_5(x)$ since each entry to be used in approximating is roughly the square of the corresponding entry in $P_4(x)$. Notice the telltale negative quantities in the third column. This indicates that the second and third roots are complex conjugates. Approximating from equations (8.1) gives

$$r_1 = -(1348 \times 10^{35}/1845 \times 10^{16})^{1/32}$$
$$\doteq -3.886,$$
$$|r_2| = |r_3| \doteq (5814 \times 10^{40}/1348 \times 10^{35})^{1/64}$$
$$\doteq 1.225,$$
$$r_4 \doteq (3434 \times 10^{27}/5814 \times 10^{40})^{1/32}$$
$$\doteq 0.386.$$

From $P(x)$ we have $-a_1/a_0 = -12/4 = -3$ so

$$\text{Re}(r_2) = \text{Re}(r_3) \doteq \frac{1}{2}\left[-3 - (-3.886 + 0.386)\right]$$
$$\doteq 0.250.$$

Computing the imaginary parts gives

$$\text{Im}(r_2) \doteq \pm\sqrt{(1.225)^2 - (0.250)^2}$$
$$\doteq \pm 1.199.$$

Therefore, the approximations are

$$\beta_1 = -3.886; \qquad \beta_2, \beta_3 = 0.250 \pm 1.199i; \qquad \beta_4 = 0.386.$$

(The actual roots are $(1 \pm \sqrt{23}\ i)/4$ and $(-7 \pm \sqrt{73})/4$.) ■

So far we have considered only the case of a single pair of complex conjugate roots whose magnitude is different from that of any real root. As in Section 7, *different arrangements of the magnitudes require individual treatment.* To indicate the approach we consider the case of *two* pairs of complex conjugate roots, call them $u_1 \pm iv_1$ and $u_2 \pm iv_2$. We assume that $|u_1 \pm iv_1| \neq |u_2 \pm iv_2|$ and that no real root has the same magnitude as any other root. As before, a look at equations (7.2) indicates that for each pair of imaginary roots there will be a column of occasionally negative coefficients for the polynomials $P_k(x)$. Also, the magnitudes $|u_1 + iv_1|$ and $|u_2 + iv_2|$ may be found in the same way as before. Once more, however, we are faced with determining real and imaginary parts. To find the real parts u_1 and u_2, we shall need two equations. One of these equations will be the same one used before, namely,

$$\sum{}' r_k + 2u_1 + 2u_2 = -\frac{a_1}{a_0}.$$

To obtain the second equation, first recall that one polynomial whose roots are the reciprocals of the roots of $P(x)$ is

$$x^n P\left(\frac{1}{x}\right) = a_n x^n + a_{n-1}x^{n-1} + \ldots + a_1 x + a_0.$$

Thus, we have

$$\sum{}' \frac{1}{r_k} + \frac{1}{u_1 \pm iv_1} + \frac{1}{u_2 \pm iv_2} = -\frac{a_{n-1}}{a_n}$$

or

$$\sum{}' \frac{1}{r_k} + \frac{2u_1}{|u_1 + iv_1|^2} + \frac{2u_2}{|u_2 + iv_2|^2} = -\frac{a_{n-1}}{a_n}.$$

With this second equation, we may now solve for u_1 and u_2, and then for v_1 and v_2 as before.

Of course, there are other cases such as repeated imaginary roots, more than two pairs of imaginary roots, real roots with the same magnitude as an imaginary pair or unequal imaginary pairs with equal magnitudes. Problems of this complexity are fortunately the exception in practice. We do wish to remark, nevertheless, that in some of these cases a helpful trick is to decrease the roots of the polynomial by some constant c to rearrange the magnitudes. (See Exercise 5 below.) Additional techniques are given in [2].

In closing we note that there are other methods of approximating imaginary roots which are efficient when complex-arithmetic functions are available (not the case with most hand-held calculators). One method

which does not require evaluating derivatives but still yields all real and imaginary roots with usually rapid convergence is Muller's method (see [3] or [9]).

EXERCISES 6.8

1. The following polynomials each have a single pair of imaginary roots. Use the method of Example 1 to approximate all roots to four decimal places.
   (a) $7x^3 - 16x^2 + 24x - 6$,
   (b) $2x^4 + 12x^3 + 29x^2 + 63x + 40$.
2. The following polynomials each have a single pair of imaginary roots. Use the method of Example 2 to approximate all roots to three decimal places.
   (a) $6x^3 - x^2 + 27x - 60$,
   (b) $24x^3 + 37x^2 - 61x + 68$,
   (c) $x^4 - 3x^3 - 15x^2 + 57x - 76$,
   (d) $4x^4 + 6x^3 - 152x^2 + 400x - 147$.
3. The polynomial $P(x) = 2x^4 + 9x^3 + 44x^2 + 14x + 150$ has two pairs of imaginary roots. Approximate them to three decimal places.
4. Suppose the (real) polynomial $P(x) = a_0x^4 + a_1x^3 + a_2x^2 + a_3x + a_4$ has roots $r_1 = r_2 = r$ and $r_3 = r_4 = r^*$ (complex conjugate roots of multiplicity 2). If $r = R\,\text{cis}(\theta)$, show that $-a_1/a_0 = 4R\cos\theta$ and $a_4/a_0 = R^4$. Find r and r^* in terms of a_0, a_1, and a_4.
5. The polynomial $P(x) = 27x^3 - 27x^2 - 36x + 64$ has one real root which has the same magnitude as its two imaginary roots. Approximate these roots to three decimal places by applying Graeffe's method to the polynomial whose roots are 1 unit more than those of $P(x)$.

*9. Speed of Convergence and Error Estimates

We close this chapter with some comments about the rates of convergence of the approximation methods considered above. The approximation method which is easiest to monitor is bisection. Indeed, since we divide the interval of search in half at each step, the error at the n-th step can be no more than $(b - a)/2^n$, where $[a, b]$ is the original interval containing the solution α. The speed of convergence for bisection is represented by the logarithmic expression of Section 2 which gives a value for the number of the step at which a prescribed accuracy is guaranteed.

In the method of linear interpolation the function under consideration is approximated by a straight line. It has already been noted (in #7 of Exercises 6.3) that if α is the solution of $F(x) = 0$ which we seek and $F'(\alpha) = 0$, then the linear interpolation method may give slow conver-

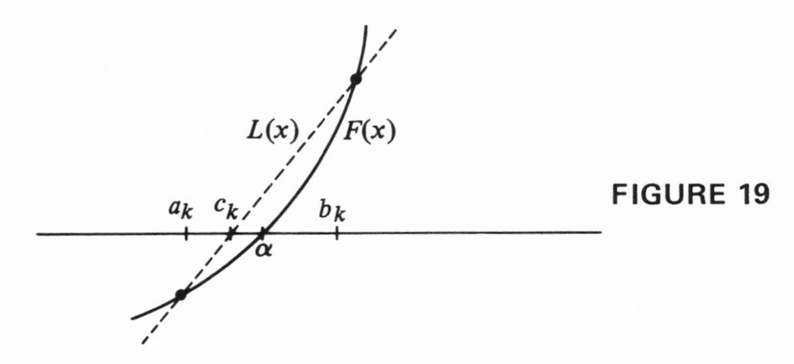

FIGURE 19

gence. So, let us assume that we are considering a "nice" function $F(x)$. That is, suppose $F(x)$ is twice differentiable in some neighborhood of α and suppose (just to be specific) that $F'(x) > 0$ and $F''(x) > 0$ in this neighborhood. Then if $[a_k, b_k]$ is an interval containing α in this neighborhood, the picture for $F(x)$ and its straight-line approximation $L(x)$ might appear as in Figure 19. Let $\triangle(u,v) = (F(v) - F(u))/(v - u)$. Then, the equation for $L(x)$ is $y - F(a_k) = \triangle(a_k, b_k) \cdot (x - a_k)$. Setting $y = 0$ and solving for x gives

$$c_k = a_k - \frac{F(a_k)}{\triangle(a_k, b_k)}\,.$$

Now, since $F(\alpha) = 0$, we find

$$\begin{aligned}\alpha - c_k &= \alpha - a_k + \frac{F(a_k)}{\triangle(a_k, b_k)}\\ &= (\alpha - a_k)\left[1 - \frac{F(\alpha) - F(a_k)}{\alpha - a_k} \cdot \frac{1}{\triangle(a_k, b_k)}\right]\\ &= (\alpha - a_k)\cdot\left(1 - \frac{\triangle(a_k, \alpha)}{\triangle(a_k, b_k)}\right).\end{aligned}$$

Hence $1 - (\triangle(a_k, \alpha)/\triangle(a_k, b_k))$ is the multiplication factor between $\alpha - a_k$ (the error in the old estimate a_k) and $\alpha - c_k$ (the error in the new estimate c_k). When a_k is "close" to α, this factor is approximately $1 - (F'(\alpha)/\triangle(a_k, b_k))$. Since b_k remains the same in each step for the function in Figure 19, the error decreases nearly "linearly," in the sense that the error at one step is roughly a constant multiple (between 0 and 1) of the error at the previous step. Other assumptions about the signs of $F'(x)$ and $F''(x)$ in a neighborhood of α lead similarly to this linear rate of convergence.

An error estimate (that is, an upper bound for the error in the k-th step) for the linear interpolation method can be obtained in the following way. Suppose $F(x)$ is as pictured in Figure 19, and let $F^{-1}(x)$ and $L^{-1}(x)$ be the corresponding inverse functions. Since $(F^{-1} - L^{-1})(0) = \alpha - c_k$ and $(F^{-1} - L^{-1})(F(a_k)) = 0$, the mean-value theorem of calculus gives us the estimate

$$|c_k - \alpha| \leq \left[\max_{F(a_k) \leq y \leq 0} |(F^{-1} - L^{-1})'(y)| \right] \left[|F(a_k)| \right].$$

Noticing that

$$(L^{-1})'(y) = \frac{1}{L'(y)} = \frac{1}{\triangle(a_k, b_k)}$$

and

$$(F^{-1})'(y) = \frac{1}{F'(F^{-1}(y))},$$

we substitute to find

$$|c_k - \alpha| \leq \left(\max_{F(a_k) \leq y \leq 0} \left| \frac{1}{F'(F^{-1}(y))} - \frac{1}{\triangle(a_k, b_k)} \right| \right) (|F(a_k)|)$$

$$= \left(\max_{a_k \leq x \leq \alpha} \left| \frac{1}{F'(x)} - \frac{1}{\triangle(a_k, b_k)} \right| \right) (|F(a_k)|)$$

$$\leq \left(\max_{a_k \leq x \leq b_k} \left| \frac{1}{F'(x)} - \frac{1}{\triangle(a_k, b_k)} \right| \right) (|F(a_k)|).$$

Because a similar argument could be given with $F(a_k)$ replaced by $F(b_k)$, the following error estimate holds

$$|c_k - \alpha| \leq \left(\max_{a_k \leq x \leq b_k} \left| \frac{1}{F'(x)} - \frac{1}{\triangle(a_k, b_k)} \right| \right) (\min\{|F(a_k)|, |F(b_k)|\}). \quad (9.1)$$

In working with Newton's method in Section 4, we assumed that the rate of convergence is usually quadratic, in the sense that the number of correct decimal digits doubles with each step. Let us now investigate why this is so. Suppose we seek the solution α for $F(x)$ which lies in an interval $[a, b]$. Suppose further that $F(x)$ has a certain amount of "smoothness" on $[a, b]$, specifically that $F(x)$ is three times continuously differentiable on $[a, b]$, and that $F'(\alpha) \neq 0$. Let $G(x) = x - (F(x)/F'(x))$. Then $G(x)$ is twice continuously differentiable, and by Taylor's theorem

$$G(x) = G(\alpha) + G'(\alpha)(x - \alpha) + \frac{1}{2} G''(\beta)(x - \alpha)^2$$

for each x in (a, b) and some β (depending on x) between x and α. Observe that

$$G'(\alpha) = 1 - \frac{F'(\alpha)\cdot F'(\alpha) - F(\alpha)\cdot F''(\alpha)}{[F'(\alpha)]^2}$$

$$= \frac{F(\alpha)F''(\alpha)}{[F'(\alpha)]^2}$$

$$= 0,$$

since $F(\alpha) = 0$ and $F'(\alpha) \neq 0$. Therefore, by substituting,

$$G(x) = G(\alpha) + \frac{1}{2} G''(\beta)(x-\alpha)^2$$

with β depending on x. Because $G''(x)$ is continuous on $[a, b]$, it is bounded (say $|G''(x)| \leq B$) on $[a, b]$. Setting $x = a_k$, the k-th Newton estimate, in the last equation then leads to

$$|G(a_k) - G(\alpha)| = |\frac{1}{2} G''(\beta)(a_k - \alpha)^2|$$

$$\leq \frac{1}{2} B(a_k - \alpha)^2.$$

Since $G(a_k) = a_{k+1}$ and $G(\alpha) = \alpha - (F(\alpha)/F'(\alpha)) = \alpha$, we finally have

$$|a_{k+1} - \alpha| \leq \frac{1}{2} B(a_k - \alpha)^2;$$

that is, the error in a_{k+1} is roughly the square of the error in a_k. If a_k is off by 10^{-3}, we expect that a_{k+1} will be off by about $(10^{-3})^2 = 10^{-6}$. This is why, when $F'(\alpha) \neq 0$, we expect quadratic convergence from Newton's method. If, in addition $F(x)$ is monotone on $[a, b]$, an error bound can be found in the same way as for linear interpolation, giving

$$|a_{k+1} - \alpha| \leq \left(\max\left| \frac{1}{F'(x)} - \frac{1}{F'(a_k)} \right|\right)(|F(a_k)|), \tag{9.2}$$

where the maximum is taken over all x between a_k and α.

Horner's method produces the decimal digits of the solution being sought one at a time, and so there is no question of rate of convergence or error bounds. The convergence in Bernoulli's method and Graeffe's method depends upon terms of the form $(r_j/r_k)^m$ approaching zero when $|r_j| < |r_k|$. Of course, the rate of this convergence depends upon the relative sizes of the roots. This is a case-by-case matter, and we have seen (in #5 of Exercises 6.7) that if two roots have nearly the same magnitude, the convergence may be quite slow.

EXERCISES 6.9

1. Let $F(x) = x^3 + x - 12$, which has a root between 2 and 3. Use equation (9.1) to obtain an error bound for the approximation obtained by one step of the linear interpolation method with $a = 2$ and $b = 3$.
2. In Example 3 of Section 3 note the approximations 0.734, 0.803, 0.831, and 0.838 to the actual solution $\alpha = 0.8408\ldots \doteq 0.841$. Compute the error in each approximation and compare the ratio of these errors to the quantity $1 - (F'(\alpha)/\Delta(1,\alpha))$ which was our estimate of the factor of linear convergence in linear interpolation.
3. The equation $F(x) = x - 2\sin x = 0$ has a solution between 1.8 and 1.9. With $a = 1.8$ use equation (9.2) to obtain an error bound for the next approximation by Newton's method.

Epilogue: Quo Vadis

As Shakespeare's Juliet put it, "Parting is such sweet sorrow." Although it will soon be time for us to part, much remains to be said and studied. In this brief final section, we can only hope to direct the reader's attention to appropriate sources for such additional study. There are many possible courses for the reader to pursue next since, as we mentioned in the introduction, we expect that this book's readers have varied backgrounds and interests. Each of the next six paragraphs describes one such course of action, via comments about texts in the annotated list of references. That list contains additional remarks concerning those texts, as well as references to a dozen articles which pertain to this book's contents.

Theory of equations: The reader may wish to compare our approach with the treatment in texts of another era. Of these, we specifically recommend [2] and [24]. In a word, [2] is quite analytic and direct, while [24] is more algebraic. Additional comments about these books, as well as remarks pertaining to other vintage texts such as [15] and [23], are given in the annotated list of references.

Number theory: In developing the second application of g.c.d. theory in Section 3 of Chapter 1, we alluded to similarities in the algebraic behavior of $\mathbf{Z}$ and $F[x]$. Although the latter has occupied us in these pages, the former certainly had its moments. (Think of $\mathbf{Z}_p$ and the rational root test, for instance.) The reader may like to see a thoroughgoing treatment of $\mathbf{Z}$ which amplifies such similarities, while also containing the few number-theoretic prerequisites for our book and going far beyond. For this purpose, many attractive, modern, undergraduate-level texts on number theory are available, and we specifically recommend [21], which was written by one of today's leading researchers in the area.

Linear algebra and differential equations: These two topics are popular and important parts of contemporary undergraduate studies. A neat, modern treatment of linear algebra may be found in [4] and/or [12]. The former is quite geometric and may be viewed as a modernization of [16]. Chapter 2 of [12] more than fits the bill as prerequisite material for our starred sections. As for differential equations, we have found it convenient to refer to [6]. Needless to say, numerous other fine texts on each of these two topics are available.

Abstract ("modern") algebra: During the fifth and sixth decades of this century, courses in abstract algebra supplanted courses in the theory of equations in many universities' programs for training future teachers and mathematics researchers. As delineated in the first 2 paragraphs of Section 5 of Chapter 1, and in the introduction, many of the concerns of abstract algebra have already been dealt with in this book; and, in many ways, the theory of equations may be the best arena in which to first encounter such topics. The reader who wishes to study abstract algebra *per se* may consult numerous fine texts, including [1], [10], and [25]. Of these, the first may be most accessible to our book's reader, proceeding, as it does, from domains to the subsequent study of more abstract systems (rings, groups, etc.). Incidentally, the first edition of [1] (in 1941) led, almost singlehandedly, to the aforementioned curricular reform in North American universities. Its European precursor, the first edition of [25], has aged gracefully, and (though terser and more demanding in its arrangement of topics) is also recommended. Finally, we recommend [10] as being typical of more recent texts which combine the readability of [1] with the efficient arrangement of abstractions in [25].

Numerical analysis: Many readers will wish to pursue the topics in Chapter 6, especially with an eye toward more refined techniques and applications to solutions of nonpolynomial equations. In this regard, we recommend [11] as a rather complete algorithmic approach to methods of approximation. Also recommended, and geared to programming use with computers (as opposed to hand-held calculators), are [3] and [9]. Be advised, however, that analytic training at the level of advanced calculus or beyond is a prerequisite for much of this material.

Other: Many readers will move on to additional mathematical studies, such as geometry, statistics, advanced calculus, real analysis, complex analysis (remember Liouville's theorem), and topology (remember profinite groups). Fortunately, there are several good texts available for each of these subjects. Other readers will pursue different sciences and arts, as varied as were their motivations for reading this book.

And now it's time.

An Annotated List of References

[1] G. Birkhoff and S. Mac Lane, *A Survey of Modern Algebra,* revised edition, Macmillan, New York, 1953. [See *Quo Vadis*?]

[2] N. B. Conkwright, *Introduction to the Theory of Equations,* Ginn, Boston, 1941. [This classic text is especially noteworthy because of its analytic proof of Newton's formulas. The reader should contrast that proof with the ones which we gave in Sections 4 and 7 of Chapter 2. In general, Conkwright's approach is much less algebraic than ours.]

[3] S. D. Conte and C. de Boor, *Elementary Numerical Analysis,* McGraw-Hill, New York, 1972. [See *Quo Vadis*?]

[4] N. Divinsky, *Linear Algebra,* Page Ficklin, Palo Alto, 1975. [In addition to being a fine text on linear algebra, this recent book contains a modern treatment of the classification of the seventeen Euclidean quadric surfaces. We referred to this geometric application in the first paragraph of Section 2 of Chapter 3.]

[5] D. E. Dobbs, Proving trig. identities to freshpersons, *Math. Assoc. of Two Year Colleges J.* **14** (1980), 39–42. [The algorithm, in Section 5 of Chapter 1, for verifying trigonometric identities was drawn from this article.]

[6] L. R. Ford, *Differential Equations,* second edition, McGraw-Hill, New York, 1955. [This text was referenced by us in Chapter 5 in regard to Legendre polynomials (see Section 2) and the oscillation theorem of Sturm (Section 4). Also see the comment in *Quo Vadis*?]

[7] J. Gerst and J. Brillhart, On the prime divisors of polynomials, *Amer. Math. Monthly* **78** (1971), 250–265. [This article, the most algebraically demanding reference that we list, contains significant applications of much of the algebraic theory in our starred sections. Its first result is the theorem of Schur, guaranteeing that an integral polynomial has root(s) modulo infinitely many prime integers, given in Section 1 of Chapter 2.]

[8] R. Gilmer, On solvability by radicals of field extensions, *Math. Ann.* **199** (1972), 263–277. [As noted in Section 5 of Chapter 3, this article compares various (inequivalent) notions of "solvability by radicals."]

[9] R. T. Gregory and D. M. Young, *A Survey of Numerical Mathematics,* Addison-Wesley, Reading, 1972. [See *Quo Vadis*?]

[10] I. N. Herstein, *Topics in Algebra*, second edition, Xerox College Publishing, Lexington, 1975. [See *Quo Vadis*?]

[11] F. B. Hildebrand, *Introduction to Numerical Analysis,* second edition, McGraw-Hill, New York, 1973. [See *Quo Vadis*?]

[12] K. Hoffmann and R. Kunze, *Linear Algebra,* second edition, Prentice-Hall, Englewood Cliffs, 1971. [As mentioned in the first paragraph of Section 3 of Chapter 3, the vector-space theory prerequisite for that section is more than adequately covered in Chapter 2 of this fine text.]

[13] F. Klein, *Famous Problems of Elementary Geometry,* English translation, second edition, Dover, New York, 1956. [As mentioned in Section 3 of Chapter 3, this historically influential account of the algebra underlying the study of constructible numbers is of special interest because of its accessible proofs that π and e are transcendental over the field of rational numbers.]

[14] A. R. Magid, Trigonometric identities, *Math. Magazine* **47** (1974), 226–227. [This article uses ring theory to obtain a nonalgorithmic proof of the result, treated via an algorithm in Section 5 of Chapter 1, that any trigonometric identity is a consequence of the Pythagorean identity, $\sin^2\theta + \cos^2\theta = 1$.]

[15] C. C. MacDuffee, *Theory of Equations,* John Wiley and Sons, New York, 1954. [As referenced in Section 6 of Chapter 5, this short, attractive text contains a particularly clear proof of Budan's theorem.]

[16] J. M. H. Olmsted, *Solid Analytic Geometry,* Appelton-Century-Crofts, New York, 1947. [Although some of this book's concerns are no longer in vogue, it contains a neat, classical approach to the classification problem for quadric surfaces which we mentioned in the first paragraph of Section 2 of Chapter 3.]

[17] M. Quagliarello, The angle trisection problem–a constructible approach, *Math. Assoc. of Two Year Colleges J.* **12** (1978), 143–149. [Without appeal to the dimension theory of vector spaces, this article presents a direct proof of the result, treated differently in Section 3 of Chapter 3, that the general angle cannot be trisected by ruler and compass.]

[18] I. Richards, An application of Galois theory to elementary arithmetic, *Advances in Math.* **13** (1974), 268–273. [This article uses Galois theory to treat generalizations of the result in Section 1 of Chapter 1 (see also Sections 3 and 5 of Chapter 3) that $\sqrt{2} + \sqrt{3}$ is irrational. Curiously enough, this theorem's first proof was obtained in 1940 by A. S. Besicovitch via a Euclidean algorithm for polynomials in several variables: the fundamental fact about degrees strikes again!]

[19] D. Rosen and J. Shallit, A continued fraction algorithm for approximating all real polynomial roots. *Math. Magazine* **51** (1978), 112–116. [As mentioned in Section 6 of Chapter 5, this article uses continued fractions (for background, see [21]) to improve the computational aspects of [24]'s technique of separation of roots via Vincent's theorem.]

[20] Z. Rubinstein, On polynomial δ-type functions and approximation by monotonic polynomials, *J. Approximation Theory* **3** (1970), 1–6. [This article contains a proof of the first of the (non-Lagrangian) interpolation results which were stated in the paragraph preceding Exercises 2.2.]

[21] H. M. Stark, *An Introduction to Number Theory,* Markham, Chicago, 1970. [This fine text contains a wealth of information, including the small amount of number theory which is prerequisite for this book and much of interest that goes beyond it. See also the comments pertaining to [18] and

the remarks in *Quo Vadis*?]

[22] I. Stewart, Gauss, *Scientific American* **237** (1977), 122–131. [This accessible article about the scientific work of Gauss was written on the occasion of Gauss's bicentenary. See the comment about it in the second paragraph of Section 6 of Chapter 3.]

[23] J. M. Thomas, *Theory of Equations*, McGraw-Hill, New York, 1938. [As referenced in the paragraph preceding Exercises 3.2, this text contains a proof that nonreal complex radicals are inevitable aspects of any solution by radicals of the *casus irreducibilis* type of real cubic polynomial.]

[24] J. V. Uspensky, *Theory of Equations*, McGraw-Hill, New York, 1948. [We heartily recommend this classic text to the reader, especially for applications and exercises. Its influence on our approach is most evident in Chapter 5. Especially noteworthy are Uspensky's proof of Rolle's theorem (contrast with ours in Section 2 of Chapter 5) and his elaboration of root separation via the theorem of Vincent. In regard to the latter, see our comment about [19].]

[25] B. L. van der Waerden, *Modern Algebra*, vol I, revised English edition, Ungar, New York, 1953. [As referenced in Section 5 of Chapter 3, this classic text contains a rederivation of Cardan's cubic formula and Ferrari's quartic formula via Lagrangian resolvents and appropriate group theory. Also see the comments in *Quo Vadis*?]

[26] C. G. Wagner, Newton's inequality and a test for imaginary roots, *Two-Year College Math. J.* **8** (1977), 145–147. [This article, written by a colleague of the authors, proves that strong logarithmic convexity guarantees the existence of nonreal roots. Contrast with our proof of this fact in Section 3 of Chapter 5.]

[27] W. C. Waterhouse, Profinite groups are Galois groups, *Proc. Amer. Math. Soc.* **42** (1974), 639–640. [If a field extension is not finite dimensional, when should it be termed "Galois"? What should be meant by its "Galois group"? Is there a relevant fundamental theorem for such objects? The answers to these questions are known and of considerable interest. As noted in Section 5 of Chapter 3, the appropriate Galois groups may be characterized topologically, and this article contains a readable proof of that characterization.]

[28] W. Wolibner, Sur un polynome d'interpolation, *Colloq. Math.* 2 (1951), 136–137. [As mentioned in the paragraph preceding Exercises 2.2, this article contains a proof of an interpolation result which is equivalent to the fundamental approximation theorem of Weierstrass.]

Additional Bibliography

The reader is encouraged to keep abreast of recent articles concerning the theory of equations. In addition to the four articles cited above ([5], [17], [19], and [26]) which have appeared after 1976, one should be aware of three additional articles which appeared in 1979 following the completion of this text. All three could be used as enrichment material for Sections 2–5 of Chapter 5.

[29] D. S. Drucker, A second look at Descartes's rule of signs, *Math. Magazine* **52** (1979), 237–238. [This article studies the number of nonreal roots of a real polynomial by analyzing the gaps in the sequence of coefficients of the polynomial.]

[30] E. Grosswald, Recent applications of some old work of Laguerre, *Amer. Math. Monthly* **86** (1979), 648–658. [This article's Theorem 6 gives a different necessary and sufficient condition for all the roots of a given complex polynomial to be real. Theorem 7 gives an alternate proof that the roots of a Legendre polynomial are all real and simple.]

[31] J. W. Pratt, Finding how many roots a polynomial has in $(0, 1)$ or $(0, \infty)$, *Amer. Math. Monthly* **86** (1979), 630–637. [This article addresses the problem described in its title by counting variations in non-Sturmian tableaux. It also indicates applications to business situations.]

Answers to Selected Problems

Exercises 1.1

1(a). Axioms (iv) and (v) each fail to be satisfied. **4(c).** For one such example, use $S = \mathbf{Q}$, $T = \mathbf{Z}$, and $a = 2$. **5(b).** $(a + a) + (b + b) = a(1 + 1) + b(1 + 1) = (a + b)(1 + 1) = (a + b)1 + (a + b)1 = (a + b) + (a + b)$. Add $-a$ on the left to get $a + b + b = b + a + b$. Next, add $-b$ on the right sides. **6(c).** The "similar result" states: "a domain S with at least three elements has characteristic 3 if and only if $(a + b)^3 = a^3 + b^3$ whenever $a \in S$ and $b \in S$." **8.** 11, 1, 10, 9, 0; 8, 4, 9, 6; $4 \cdot m$ is never 1 in $\mathbf{Z}_{12}$.

Exercises 1.2

1(first, third, fifth, and seventh). $-\frac{1}{2} + 7i$; $(\pi + 2) + (-10)i$; $\frac{165}{8} + 6i$; $\frac{36}{45} + \frac{33}{45}i$. **2(first, fifth).** $r = 7.0178344, \theta = 1.6421037$ radians; $r = 21.480005$, $\theta = 0.2830958$. **3(b).** $\cos^2(\theta) + \sin^2(\theta) = 1$. **4(first).** $|z^{-1}| = |z|^{-1}$; $\arg(z^{-1}) = 2\pi - \arg(z)$ if z is not real, and $\arg(z^{-1}) = 0$ if z is real. **5(first).** $\cos(4\theta) = \cos^4\theta - 6\cos^2(\theta)\sin^2(\theta) + \sin^4\theta$. **6(b, second).** $r = 1.0184081$, $\theta = (7\pi/76) + (2\pi k/19)$ for $k = 0, 1, \ldots, 18$. **7(a).** 3, $\frac{-3}{2}$. **7(c).** $\frac{-1}{2} \pm (i/\sqrt{2})$. **10(a, first and second).** $-1 - i, -1 + i$.

Exercises 1.3

1(b). $2x^3 - (7 + 2\pi)x^2 + (7\pi + 18)x - 63$. **2(b).** $q(x) = x^2$, $r(x) = 2$. **2(c).** $q(x) = \frac{2}{3}$, $r(x) = \frac{-5}{3}x$. **3(b).** Observe that $f^{-1} \neq 0$. Then deduce from $\deg(f) + \deg(f^{-1}) = \deg(1) = 0$ that $\deg(f) = 0$. **5(first and second).** 1, $x^2 + x$. **7(c).** Yes. **8.** Observe, for p prime, that the binomial coefficient $\binom{p}{i}$ is an integral multiple of p whenever $1 \leq i \leq p - 1$. **9(a).** $(2x^2 - 7)^3(x^2 + 1)^2$. **9(e).** x^3. **10(a).** Observe that the integer g.c.d. of 4 and 7 is 1. Use the equations $7 = 1(4) + 3$, $4 = 1(3) + 1$ (and $3 = 3(1)$) to deduce $1 = 2(4) - 1(7)$, whence $4^{-1} = 2$ in $\mathbf{Z}_7$. **11.** $\text{cis}(k\pi/4)$ for $k = 1, 3, 5$, and 7. **12(last part).** $\text{char}(S) = 2, 5$.

Exercises 1.5

2(a, second). $(-y^2)x^3 + (2y^3 + 7)x^2 + (-9y + \pi)$.

Exercises 2.1

1(a, fourth and sixth). $2 - 2i$; $x^3 + (-7 + i)x^2 + (16 - 5i)x + (-10 + 6i)$. **1(b, second).** 1. **4(a).** 1. **4(c).** 4. **6(a).** $x^3 - 24x^2 + 191x - 502$. **6(b).** $x^3 - 9x + 54$. **6(f).** $4x^3 - x^2 + 2x - 1$. **8(a).** $f(0) = 7$ and $f(1) = 3$ are each odd. **9(a).** One generalization states that if n is an integer which is not divisible by the prime p, then for $f(x) = nx + 1$, the reduction $\overline{f}(x)$ has a root in $\mathbf{Z}_p$.

Exercises 2.2

1(a). $\frac{1}{2}x^2+4$. **2(a).** Same as in #1(a). **3(b).** $\bar{2}$. **5.** Lagrange interpolation gives a constructive proof. For an alternate approach, observe, in case F has precisely n elements, that polynomials in $F[x]$ of degree at most $n-1$ induce n^n distinct functions $F \to F$.

Exercises 2.3

1(b). $49(2^2-3)^{48}(2 \cdot 2) = 196$. **4(a).** 2. **4(c).** The multiplicity is 0. **5(first).** No. **6(second).** $g_1 = x-1$, $g_2 = x-3$, and $g_n = 1$ for each $n \geq 3$. **7.** One generalization asserts that all the roots of a real cubic polynomial $f(x)$ are simple if $f(x)$ has a nonreal root.

Exercises 2.4

1(a). $-2x^4+26x^3-\frac{197}{2}x^2+102x-90$. **1(b).** $2x^4-x^3-113x^2-344x-144$. **2(a).** Roots are $2 \pm i$ and -3; $a_2=-7$ and $a_3=15$. **3(a).** $s_2/s_3=\frac{1}{2}$. **3(c).** $s_1s_2-3s_3=\frac{9}{4}$. **7(a).** 6, 16, 50, 168, 1748/3, 6148/3, 21760/3, 231712/9. **7(b).** 2.66667, 3.12500, 3.36000, 3.46825, 3.51716, 3.53936, 3.54951. **7(c).** 4.00000, 3.60021, 3.55908.

Exercises 2.5

2(a, second). $s_2s_3-3s_1s_4+5s_5$.

Exercises 2.6

3(second, third, and fifth). **C**, **C**, **Q**. **5(a).** Examples of such include 3, 11, 19, 27, 35, and 43. **5(b, first and third).** $\bar{1}$, $\bar{3}$.

Exercises 2.7

3. One such example is $\mathbf{Z}_2((x))$. **5(a, second and sixth).** 21, 1287. **5(c).** 35. **5(d, first).** Purchase three split-foyers and one mobile home.

Exercises 3.1

1(fourth and fifth). $-2, -1, -1$; -0.6823278, $0.3411639 \pm 1.1615413i$. **4.** It is not meaningful to use Cardan's formula if $p=0$, for α is then 0. In case $p \neq 0$ $(=q)$, Cardan's formula is valid.

Exercises 3.2

1(fourth and fifth). $\triangle = 0$ (case 1), $\triangle = 31 > 0$ (case 2). **3.** $\triangle = 0$; multiplicity is 2.

Exercises 3.3

1(a). Two examples of such complex numbers are π and ei. **2(b, second and fifth).** 3, 2. **6(a).** 15 is not of the form p^n. **7(last part).** One such example is given by $F=\mathbf{Q}$ and $K=\mathbf{C}$. **9(e, last part).** The minimum polynomial of $\cos(2\pi/5)$ over $\mathbf{Q}$ is x^2+2x-4.

Exercises 3.4

1(first). $x^3 + 24x^2 - 128x - 64$. **2(first).** $2x^3 - 18x^2 - 90x + 810$.

Exercises 3.5

1(second, fifth). S_3, S_2 (or equivalently, the additive structure of $\mathbf{Z}_2$). **3(c, last part).** Find m so that τ is a primitive m-th root of 1, thus reducing to the first case. **5(first).** Observe that if r and s are respectively roots of distinct irreducible factors of $f(x)$, then no automorphism σ can send r to s. **6.** Use the rational root $r = -1$ to initiate the factoring. **7(c).** $n = 3$.

Exercises 4.1

1. One such example is $12x^3 + 15x^2 - 42x + 6$. **3(a).** There are none. **4.** See the text's fourth and fifth applications of the rational root test. **5(a).** $x^4 - 18x^2 + 25$. **6(first).** Use $p = 3$.

Exercises 4.2

1. The first and second polynomials are primitive; the third is not primitive. **3.** $2, -2, 3, -3, x - 3, x^3 - x + 1$. **6(last part).** Reducible if $p = 2$.

Exercises 4.3

3. Irreducible, regardless of which n is considered.

Exercises 5.1

2. $x^8 + 6x^6 + x^4 - 24x^2 + 16$. **3(first, fourth).** $+, +; -, +$. **4(first).** Isolation is effected by the intervals $(-1, 0)$ and $(0, 3)$; more sharply, by the intervals $(-1, 0)$ and $(2, 3)$.

Exercises 5.2

1(second). Even. **2(second).** Even. **3.** Observe that the derivative, $4x^3 - 4x$, has no roots in any of the given four *open* intervals. **6.** 4, 6, 8.

Exercises 5.3

1. $x^4 - x^2 + x - 2$ has, according to Descartes's rule of signs, precisely one negative root and either one or three positive roots. 0 is not a root. **4.** All but the last. Yes.

Exercises 5.4

1. The real root of $x^5 + 2x^3 - x^2 + x - 1$ is in the interval $(0, 1)$. **2.** The real roots of $x^3 - 2x^2 - x + 2$ are isolated by use of the intervals $(-2, 0)$, $(0, \frac{3}{2})$, and $(\frac{3}{2}, 3)$. **6.** There is but one real root, in the interval $(-2, -1)$.

Exercises 5.5

3. (4) is used to obtain δ in case 2; and (5) is used to guarantee $m < \infty$ in case 1.

Exercises 6.1

1(a). $x^4 - 2x^3 + 4x^2 - 8x + 16$. **2(last part).** -2.45. **3(last part).** $1.09 \pm 1.59i$. **4.** $P^{(1)}(-1.1) = 15.29$, $P^{(2)}(-1.1) = -23.2$.

Exercises 6.2

2. -1.585, 2.820. **3.** 1.18. **5.** 1.896 radians. **7.** $a = 2$, $b = (\log_{10}2)^{-1}$. **8(c).** One such example is $(x - \frac{1}{4})(x - \frac{3}{4})^2$.

Exercises 6.3

2. -1.802, -0.445, 1.247. **3.** 0.834, -1.677. **5.** 8.8097. **6.** 41.986 feet and 53.986 feet. **7.** $c = 0.8$, $c_1 = 0.79$, $c_2 = 0.78$.

Exercises 6.4

1. -10.3837818. **3.** 0.84788566, 1.92909505. **5.** 0.7339. **8.** 257 days. **9.** 1, 0.9, 0.85, 0.8410, 0.84089643.

Exercises 6.5

1. $4x^3 + 130x^2 + 1000x + 2000$. **3.** 2.7594863794. **5.** 2.2064262782. **6.** No.

Exercises 6.6

1. 7.0225; 7.0236. **2(last part).** ± 2.2347. **4.** 1, 0, undefined, 1, 0, undefined. **6.** $\frac{8}{3}$ if k is even; 24 if k is odd. **7.** One such example is $(x^2 + 2)(x^2 + 5)$.

Exercises 6.7

1(a). 5.000, -3.000, 2.000. **1(c).** 4.500, 0.333. **1(e).** -5.667, -0.333. **1(f).** -4.828, ± 3.162, 0.828. **4.** -8.997, 4.646, 3.216, 2.383, -1.236. **5.** $m = 11$.

Exercises 6.8

1(a). 0.3031, $0.9913 \pm 1.3583i$. **2(b).** -2.807, $0.633 \pm 0.781i$. **2(d).** -7.942, $3.002 \pm 1.235i$, 0.439. **5.** $1.167 \pm 0.646i$, -1.333.

Exercises 6.9

1. 0.054. **2.** Approximate factor 0.243. **3.** 0.012.

Appendix

The programs in this appendix will solve real polynomial equations and perform a few other related tasks (e.g., Lagrange interpolation) using the methods in the text.

The methods we have chosen to concentrate on are bisection, regula falsi, and Ferrari's method. We present a regula-falsi program to find real roots, and then later modify it into a program to find complex roots.

We are working toward the following goal: a program to find all roots of an arbitrary polynomial. Theoretically (ignoring round-off error build up for one thing) this goal can be realized by combining a regula-falsi program with deflation methods. Such a combined program is presented as Exercise 10, so that exercise can be considered as the goal of the appendix.

Ferrari's method is presented for several reasons. The main reason being that Ferrari's method always works. (Regula falsi, in contrast, may not always work. The method requires the user to make some initial guesses. If these are unfortunately chosen, the method won't produce an answer and the user will have to go back an choose new initial guesses.)

Cardan's method (to find all 3 roots of a real cubic polynomial) also always works, but then again so does a combination of bisection, deflation, and the quadratic formula. Both of these techniques are presented as exercises and the reader is encouraged to try the relevant exercises, compare results, and decide for himself which technique he prefers for use on cubic polynomials.

We have attempted to keep the programs as simple as possible, with few frills, so as to make them easier to read and understand. Many of the missing frills pertain to screen display. For instance, complex numbers such as $-6+7i$ and $4i$ will appear on-screen as $-6 \quad 7$ and $0 \quad 4$. A polynomial such as $3x^2+5$ will appear on-screen as $3 \quad 0 \quad 5$. In writing the programs we had in mind a user who does not object to displays of this kind. We do, however, indicate what some of the missing frills are, and — in some cases — what can be done to add them to the programs.

One further characteristic of the user we had in mind is that he is not malicious or careless about data entry. When asked for the degree of a polynomial, he will not enter a noninteger or a string of letters; nor will he enter 0 for the leading coefficient of a polynomial. Either of these data entries will cause the programs to fail. Program failure due to malicious or careless data entry can be prevented, but to keep the programs simple, we have not attempted to do so. (The second problem is easily handled with a repeat ... until loop which repeats the request to enter the leading coefficient until such time as the number entered is nonzero. The first problem is more difficult, but subroutines can be written to screen every user entry, to decide whether or not these entries are of the right type (e.g., an integer for the degree of a polynomial), to pass them on to the rest of the program if they are, or to sound the buzzer and repeat the request for data entry if they are not.)

The programs are written in Pascal.

GENERAL

The programs can handle polynomials up to degree 20. This is an arbitrary (and changeable) limit, but one needs to specify some limit to be able to use arrays to represent polynomials.

It will be convenient to have the degree of a polynomial coded into the array. This will be done by starting the array at -1 rather than at 0. So for a polynomial $f(x)$, $f[-1]$ will denote its degree, while $f[0], f[1], \ldots$ will denote its coefficients. Example: for $f(x) = 2x^4 - 7x + 1$, $f[-1]$ is 4; $f[0]$ is 2; $f[1]$ is 0 (the coefficient of x^3); $f[2]$ is 0; $f[3]$ is -7; and $f[4]$ is 1. We'll use the type declaration:

```
type poly = array[-1..20] of real;
```

for polynomials.

We'll use the following program fragment for polynomial input:

```
write('enter the degree '); readln(n);
f[-1]:=n; writeln;
writeln('enter, when prompted, the coefficients ');
for i:=0 to n do
  begin
  write('coefficient of x super ',n-i,' '); readln(f[i]);
  end;
writeln;
```

The first line asks the user for the degree of the polynomial. Then `f[-1]:=n;` assigns that number as the degree of $f(x)$. The remaining lines ask the user to input, in this order, the coefficient of x^n, the coefficient of $x^{n-1}, \ldots$, the coefficient of x, the constant term.

Once we've inputed a polynomial $f(x)$, one of the first things we'll frequently want to do is change it to its monic associate. This is easily done by dividing all the coefficients by the leading coefficient $f[0]$. The program fragment

```
y:=f[0];
for i:=0 to n do f[i]:=f[i]/y;
```

will do this. Note that we've changed the polynomial, but it keeps its old name. For example the above fragment changes $f(x) = 2x^4 - 7x + 1$ to its monic associate $f(x) = x^4 - 3.5x + 0.5$.

We'll need a subroutine to evaluate polynomials. Suppose we want to evaluate $f(x) = a_0x^n + a_1x^{n-1} + \ldots + a_{n-1}x + a_n \in \mathbf{R}[x]$ at the real number r. We can avoid powers by using nested multiplication: a second-degree polynomial is considered to be $(a_0x + a_1)x + a_2$, a third-degree polynomial is considered to be $((a_0x + a_1)x + a_2)x + a_3$, a fourth-degree polynomial is considered to be $(((a_0x + a_1)x + a_2)x + a_3)x + a_4$, etc. E.g., repeating the example on page 159, $2x^4 + 3x^2 - 5x + 7 = (((2x + 3)x + 0)x - 5)x + 7$. We'll use the following subroutine:

```
function eval(f:poly; r:real):real;
var j,n:integer;
    y:real;
begin
n:=round(f[-1]);
y:=f[0];
for j:=1 to n do y:=y*r+f[j];
eval:=y;
end;
```

The first thing the function eval does is to set n equal to the degree of the polynomial. This is done with the command `n:= round(f[-1])` (the function round is necessary to convert the real number $f[-1]$ — it's a real number because it's part of the array of real numbers f — to the integer $f[-1]$). Next it initially defines a number y to be $f[0]$, i.e., the leading coefficient of the polynomial. It then redefines (using the command `y:=y*r+f[j]`) the new value of y to be the old value times r, plus $f[j]$. This being done for all j from 1 to n, we wind up with $(((f[0]r + f[1])r + f[2])r + f[3]) + \ldots + f[n]$, i.e., with $f(r)$.

One of the uses of evaluation is, of course, to help us decide whether or not a candidate for a root of a polynomial really is a root. Since we're dealing with real numbers here, it is best to allow a little leeway (how much will depend upon the requirements of the user and/or the degree of precision available from the computer). Let's say that a number r is a root of $f(x)$ if and only if $|f(r)| < 0.000001$. The parameter 0.000001 is arbitrary and can be changed.

BISECTION (Section 6.2)

Recall from Section 6.2 that bisection is a root-finding technique that depends on the intermediate-value theorem. Given a polynomial $f(x) \in \mathbf{R}[x]$, if we can find real numbers a and b such that a evaluates negatively (i.e., $f(a) < 0$) and b evaluates positively ($f(b) > 0$), then the intermediate-value theorem guarantees existence of a root of $f(x)$ in the interval (a, b), and bisection will find such a root.

If our polynomial $f(x)$ is monic then a number b that evaluates positively must exist. This follows from the limit formula $\lim_{x\to\infty} f(x) = \infty$ on page 134. But this formula is nonconstructive; we'd like to actually produce a number b for use in a computer program. Here's one way:

THEOREM *Let $f(x) = x^n + a_1x^{n-1} + \ldots + a_{n-1}x + a_n \in \mathbf{R}[x]$ be a monic polynomial. Let $G = \max\{|a_1|, |a_2|, \ldots, |a_n|\}$. Then $b > G + 1$ implies $f(b)$ is positive.*

Proof: Suppose $b > G + 1$. Then $G < b - 1$, so $b - 1$ is positive since $0 \leq G$. Dividing by $b - 1$, we get $\frac{G}{b-1} < 1$; hence $1 - \frac{G}{b-1}$ is positive.

Since $0 \leq G$, $1 \leq G + 1 < b$ and consequently $1 < b^n$; thus $b^n - 1$ is positive. Multiplying the positive numbers $1 - \frac{G}{b-1}$ and $b^n - 1$ we get $b^n - 1 - \frac{G(b^n-1)}{b-1}$ is positive. Hence so is $b^n - \frac{G(b^n-1)}{b-1}$.

Since $\frac{b^n-1}{b-1} = b^{n-1} + b^{n-2} + \ldots + b + 1$, we have $0 < b^n - G(b^{n-1} + b^{n-2} + \ldots + b + 1)$ and hence

$$G(b^{n-1} + b^{n-2} + \ldots + b + 1) < b^n. \tag{1}$$

Now $-a_i \leq |a_i| \leq G$ for all i, $1 \leq i \leq n$. And since b is positive, so is b^{n-i}. Thus $-a_ib^{n-i} \leq Gb^{n-i}$ for all such i. Adding gives us $-a_1b^{n-1} - a_2b^{n-2} - \ldots - a_{n-1}b - a_n \leq G(b^{n-1} + b^{n-2} + \ldots + b + 1)$. That and (1) yield $-a_1b^{n-1} - a_2b^{n-2} - \ldots - a_{n-1}b - a_n < b^n$; hence $0 < f(b)$.

We now have a number b (for definiteness, say $b = G+1.1$) that evaluates positively. If we can find a number a that evaluates negatively, then we're in business; bisection will work. If not, bisection won't get off the ground.

In the general case, such a number a need not exist. For instance, suppose all the roots of a polynomial (e.g., x^2+2x+2) are nonreal complex; i.e., the polynomial doesn't have any real roots. If there were a number a that evaluated negatively, then the intermediate-value theorem applied to a and b (we know b exists from the above) would produce a real root c. Hence since c doesn't exist, neither does a.

But there are several special cases where a must exist and hence where bisection will work. Two such cases are (i) $\deg(f)$ odd and (ii) constant term (i.e., a_n) negative.

For odd-degree polynomials, we can use the following:

THEOREM *Let $f(x) = x^n + a_1x^{n-1} + \ldots + a_{n-1}x + a_n \in \mathbf{R}[x]$ be a monic polynomial of odd degree (i.e., n is odd). Let $G = \max\{|a_1|, |a_2|, \ldots, |a_n|\}$. Then $r < -(G + 1)$ implies $f(r)$ is negative.*

Proof: Define a new polynomial $g(x) = x^n + b_1 x^{n-1} + \ldots + b_{n-1}x + b_n$ where $b_i = a_i$ if i is even, while $b_i = -a_i$ if i is odd. Note that $g(x)$ is monic. Also note that $\max\{|b_1|, |b_2|, \ldots, |b_n|\}$ is the same as $\max\{|a_1|, |a_2|, \ldots, |a_n|\}$, i.e., G. By the first theorem then, $y > G + 1$ implies $g(y) > 0$.

Suppose $r < -(G+1)$. Then $G + 1 < -r$ so $g(-r) > 0$. But $g(-r) = -f(r)$ ($g(-r) = (-r)^n - a_1(-r)^{n-1} + \ldots + a_{n-1}(-r) - a_n$ which is $-r^n - a_1 r^{n-1} - \ldots - a_{n-1}r - a_n$ since n is odd). So $-f(r) > 0$, i.e., $f(r)$ is negative.

Taking (for definiteness) $a = -(G+1.1)$, we get $f(a) < 0$ as we wanted. The other special case mentioned above ($a_n < 0$) is even easier: take $a = 0$.

The program below decides whether or not a polynomial satisfies either of the special cases. If so, it automatically computes a root. If not, it asks the user to supply a number that evaluates negatively. The user may or may not know such a number (as explained above, such a number may not even exist). Any number a can be entered here. The program will evaluate the polynomial at a and display the result. If $f(a)$ is indeed negative, the program computes a root by bisection. If $f(a)$ is positive, nothing will happen; the program just ends. (The program can be modified to allow the user more than one try at a number that evaluates negatively.) The remaining logical possibility (slim as it is) is that $f(a) = 0$; i.e., the user, guessing at a number that evaluates negatively, stumbles on a root. The program in this case will say so.

Here's the program:

```
program bisection;

type poly = array[-1..20] of real;                    {line 3}

var G,y,a,b:real;
    i,n:integer;
    f:poly;

function eval(f:poly; r:real):real;
var j,n:integer;
    y:real;
begin
n:=round(f[-1]);
y:=f[0];
for j:=1 to n do y:=y*r+f[j];
eval:=y;
end;
```

```
function bis(f:poly; b,a:real):real;
var c,d,m,y:real;
begin
c:=b; d:=a;
repeat
m:=(c+d)/2; y:=eval(f,m);
if y>0 then c:=m;                                    {line 25}
if y<0 then d:=m;
until (abs(y)<0.000001) or (abs(c-d)<0.000001);      {line 27}
bis:=m;
end;

begin                                                {line 31}
write('enter the degree '); readln(n); f[-1]:=n; writeln;
writeln('enter, when prompted, the coefficients ');
for i:=0 to n do
  begin
  write('coefficient of x super ',n-i,' '); readln(f[i]);
  end;
writeln;
y:=f[0];                                             {line 39}
for i:=0 to n do f[i]:=f[i]/y;
G:=abs(f[1]);                                        {line 41}
for i:=2 to n do
  begin
  if G<abs(f[i]) then G:=abs(f[i]);
  end;
b:=G+1.1;                                            {line 46}
if (n mod 2)=1 then writeln(bis(f,b,-b):6:6);        {line 47}
if ((n mod 2)=0) and (f[n]<0) then writeln(bis(f,b,0):6:6);
if ((n mod 2)=0) and (f[n]>=0) then
  begin
  write('enter a real number a such that f(a) < 0 ');
  readln(a); y:=eval(f,a);
  writeln('f(',a:6:6,')=',y:6:6);                    {line 53}
  if abs(y)<0.000001 then                            {line 54}
    writeln('CONGRATULATIONS! Your guess is a root');
  if (abs(y)>=0.000001) and (y<0) then               {line 56}
    writeln(bis(f,b,a):6:6);                         {line 57}
  end;                                               {line 58}
end.
```

The function eval was explained in §General. The second function, bis, does the bisection. The arguments of the function are, in this order, a polynomial $f(x)$, a real number b that evaluates positively, and a real number a that evaluates negatively. First bis sets c equal to b, d equal to a, and m equal to the average of c and d. If $f(m) > 0$, it replaces c by m (Line 25); if $f(m) < 0$, it replaces d by m. It keeps doing this (averaging c and d, and replacing one or the other according to how the average evaluates) until either $|f(m)| < 0.000001$ (which, recall, is our criterion for m being a root) or else until c and d are so close (namely, $|c-d| < 0.000001$) that any two numbers in the interval (c,d) are virtually the same; and hence $m = (c+d)/2$ will be virtually the same as the root we're looking for.

The main program begins on Line 31. Lines 32–38 handle the polynomial input. Lines 39 and 40 convert the polynomial to its monic associate. Lines 41–45 compute G by initially defining G to be $|f[1]|$ then comparing, for $i = 2, \ldots, n$, G with $|f[i]|$, leaving G unchanged if $|f[i]| \leq G$, or changing G to $|f[i]|$ if $G < |f[i]|$. We wind up with $G = \max\{|f[1]|, \ldots, |f[n]|\}$ as we wanted. Line 46 sets b equal to $G + 1.1$.

Lines 47–49 decide whether or not our polynomial has odd degree or negative constant term. If it has odd degree, the program bisects using $a = -b$ (i.e., $a = -(G + 1.1)$) and displays the root returned by the function bis. If $f(x)$ has even degree and negative constant term, the program bisects using $a = 0$. In the remaining case, the program asks the user to input a number that evaluates negatively.

Sample Runs

(1) For the polynomial $x^5 + 7x^4 + 4x^3 + 28x^2 + 1155x + 1125$ we get the root -1.000000.

(2) For the polynomial $x^4 + 3x^3 + x^2 + x - 63$ we get the root 2.206427.

(3) For the polynomial $4x^6 - 5x^5 + 4x^4 - 3x^3 + 7x^2 - 7x + 1$ we're asked for a number that evaluates negatively. If we enter 0.5, we'll see that $f(0.5) = -0.242188$ and the program will go on to compute the root 0.902474.

(4) For the polynomial $f(x) = x^3 - 0.6x^2 + 0.12x - 0.008$ we get the root 0.199219. Since $f(x) = (x - 0.2)^3$, which has 0.2 as a triple root, our answer is not especially accurate. This illustrates the difficulty involved when a polynomial has multiple roots; namely that a number fairly far off from the true value of a root might evaluate close enough to zero so as to fool a computer program into thinking it really is a root. In our case $0.199219 - 0.2$ is $-781 \cdot 10^{-6}$ so $|f(0.199219)|$ is $(781 \cdot 10^{-6})^3$. That's less than $(8 \cdot 10^{-4})^3 = 512 \cdot 10^{-12}$, well under our close-to-zero parameter 0.000001. (For a real horror show, try the program on $g(x) = x^5 - x^4 + 0.4x^3 - 0.08x^2 + 0.008x - 0.00032$ which is $(x - 0.2)^5$.) We'll come back to the polynomial $f(x)$ and get a better answer in Exercise 6.

Program Modifications

(1) When the program doesn't automatically compute a root (i.e., degree even and constant term positive) the user may want more than one try at inputing a number that evaluates negatively. Replacing Lines 51–53 by the following, for example, will allow 6 tries:

```
writeln('enter a real number a such that f(a) < 0 (6 tries)');
i:=0;
repeat
i:=i+1;
write('try number ',i,' '); read(a); y:=eval(f,a);
writeln('       f(',a:6:6,')=',y:6:6);
until (y<0) or (abs(y)<0.000001) or (i=6);
```

(2) The program as written can find only one root of a given polynomial, this despite the fact that the polynomial may have several real roots amenable to bisection. To allow the user to input both a and b (and thereby find a root between them) replace Lines 39–58 with:

```
write('enter a real number a such that f(a) < 0 ');
readln(a); y:=eval(f,a);
writeln('f(',a:6:6,')=',y:6:6);
if abs(y)<0.000001 then
  writeln('CONGRATULATIONS! Your guess is a root');
if (abs(y)>=0.000001) and (y<0) then
  begin
  write('enter a real number b such that f(b) > 0 ');
  readln(b); y:=eval(f,b);
  writeln('f(',a:6:6,')=',y:6:6);
  if abs(y)<0.000001 then
    writeln('CONGRATULATIONS! Your guess is a root');
  if (abs(y)>=0.000001) and (y>0) then
    writeln(bis(f,b,a):6:6);
  end;
```

(3) The following parameters are arbitrary and can be changed: the polynomial degree limit 20 (Line 3); the close-to-zero parameter 0.000001 (Lines 27, 54, 56); the real number display parameter 6:6 (Lines 47, 48, 53, 57); and 1.1 (Line 46; this can be changed to anything larger than 1).

REGULA FALSI — REAL ROOTS (Section 6.3)

Regula falsi is an iterative method that begins with two real numbers x_1 and x_2; uses these to manufacture a third real number x_3; uses x_2 and x_3 to manufacture x_4; etc.

In more detail, we have a polynomial $f(x)$ and we make two guesses x_1 and x_2 as to a root. If these are intelligent guesses (i.e., numbers that we know are close to a root), all the better. But this is not necessary; the guesses can be completely wild and need not even depend on the particular polynomial; we just need two numbers to get started. Next the two points $(x_1, f(x_1))$ and $(x_2, f(x_2))$ are connected with a straight line. (See Figure 8 on Page 166 in which a represents x_1 and b represents x_2. The situation depicted by this example — that $(a, f(a))$ and $(b, f(b))$ fall on different sides of the x-axis — while desirable, is not necessary. We repeat that the first two guesses can be wild.) Define $(x_3, 0)$ to be the point of intersection of this straight line and the x-axis (labelled c in Figure 8). Then repeat: connect $(x_2, f(x_2))$ and $(x_3, f(x_3))$ with a straight line and call the intersection point of this line and the x-axis $(x_4, 0)$; etc.

If the method works — and as you will be able to see from trying the program, there are plenty of polynomials on which it works nicely — two sequences will converge: $f(x_1), f(x_2), f(x_3), \ldots$ will converge to 0 and $x_1, x_2, x_3, \ldots$ will converge to a root of $f(x)$.

On the other hand the method may fail to work for any of several reasons. In addition to the obvious problem that there may not be any real roots of the polynomial (all roots may be complex) and hence $x_1, x_2, x_3, \ldots$ can't possibly converge to a root, some subtler things can go wrong. (Two such are (i) the sequence $x_1, x_2, x_3, \ldots$ may oscillate, e.g., 6,7,6,7, ... and thus may not converge, or (ii) at any stage in the game — including the two initial guesses — we may have a number x_{k+1} such that $f(x_{k+1}) = f(x_k)$; in this case the straight line connecting $(x_k, f(x_k))$ and $(x_{k+1}, f(x_{k+1}))$ will be horizontal and thus fail to intersect the x-axis, giving us no way to continue.)

To find the number x_3 (given x_1 and x_2) we use the equation from analytic geometry for a nonvertical straight line: $y = mx + b$. Here m denotes the slope and b the y-intercept. Since $(x_2, f(x_2))$ is on this straight line, $f(x_2) = mx_2 + b$; hence $b = f(x_2) - mx_2$. Since $(x_3, 0)$ is on this line, $0 = mx_3 + b$. Thus $mx_3 = -b = mx_2 - f(x_2)$, so $x_3 = x_2 - \frac{f(x_2)}{m}$. The slope m is found from $(x_1, f(x_1))$ and $(x_2, f(x_2))$; namely $m = \frac{f(x_2)-f(x_1)}{x_2-x_1}$. Thus $x_3 = x_2 - \frac{(x_2-x_1)f(x_2)}{f(x_2)-f(x_1)}$. Continuing the iterations, we have in general

$$x_{i+1} = x_i - \frac{(x_i - x_{i-1})f(x_i)}{f(x_i) - f(x_{i-1})}$$

for all $i \geq 2$.

The following program computes and displays 20 iterations, i.e., $x_3, x_4, \ldots, x_{22}$, and their corresponding evaluations $f(x_3), \ldots, f(x_{22})$. Twenty will usually be sufficient to either produce a root or to convince the user that the method isn't working (that $f(x_1), f(x_2), \ldots$ is not converging to 0). If — for a particular problem — twenty fails to convince, the program can always be rerun using x_{21} and x_{22} as the initial guesses. (Then too the program can easily be modified, with 20 changed to something larger.)

Either of two events will cause the program to stop before it does all 20 iterations: (1) if it finds a root, it will say so and stop iterating, or (2) if $x_{k+1} = x_k$ for some k, it will say so and stop iterating.

Here's the program:

```
program regulafalsi(real);

type poly = array[-1..20] of real;                             {line 3}

var x1,x2,x3,y1,y2,y3:real;
    i,n:integer;
    f:poly;
    stop:boolean;

function eval(f:poly; r:real):real;
var j,n:integer;
    y:real;
begin
n:=round(f[-1]);
y:=f[0];
for j:=1 to n do y:=y*r+f[j];
eval:=y;
end;

begin                                                          {line 20}
write('input the degree '); readln(n); f[-1]:=n; writeln;
writeln('enter, when prompted, the coefficients ');
for i:=0 to n do
  begin
  write('coefficient of x super ',n-i,' '); readln(f[i]);
  end;
writeln;
write('enter a first guess ');                                 {line 28}
readln(x1); y1:=eval(f,x1);
write('enter a second guess ');
readln(x2); y2:=eval(f,x2); writeln;
writeln(x1:6:6,' ',y1:6:6);                                    {line 32}
writeln(x2:6:6,' ',y2:6:6);                                    {line 33}
if (abs(y1)<0.000001) or (abs(y2)<0.000001) then               {line 34}
  begin
  write('CONGRATULATIONS! Your guess ');
  if abs(y1)<0.000001 then write(x1:6:6) else write(x2:6:6);
```

```
  writeln(' is a root ');
  end
else                                                       {line 40}
  begin
  i:=1; stop:=false;                                       {line 42}
  repeat                                                   {line 43}
  if y2=y1 then
    begin
    stop:=true;                                            {line 46}
    write('The method fails:  two x values ');
    write('have the same evaluation.  You can try again ');
    writeln('with different first and second guesses.');
    end
  else                                                     {line 51}
    begin
    x3:=x2-(x2-x1)*y2/(y2-y1); y3:=eval(f,x3);             {line 53}
    writeln(x3:6:6,' ',y3:6:6);                            {line 54}
    if abs(y3)<0.000001 then                               {line 55}
      begin
      writeln; writeln(x3:6:6,' is a root');               {line 57}
      stop:=true;
      end;
    x1:=x2; x2:=x3; y1:=eval(f,x1); y2:=eval(f,x2);        {line 60}
    i:=i+1;                                                {line 61}
    end;
  until (i>20) or (stop=true);                             {line 63}
  end;
end.
```

Variables **x1** and **x2** are the initial guesses x_1 and x_2, and variables **y1** and **y2** are $f(x_1)$ and $f(x_2)$. Variable **x3** is used generically to represent any of $x_3, x_4, \ldots x_{22}$ and **y3** represents the evaluation of $f(x)$ at **x3**. Variable **stop** is a flag which will be used to stop the iteration for either of the two reasons given above.

The main program begins on Line 20. Lines 21–27 handle the polynomial input. Lines 28–33 handle the input of the initial guesses, evaluate the polynomial at these initial guesses, and display the evaluations. Line 34 tests whether either initial guess is a root; if so, Lines 35–39 tell the user he's found a root and the program ends there.

If not, the program continues with Line 40. Line 42 starts the iteration count and sets the variable **stop** to be false (which is to say that at this point — just prior to starting the iteration process — we do not yet want to stop it).

Lines 43 and 63 define a repeat loop for the iterations: we keep computing iterations until (i) 20 such have been computed, or (ii) we've found a root, or (iii) we can't go any further owing to $f(x_{k+1}) = f(x_k)$ for some k.

If **x1** and **x2** have the same evaluation, Lines 46–49 will say so on-screen and will set the variable **stop** equal to true (and thereby stop the iterations). If not, we continue with Line 51. Lines 53 and 54 compute **x3**, evaluate the polynomial at **x3**, and display both numbers (**x3** in the first column on the screen, the evaluation in the second column). Line 55 tests **x3** for being a root. If it is, Line 57 will say so on-screen and Line 58 will set the variable **stop** equal to true.

Line 60 now sets **x1** to be **x2**, sets **x2** to be **x3**, and sets **y1, y2** to be the evaluations of $f(x)$ at **x1, x2**. Line 61 advances the iteration count.

Sample Runs

(1) For the polynomial $3x^4 - 12x^3 - 39x^2 + 84x + 126$ and the initial guesses 1,2 we get the root 2.645751.

(2) For the polynomial $x^4 - 16x^3 + 78x^2 - 412x + 624$ and the initial guesses 1,2 we luck out; our 2nd guess is a root.

(3) For the polynomial $x^3 + x^2 - 10x + 13$ and the initial guesses 1,2 the program tells us that $f(1) = 5 = f(2)$, so the method won't work with these initial guesses. Retrying with guesses 3,4, we get the root −4.157447.

(4) For the polynomial $x^6 + x + 1$ and the initial guesses 1,2, we get a sequence whose evaluations clearly do not converge to zero. So the method fails for this polynomial and these initial guesses (as well it should — all roots of this polynomial are nonreal complex).

Program Modifications

(1) If any iteration (or even initial guess) evaluates negatively, then a real root must exist and bisection can find it (assuming the leading coefficient of the polynomial is positive). You may want to modify the program to automatically go into a bisection if the evaluation of any iteration turns up negative.

(2) Arbitrary, changeable parameters: the polynomial degree limit 20 (Line 3); the close-to-zero parameter 0.000001 (Lines 34, 37, 55); the real number display parameter 6:6 (Lines 32, 33, 54, 57); and the number of iterations 20 (Line 63). For the latter, note that not every iteration need be displayed. For instance, you may want to change 20 to 200 and display every 10th iteration (replace Line 54 by "`if (i mod 10)=0 then writeln(x3:6:6,' ',y3:6:6);`").

DEFLATION

Once we've managed to find a root of a given polynomial, we might want to find another root. (Not inconceivably, we might want to find all the roots.) Recall from the factor theorem on Page 39 that if r is a root of $f(x)$, then $x - r$ divides $f(x)$, i.e., there exists a polynomial $q(x)$ such that $f(x) = q(x)(x - r)$. Now note that any root of $q(x)$ will also be a root of $f(x)$. This gives us a strategy for finding further roots of $f(x)$: divide $f(x)$ by $x - r$ and look for a root of the quotient polynomial. There is a fancy term for this strategy; it's called *deflating* $f(x)$ by the root r.

Starting with the polynomial $f(x) = a_0x^n + a_1x^{n-1} + \ldots + a_{n-1}x + a_n$, the coefficients of the deflated polynomial $q(x) = b_0x^{n-1} + b_1x^{n-2} + \ldots + b_{n-2}x + b_{n-1}$ are given by

$$b_0 = a_0 \quad \text{and} \quad b_i = a_i + rb_{i-1} \quad \text{for } 1 \leq i \leq n-1.$$

(The easiest way to see this is via multiplication. Multiply $b_0x^{n-1} + b_1x^{n-2} + \ldots + b_{n-2}x + b_{n-1}$ by $x - r$ and compare coefficients with those of $f(x)$.)

Here is a program to deflate the polynomial $f(x) \in \mathbf{R}[x]$ by a real root r:

```
program deflate;

type poly = array[-1..20] of real;                        {line 3}

var f:poly;
    i,n:integer;
    r:real;

procedure def(var f:poly; r:real);                        {line 9}
var q:poly;
    i,n:integer;
begin
n:=round(f[-1]); q[-1]:=n-1;                              {line 13}
q[0]:=f[0];                                               {line 14}
for i:=1 to n-1 do q[i]:=f[i]+r*q[i-1];
f:=q;                                                     {line 16}
end;

begin                                                     {line 19}
write('enter the degree '); readln(n); f[-1]:=n; writeln;
writeln('enter, when prompted, the coefficients ');
for i:=0 to n do
begin
write('coefficient of x super ',n-i,' '); readln(f[i]);
end;
writeln;                                                  {line 26}
```

```
write('enter a root '); readln(r); writeln;          {line 27}
def(f,r);                                            {line 28}
for i:=0 to n-1 do write(f[i]:6:6,' ');              {line 29}
end.
```

The actual deflation is accomplished by the procedure def. This procedure takes a polynomial $f(x)$ and a root r and changes the polynomial to the deflated polynomial $\frac{f(x)}{x-r}$. The polynomial retains its name (f), but its coefficients change (as does its degree). The Pascal var designation in Line 9 indicates that the polynomial will be changed by the procedure. Line 13 sets n equal to the degree of the polynomial (the function round is necessary to convert the real number $f[-1]$ to an integer). Line 13 also sets the degree of the polynomial $q(x)$ equal to $n-1$. Lines 14 and 15 define the coefficients of $q(x)$ as per the above display. And finally Line 16 sets $f(x)$ equal to $q(x)$.

The main program starts on Line 19. Lines 20–26 handle the polynomial input. Line 27 inputs the root. Line 28 sends the polynomial and its root to the deflation procedure and Line 29 displays the deflated polynomial on-screen (by listing its coefficients $a_0, a_1, a_2, \ldots$ in that order).

Note: Deflation should be used with some caution: if several deflations are performed on a polynomial, there are likely to be roundoff errors in the coefficients of the deflated polynomial; so any root-finding method which produces its own roundoff errors might get you further away from the true value of a root than is desirable.

Sample Runs

(1) For the polynomial $8x^4 - 52x^2 + 32x + 5$ and the root 0.80277564, we get $8x^3 + 6.422205x^2 - 46.84441x - 5.605551$.

(2) For the polynomial $x^3 - 20x^2 - 558.7691x + 1193.926$ and the root -17, we get $x^2 - 37x + 70.2309$.

(3) For the polynomial $x^{15} - 1$ and the root 1, we get $x^{14} + x^{13} + x^{12} + x^{11} + x^{10} + x^9 + x^8 + x^7 + x^6 + x^5 + x^4 + x^3 + x^2 + x + 1$.

Program Modifications

(1) For a fancier on-screen display of the polynomial (one that will include the indeterminate x and will suppress terms whose coefficients are zero) add the following procedure and change Line 29 to "`fancy(f);`":

```
procedure fancy(f:poly);
var i,j,n:integer;
begin
writeln;
n:=round(f[-1]);
for i:=0 to n do
```

```
  begin
  if f[i]<>0 then
    begin
    write(f[i]:6:6);
    if i<n then write('x');
    gotoxy(wherex,wherey-1);
    if n-i<>0 then write(n-i);
    gotoxy(wherex,wherey+1);
    j:=i;
    repeat
    j:=j+1
    until (f[j]<>0) or (j=n);
    if (i<>n) and (f[j]>0) then write('+');
    end;
  end;
end;
```

(2) Arbitrary, changeable parameters: the polynomial degree limit 20 (Line 3); and the real number display parameter 6:6 (Line 29).

COMPLEX ARITHMETIC

We'll consider complex numbers to be pairs of real numbers. We can then represent complex numbers by arrays of real numbers. Thus:

```
type pair = array[1..2] of real;
```

For instance, the complex number $z = 3 + 7i$ is represented by $z[1] = 3$ and $z[2] = 7$, where the variable z has been declared to be a pair.

We'd like to have functions to handle complex addition, multiplication, etc, but since Pascal doesn't allow an array type to be the result (i.e., output) of a function, we'll use procedures instead. (It's possible to convert each of these procedures into two functions — outputting respectively the real part and complex parts of the result — but it doesn't seem worth it.)

To handle complex addition we'll use the following:

```
procedure plus(s,t:pair; var sum:pair);
begin
sum[1]:=s[1]+t[1]; sum[2]:=s[2]+t[2];
end;
```

This procedure adds the pairs s and t coordinatewise (Page 12, Line 7), and calls the result sum; s,t, and sum are complex numbers (i.e., variables of the type pair). The Pascal var designation before sum indicates that the variable sum will be changed by the procedure. It isn't necessary to define the variable sum before calling the procedure; we can leave it undefined and the procedure will set it equal to $s + t$ as we want it to. Note that the variables s and t, which are not preceded by the Pascal var designation, will not be changed by the procedure. Example: the fragment

```
a[1]:=3; a[2]:=7;
b[1]:=2; b[2]:=-6;
plus(a,b,c);
```

will add $3 + 7i$ and $2 - 6i$. The result is an array called c, where $c[1]$ is 5 and $c[2]$ is 1.

Complex multiplication is analogous (see the definition on Lines 9 and 10 of Page 12):

```
procedure times(s,t:pair; var prod:pair);
begin
prod[1]:=(s[1]*t[1])-(s[2]*t[2]);
prod[2]:=(s[1]*t[2])+(s[2]*t[1]);
end;
```

To divide by the nonzero complex number $e = a + bi$, we'll take the multiplicative inverse of e and then multiply (using the above multiplication procedure). Recall (2nd display on Page 12) that $\frac{1}{e} = \frac{a}{a^2+b^2} - \frac{b}{a^2+b^2}i$. We could use a procedure to invert e; and indeed for complicated programs involving many inversions it would make sense to have such a procedure. For the two programs we have in mind though, which would call such a procedure just one time each, it's just as easy to use the above formula in the main part of the program.

The procedure we'll use to evaluate a polynomial $f(x) \in \mathbf{R}[x]$ at a complex number r is just an analog of the function eval we've been using for real numbers r. Here's the procedure:

```
procedure eval(f:poly; r:pair; var result:pair);
var j,n:integer;
    y,s,t:pair;
begin
n:=round(f[-1]);
y[1]:=f[0]; y[2]:=0;
for j:=1 to n do
  begin
  times(y,r,s);
  t[1]:=f[j]; t[2]:=0;
  plus(s,t,y);
```

```
  end;
result:=y;
end;
```

It begins (as does the real-valued analog) by initially defining y to be the leading coefficient of the polynomial. But this time y is complex, so this definition takes the form $y[1] := f[0]$; $y[2] := 0$. It then redefines y to be the old value of y times r, plus $f[j]$. (The multiplication is done by the command **times(y,r,s)**; i.e., y gets multiplied by r and the result is called s. It then sets (the complex number) t equal to (the real number) $f[j]$, then adds t to s and calls the result y.) This being done for all j from 1 to n, we wind up with $f(r)$.

We'll need one more arithmetic operation for the Ferrari-method program, namely square roots. The following finds a square root of the complex number z. (Once we know one square root, the other is just -1 times the first). For the complex number z in polar form, $z = |z|(\cos\theta + (\sin\theta)i)$, recall (Page 14, 7 lines from the bottom) that $\sqrt{|z|}(\cos\frac{\theta}{2} + (\sin\frac{\theta}{2})i)$ is a square root of z. The following subprogram puts z in polar form, computes θ, and then computes the above-mentioned square root:

```
function arccos(x:real):real; {0 <= x and x <= 1}
const pi = 3.14159265;
var y:real;
begin
if x=0 then y:=pi/2;
if x<>0 then y:=arctan(sqrt(1-x*x)/x);
arccos:=y;
end;

procedure sr(var z:pair);                                  {line 10}
const pi = 3.14159265;
var r,theta,x,y:real;
begin
x:=z[1]; y:=z[2]; r:=sqrt(x*x+y*y);                        {line 14}
if (x<>0) or (y<>0) then                                   {line 15}
  begin
  if (x>=0) and (y>=0) then theta:=arccos(x/r);            {line 17}
  if (x<=0) and (y>=0) then theta:=pi-arccos(-x/r);
  if (x<=0) and (y<=0) then theta:=pi+arccos(-x/r);
  if (x>=0) and (y<=0) then theta:=2*pi-arccos(x/r);
  z[1]:=sqrt(r)*cos(theta/2);                              {line 21}
  z[2]:=sqrt(r)*sin(theta/2);                              {line 22}
  end;
end;
```

The function computes the arccosine of a real number x in the interval $[0,1]$. Since Pascal has a built-in arctangent function, but no arccosine function, we express arccosines in terms of arctangents using the identity

$$\arccos(x) = \arctan\left(\frac{\sqrt{1-x^2}}{x}\right)$$

valid for all x in $[0,1]$ except $x = 0$. The function simply sets arccos(0) to be $\pi/2$ and sets $\arccos(x)$, for $x \neq 0$, to be $\arctan(\sqrt{1-x^2}/x)$.

The procedure computes the square root. The Pascal var designation in Line 10 indicates that the argument z will be changed by the procedure (changed to its square root). The variables are x and y for the real and imaginary parts of z; and r and theta for $|z|$ and θ. Line 14 defines x, y, and r. Line 15 continues the procedure provided that $x \neq 0$ or $y \neq 0$; i.e., provided that $z \neq 0$. (For $z = 0$, the procedure leaves z unchanged; thus it returns 0 for a square root of 0.)

Suppose z is in the first quadrant; i.e., both $x, y \geq 0$. Then $\cos\theta = x/r$, so $\theta = \arccos(x/r)$.

In the remaining three cases, let δ denote the angle between the x-axis and the line from the origin to z. It will be helpful to draw diagrams for the three cases. Suppose z is in the second quadrant. Then $\delta + \theta = \pi$. We have $\cos(\delta) = |x|/r = -x/r$ (since $x \leq 0$ in this quadrant), so $\cos(\pi - \theta) = -x/r$, so $\theta = \pi - \arccos(-x/r)$. Suppose z is in the third quadrant. This time $\theta = \pi + \delta$ and again $\cos(\delta) = -x/r$. So $\theta = \pi + \arccos(-x/r)$. Finally, suppose z is in the fourth quadrant. This time $\delta + \theta = 2\pi$. That and $\cos(\delta) = x/r$ give us $\theta = 2\pi - \arccos(x/r)$.

Lines 17–20 define theta using the formulas in the above four cases. Finally Lines 21,22 set the real and complex parts of the square root equal to $|z|\cos(\theta/2)$ and $|z|\sin(\theta/2)$.

REGULA FALSI — COMPLEX ROOTS

The regula falsi technique can also find complex roots. Indeed, this is probably the most useful program of this appendix.

The technique is virtually the same as in the real-roots case: start with 2 initial guesses x_1 and x_2 (but this time x_1 and x_2 are complex) then compute iterations $x_3, x_4, \ldots$ using

$$x_{i+1} = x_i - \frac{(x_i - x_{i-1})f(x_i)}{f(x_i) - f(x_{i-1})}.$$

If the method works, then $f(x_1)$, $f(x_2)$, $f(x_3), \ldots$ will converge to 0 and $x_1, x_2, x_3, \ldots$ will converge to a root.

Here's the program:

```
program regulafalsi(complex);

type poly = array[-1..20] of real;                    {line 3}
     pair = array[1..2] of real;

var x1,x2,x3,y1,y2,y3,p,q,r:pair;
    i,n:integer;
    f:poly;
    w:real;
    stop:boolean;

procedure plus(s,t:pair; var sum:pair);
begin
sum[1]:=s[1]+t[1]; sum[2]:=s[2]+t[2];
end;

procedure times(s,t:pair; var prod:pair);
begin
prod[1]:=(s[1]*t[1])-(s[2]*t[2]);
prod[2]:=(s[1]*t[2])+(s[2]*t[1]);
end;

procedure eval(f:poly; r:pair; var result:pair);
var j,n:integer;
    y,s,t:pair;
begin
n:=round(f[-1]);
y[1]:=f[0]; y[2]:=0;
for j:=1 to n do
  begin
  times(y,r,s);
  t[1]:=f[j]; t[2]:=0;
  plus(s,t,y);
  end;
result:=y;
end;

begin                                                 {line 38}
write('enter the degree '); readln(n); f[-1]:=n; writeln;
writeln('enter, when prompted, the coefficients ');
for i:=0 to n do
  begin
  write('coefficient of x super ',n-i,' '); readln(f[i]);
  end;
```

```
writeln;
write('enter a first guess:  real part '); readln(x1[1]);
write('                      imag part '); readln(x1[2]);
eval(f,x1,y1);
write('enter a second guess:  real part '); readln(x2[1]);
write('                       imag part '); readln(x2[2]);
eval(f,x2,y2); writeln;
write(x1[1]:6:6,' ',x1[2]:6:6);                           {line 52}
writeln(' ',y1[1]:6:6,' ',y1[2]:6:6);                     {line 53}
write(x2[1]:6:6,' ',x2[2]:6:6);                           {line 54}
writeln(' ',y2[1]:6:6,' ',y2[2]:6:6);                     {line 55}
if (abs(y1[1]*y1[1]+y1[2]*y1[2])<0.000001)                {line 56}
  or (abs(y2[1]*y2[1]+y2[2]*y2[2])<0.000001) then         {line 57}
  begin
  write('CONGRATULATIONS! Your guess ');
  if (abs(y1[1]*y1[1]+y1[2]*y1[2])<0.000001) then         {line 60}
    write(x1[1]:6:6,' ',x1[2]:6:6)                        {line 61}
  else write(x2[1]:6:6,' ',x2[2]:6:6);                    {line 62}
  writeln(' is a root');
  end
else
  begin
  i:=1; stop:=false;
  repeat
  if (y2[1]=y1[1]) and (y2[2]=y1[2]) then
    begin
    stop:=true;
    write('The method fails:  two x values ');
    write('have the same evaluation.  You can try again ');
    writeln('with different first and second guesses');
    end
  else
    begin
    p[1]:=x2[1]-x1[1]; p[2]:=x2[2]-x1[2];                 {line 78}
    times(p,y2,q);
    p[1]:=y2[1]-y1[1]; p[2]:=y2[2]-y1[2];                 {line 80}
    w:=p[1]*p[1]+p[2]*p[2]; p[1]:=p[1]/w; p[2]:=-p[2]/w;
    times(q,p,r);                                         {line 82}
    x3[1]:=x2[1]-r[1]; x3[2]:=x2[2]-r[2]; eval(f,x3,y3);
    write(x3[1]:6:6,' ',x3[2]:6:6);                       {line 84}
    writeln('           ',y3[1]:6:6,' ',y3[2]:6:6);       {line 85}
    if abs(y3[1]*y3[1]+y3[2]+y3[2])<0.000001 then         {line 86}
      begin
      writeln;
```

```
      writeln(x3[1]:6:6,' ',x3[2]:6:6,' is a root'); {line 89}
      stop:=true;
      end;
    x1:=x2; x2:=x3; eval(f,x1,y1); eval(f,x2,y2);
    i:=i+1;
    end;
  until (i>20) or (stop=true);                       {line 95}
  end;
end.
```

Variables **x1, x2, x3, y1, y2,** and **y3** are analogs of their namesakes in the program regulafalsi(real). (Here they are pairs rather than real numbers.) Variables **p, q,** and **r** will be used for various pairs (i.e., complex numbers) where needed. Variable **w** will be used for real numbers where needed.

The four procedures were explained in §Complex Arithmetic. The main program starts on Line 38, and is analogous to the main program of regulafalsi(real). The only part that might require some explanation is Lines 78–83. This 6-line section is the analog of Line 53 in regulafalsi(real). It serves to define x_3 as

$$x_3 = x_2 - \frac{(x_2 - x_1)f(x_2)}{f(x_2) - f(x_1)}.$$

Line 78 defines p to be $x_2 - x_1$. Line 79 multiplies p by y_2 $(= f(x_2))$ and calls the result q. At this point q is the numerator of the fraction in the above display. Line 80 sets p equal to $y_2 - y_1$ $(= f(x_2) - f(x_1))$. Line 81 changes p to $1/p$ using the formula for inversion of a complex number given in the second display on Page 12. So at this stage $p = \frac{1}{f(x_2)-f(x_1)}$. Line 82 multiplies q by p and calls the result r. Finally, Line 83 defines x_3 to be $x_2 - r$, then evaluates the polynomial at x_3 and calls the evaluation y_3.

Sample Runs

(1) For the polynomial $x^4 - 8x^3 + 39x^2 - 62x + 51$ and initial guesses $1 + i$, $2 + 2i$, we get $0.994678 + 1.024879i$ is a root.

(2) For the polynomial $x^6 + x + 1$ and initial guesses $1 + i$, $2 + 2i$, we get $0.945402 + 0.611837i$ is a root.

(3) For the polynomial $3x^4 - 12x^3 - 39x^2 + 84x + 126$ and initial guesses $1 + i$, $2 + 2i$, we get 2.645751 is a root.

(4) For the polynomial $x^4 + 3x^3 + x^2 + x - 63$ and initial guesses $1 + i$, $2+2i$, we get $x_{21} = -0.677692-2.621097i$ and $x_{22} = -0.660265-2.629975i$. Continuing with these as initial guesses (i.e., rerunning the program) we get $-0.661198 - 2.629485i$ is a root.

Program Modifications

(1) For a fancier on-screen display of complex numbers, add the procedure

```
procedure fancy2(c:pair);
var x,y:real;
begin
x:=c[1]; y:=c[2];
if (x=0) and (y=0) then write('0');
if (x=0) and (y=1) then write('i');
if (x=0) and (y=-1) then write('-i');
if (x=0) and (y<>0) and (y<>1) and (y<>-1) then
  write(y:6:6,'i');
if (x<>0) and (y=0) then write(x:6:6);
if (x<>0) and (y=1) then write(x:6:6,'+i');
if (x<>0) and (y=-1) then write(x:6:6,'-i');
if (x<>0) and (y>0) and (y<>1) then
  write(x:6:6,'+',y:6:6,'i');
if (x<>0) and (y<0) and (y<>-1) then write(x:6:6,y:6:6,'i');
end;
```

Change Line 52 to "**fancy2(x1); write('   '); fancy2(y1); writeln;**" and make similar changes in Lines 53–55, 61, 62, 84, 85, and 89.

(2) Arbitrary, changeable parameters: the polynomial degree limit 20 (line 3); the close-to-zero parameter 0.000001 (Lines 56, 57, 60, 86); the real number display parameter 6:6 (Lines 52–55, 61, 62, 84, 85, 89); and the number or iterations 20 (Line 95).

FERRARI'S METHOD (Section 3.4)

Let's review the method (see Page 101). We have a fourth-degree monic polynomial $f(x) = x^4 + f_1x^3 + f_2x^2 + f_3x + f_4$ and we seek numbers a, b, c, and d such that $f(x)+(ax+b)^2 = (x^2+cx+d)^2$ (c and d can be assumed to be real, but a and b might be complex). Equating corresponding coefficients leads to the set of equations

$$f_1 = 2c, \tag{1}$$

$$f_2 + a^2 = c^2 + 2d, \tag{2}$$

$$f_3 + 2ab = 2cd, \tag{3}$$

$$f_4 + b^2 = d^2. \tag{4}$$

Set $c = f_1/2$, so (1) is immediately satisfied. Plugging $c = f_1/2$ into (3), we get $f_1 d - f_3 = 2ab$, so $(f_1 d - f_3)^2 = 4a^2b^2 = 4(\frac{f_1^2}{4} + 2d - f_2)b^2$ by (2). That and (4) give us

$$(f_1 d - f_3)^2 = (f_1^2 + 8d - 4f_2)(d^2 - f_4). \tag{5}$$

Expanding and simplifying leads to the resolvent cubic polynomial

$$d^3 - \frac{f_2}{2}d^2 + \left(\frac{f_1 f_3}{4} - f_4\right) d + \frac{4f_2 f_4 - f_1^2 f_4 - f_3^2}{8} = 0. \tag{6}$$

Note that the above simplification is reversible; i.e., Equation (5) is equivalent to Equation (6).

Find a real root d of the resolvent cubic polynomial. (This is always possible since 3 — the degree of that polynomial — is odd.) Thus Equation (5) will also hold. Define a to be

$$a = \sqrt{c^2 - f_2 + 2d} \tag{7}$$

(use either square root if the number under the radical sign is negative). Thus (2) is satisfied.

Define b as

$$b = \begin{cases} \sqrt{d^2 - f_4} & \text{if } a = 0 \\ \frac{f_1 d - f_3}{2a} & \text{if } a \neq 0 \end{cases} \tag{8}$$

(again, use either square root if the number under the radical sign is negative). It's not hard to show (using (5)) that with this definition of b, both (3) and (4) are satisfied.

We now have $f(x) + (ax + b)^2 = (x^2 + cx + d)^2$ so

$$\begin{aligned} f(x) &= (x^2 + cx + d)^2 - (ax + b)^2 \\ &= [x^2 + (c - a)x + (d - b)][x^2 + (c + a)x + (d + b)]. \end{aligned}$$

Find roots of the two quadratic polynomials on the right-hand side. The four roots produced are the roots of $f(x)$.

The following program computes the resolvent cubic polynomial, finds a real root d of this polynomial by bisection, computes a and b, then finds roots of the two quadratic polynomials by the quadratic formula.

```
program ferrari;

type poly = array[-1..20] of real;                        {line 3}
     pair = array[1..2] of real;

var y,G,c,d:real;
    i:integer;
    f,res:poly;
    a,b,s,t,root1,root2:pair;

function eval(f:poly; r:real):real;
var j,n:integer;
    y:real;
begin
n:=round(f[-1]);
y:=f[0];
for j:=1 to n do y:=y*r+f[j];
eval:=y;
end;

function bis(f:poly; b,a:real):real;
var c,d,m,y:real;
begin
c:=b; d:=a;
repeat
m:=(c+d)/2; y:=eval(f,m);
if y>0 then c:=m;
if y<0 then d:=m;
until (abs(y)<0.000001) or (abs(c-d)<0.000001);       {line 29}
bis:=m;
end;

function arccos(x:real):real; {0 <= x and x <= 1}
const pi = 3.14159265;
var y:real;
begin
if x=0 then y:=pi/2;
if x<>0 then y:=arctan(sqrt(1-x*x)/x);
arccos:=y;
end;

procedure sr(var z:pair);
const pi = 3.14159265;
var r,theta,x,y:real;
```

```
begin
x:=z[1]; y:=z[2]; r:=sqrt(x*x+y*y);
if (x<>0) or (y<>0) then
  begin
  if (x>=0) and (y>=0) then theta:=arccos(x/r);
  if (x<=0) and (y>=0) then theta:=pi-arccos(-x/r);
  if (x<=0) and (y<=0) then theta:=pi+arccos(-x/r);
  if (x>=0) and (y<=0) then theta:=2*pi-arccos(x/r);
  z[1]:=sqrt(r)*cos(theta/2); z[2]:=sqrt(r)*sin(theta/2);
  end;
end;

procedure quad(s,t:pair; var root1,root2:pair);
var p:pair;
begin
p[1]:=s[1]*s[1]-s[2]*s[2]-4*t[1];                          {line 60}
p[2]:=2*s[1]*s[2]-4*t[2];
sr(p);                                                     {line 62}
root1[1]:=(-s[1]+p[1])/2; root1[2]:=(-s[2]+p[2])/2;        {line 63}
root2[1]:=(-s[1]-p[1])/2; root2[2]:=(-s[2]-p[2])/2;
end;

begin                                                      {line 67}
f[-1]:=4; res[-1]:=3;                                      {line 68}
writeln('enter, when prompted, the coefficients ');
for i:=0 to 4 do
  begin
  write('coefficient of x super ',4-i,' '); readln(f[i]);
  end;
writeln;
y:=f[0];                                                   {line 75}
for i:=0 to 4 do f[i]:=f[i]/y;
res[0]:=1; res[1]:=-f[2]/2; res[2]:=(f[1]*f[3])/4-f[4];
res[3]:=(4*f[2]*f[4]-f[1]*f[1]*f[4]-f[3]*f[3])/8;
G:=abs(res[1]);                                            {line 79}
for i:=2 to 3 do
  begin
  if G<abs(res[i]) then G:=abs(res[i]);
  end;
d:=bis(res,G+1.1,-(G+1.1));                                {line 84}
c:=f[1]/2;                                                 {line 85}
y:=c*c-f[2]+2*d; a[1]:=y; a[2]:=0;                         {line 86}
sr(a);                                                     {line 87}
if y=0 then
```

```
  begin
  b[1]:=d*d-f[4]; b[2]:=0; sr(b);                        {line 90}
  end
else
  begin
  y:=(f[1]*d-f[3])/2; y:=y/(a[1]*a[1]+a[2]*a[2]);        {line 94}
  b[1]:=y*a[1]; b[2]:=-y*a[2];
  end;
s[1]:=c-a[1]; s[2]:=-a[2];                               {line 97}
t[1]:=d-b[1]; t[2]:=-b[2];
quad(s,t,root1,root2);                                   {line 99}
writeln(root1[1]:6:6,' ',root1[2]:6:6);                  {line 100}
writeln(root2[1]:6:6,' ',root2[2]:6:6);                  {line 101}
s[1]:=c+a[1]; s[2]:=a[2];                                {line 102}
t[1]:=d+b[1]; t[2]:=b[2];
quad(s,t,root1,root2);                                   {line 104}
writeln(root1[1]:6:6,' ',root1[2]:6:6);                  {line 105}
writeln(root2[1]:6:6,' ',root2[2]:6:6);                  {line 106}
end.
```

The variables `a, b, c, d` are the numbers Ferrari's method produces (`a` and `b` are pairs; `c` and `d` are real). Variable `G` has the same meaning as in the bisection Variable `y` will be used for various real numbers where needed. Variable `res` will denote the resolvent polynomial. Variables `s` and `t` will denote the coefficients of the quadratic factors (i.e., $s = c \pm a$ and $t = d \pm b$). Variables `root1` and `root2` will denote roots of the quadratic factors.

The three functions and the procedure sr have been explained previously. (Note that eval is the function that evaluates polynomials at real numbers. It will only be used on the resolvent polynomial, not on the quartic polynomial we're trying to solve.)

The procedure quad uses the quadratic formula to find both roots of monic polynomials x^2+sx+t, where s and t are complex. The two roots are represented by the variables `root1` and `root2`. Lines 60 and 61 define p to be s^2-4t (by defining the real and imaginary parts of p to be, respectively, $\text{Re}(p) = \text{Re}(s)^2 - \text{Im}(s)^2 - 4\text{Re}(t)$ and $\text{Im}(p) = 2\text{Re}(s)\,\text{Im}(s) - 4\text{Im}(t)$). Line 62 changes p to one of its square roots. Lines 63 and 64 then define root1 and root2 to be root1 $= \frac{-s+p}{2}$ and root2 $= \frac{-s-p}{2}$ as the quadratic formula dictates.

The main program begins on Line 67. Line 68 sets the degrees of $f(x)$ and res equal to 4 and 3 respectively. Lines 69–74 handle input of $f(x)$ and Lines 75, 76 change it to its monic associate. Lines 77,78 define the coefficients of the resolvent polynomial, as given in (6). Lines 79–84 find a real root of the resolvent polynomial using bisection and call it d. Line 85 defines c. Line 86 defines y to be $c^2 - f_2 + 2d$ and (temporarily) defines a to be the complex number equal to the real number y. Line 87 changes a to one of its square roots so that a has now been computed as per (7).

If $a = 0$ (which is equivalent to $y = 0$), Line 90 defines b as one of the square roots of $d^2 - f_4$ as (8) requires. If $a \neq 0$, Lines 94,95 define b as $\frac{f_1 d - f_3}{2a}$. (Since $\frac{1}{a} = \frac{\text{Re}(a)}{|a|^2} - \frac{\text{Im}(a)}{|a|^2} i$ (see Page 12, second display), $\frac{f_1 d - f_3}{2a}$ will be $\frac{f_1 d - f_3}{2|a|^2}\text{Re}(a) - \frac{f_1 d - f_3}{2|a|^2}\text{Im}(a)i$; Line 94 defines y to be $\frac{f_1 d - f_3}{2|a|^2}$, and Line 95 defines b to be $y\,\text{Re}(a) - y\,\text{Im}(a)i$.) Lines 97–101 send $x^2 + (c-a)x + d + b$ to the procedure quad, which finds both roots, and then displays these two roots. Lines 102–106 do the same with $x^2 + (c+a)x + d + b$.

Sample Runs

(1) For the polynomial $2x^4 - 4x^3 - 10x^2 + 20x - 6$ (the example on Page 100) we get the roots -2.302776, 0.381966, 1.302776, and 2.618034. (Cf the exact roots given on page 101.)

(2) For the polynomial $x^4 + 3x^3 + x^2 + x - 63$ we get the roots 2.206428, -3.884033, and $-0.661198 \pm 2.629487i$. (We found the root 2.206427 in a sample run of the bisection program. The bisection answer is the better of the two. We found the root $-0.661198 - 2.629485i$ in a sample run of the program regulafalsi(complex). The regula falsi answer is the better of the two.)

(3) For the polynomial $x^4 - 8x^3 + 39x^2 - 62x + 51$ we get the roots $3.005322 \pm 3.996369i$ and $0.994678 \pm 1.024879i$. (A sample run of program regulafalsi(complex) produced the root $0.994678 + 1.024879i$.)

Program Modifications

(1) If you want to see the resolvent cubic and its root d, add "`for i:=0 to 3 do write(res[i]:6:6,' '); writeln; writeln(d:6:6); writeln;`" between Lines 84 and 85.

(2) You may consider it more elegant to declare the variable root as `array[1..4] of pair`; so that root[i] represents the ith root, $1 \leq i \leq 4$. This will require some minor changes in Lines 99–104.

(3) Arbitrary, changeable parameters: the polynomial degree limit 20 (Line 3); the close-to-zero parameter 0.000001 (Line 29); and the real number display parameter 6:6 (Lines 100, 101, 105, 106).

DEFLATION BY COMPLEX ROOTS

Suppose we have a polynomial $f(x) \in \mathbf{R}[x]$ and we've found a nonreal complex root $r = c + di$. We know that the complex conjugate $r^* = c - di$ is also a root of $f(x)$ (Page 16). Thus both $x - r$ and $x - r^*$ divide $f(x)$. Furthermore, their product $(x-r)(x-r^*)$ also divides $f(x)$ because $x - r$ and $x - r^*$ are relatively prime (since $d \neq 0$, $r^* - r \neq 0$ and so

$u(x-r)+v(x-r^*)=1$ for the complex numbers $u=\frac{1}{r^*-r}$ and $v=\frac{-1}{r^*-r}$). Consequently we can deflate $f(x)$ by the two roots r, r^* simultaneously by dividing it by the quadratic polynomial $(x-r)(x-r^*)$ which, after simplification, is $x^2-2cx+c^2+d^2$.

Denote the deflated polynomial by $q(x)=b_0x^{n-2}+b_1x^{n-3}+\ldots+b_{n-3}x+b_{n-2}$ (since $f(x)=q(x)(x^2-2cx+c^2+d^2)$, deg($q$) is 2 less than deg($f$)). The coefficients are given by

$$b_0=a_0, \qquad b_1=a_1+2cb_0, \quad \text{and}$$

$$b_i=a_i+2cb_{i-1}-(c^2+d^2)b_{i-2} \quad \text{for } 2\le i\le n-2.$$

(To see this, multiply $b_0x^{n-2}+b_1x^{n-3}+\ldots+b_{n-3}x+b_{n-2}$ by $x^2-2cx+c^2+d^2$ and compare coefficients to those of $f(x)$.)

Here's the program. It's analogous to the previous program that deflated by a real root. (And as with that program, it can be modified to give a fancier on-screen display of the polynomial.)

```
program deflate2;

type poly = array[-1..20] of real;
     pair = array[1..2] of real;

var f:poly;
    i,n:integer;
    r:pair;

procedure def2(var f:poly; r:pair);
var q:poly;
    y:real;
    i,n:integer;
begin
n:=round(f[-1]); q[-1]:=n-2;
y:=r[1]*r[1]+r[2]*r[2];
q[0]:=f[0]; q[1]:=f[1]+2*r[1]*q[0];
for i:=2 to n-2 do q[i]:=f[i]+2*r[1]*q[i-1]-y*q[i-2];
f:=q;
end;

begin
write('enter the degree '); readln(n); f[-1]:=n; writeln;
writeln('enter, when prompted, the coefficients ');
for i:=0 to n do
  begin
  write('coefficient of x super ',n-i,' '); readln(f[i]);
```

```
  end;
writeln;
write('enter a complex root:  real part '); readln(r[1]);
write('                       imag part '); readln(r[2]);
writeln;
def2(f,r);
for i:=0 to n-2 do write(f[i]:6:6,' ');
end.
```

Sample Runs

(1) For the polynomial $5x^6 - 38x^5 + 166.2x^4 - 413.12x^3 + 684.4x^2 - 648x + 360$ and the root $1.2 + 1.6i$, we get $5x^4 - 26x^3 + 83.8x^2 - 108x + 90$.

(2) For the polynomial $x^6 - 2x^5 - 3x^4 - 7x^3 + 14x^2 + 15x + 18$ and the root $-1 + 1.414214i$, we get $x^4 - 4x^3 + 1.999999x^2 + 1.000007x + 5.999986$. (The exact value of the root is $-1 + \sqrt{2}i$ and the exact value of the deflated polynomial is $x^4 - 4x^3 + 2x^2 + x + 6$.)

THE DIVISION ALGORITHM (Section 1.3)

The division algorithm (see Page 21) starts with two polynomials $f(x)$ of degree n and $g(x)$ of degree m and produces two more polynomials $q(x)$ and $r(x)$ such that $f(x) = q(x)g(x) + r(x)$ and $\deg(r) < m$.

The computer implementation we'll use is close to the proof on Page 21; namely, we'll initially define $r(x)$ to be $f(x)$, then redefine $r(x)$ to be

$$r(x) - \frac{\text{leading coefficient of } r(x)}{\text{leading coefficient of } g(x)} x^{\deg(r)-m} g(x)$$

and keep on doing this until the degree of $r(x)$ is less than m. (For the example on Page 20, $r(x)$ is successively $7x^5 - 2x^3 + 9$, $-14x^4 + 26x^3 + 9$, $54x^3 - 56x^2 + 9$, $-164x^2 + 216x + 9$, and $544x - 647$). The final value of $r(x)$ ($544x - 647$ in our example) is then the remainder polynomial. Along the way, we'll keep track of the various numbers $\frac{\text{leading coefficient of } r(x)}{\text{leading coefficient of } g(x)}$ (7/2, -7, 27, -82 in our example) for they are the coefficients of the quotient polynomial $q(x)$.

It'll be useful to have a procedure which can look at a polynomial, decide whether or not its leading coefficient is really nonzero, and, if necessary, bring the degree down and redefine the coefficients until the leading coefficient really is nonzero. We'll call this trimming the polynomial. (Again using the example on Page 20, notice what happened after $7x^5 + 14x^4 - 28x^3$ was subtracted from $7x^5 - 2x^3 + 9$. We were left with $0x^5 - 14x^4 + 26x^3 + 9$, a purported 5th-degree polynomial which is really a 4th-degree polynomial. Trimming will render it as $-14x^4 + 26x^3 + 9$.)

Here's the program:

```
program divalgorithm;

type poly = array[-1..20] of real;                          {line 3}

var y:real;
    i,j,n,m:integer;
    f,g,q,r:poly;

procedure trim(var r:poly);
var j:integer;
begin
repeat                                                      {line 12}
if r[0]=0 then                                              {line 13}
  begin
  r[-1]:=r[-1]-1;                                           {line 15}
  for j:=0 to round(r[-1]) do r[j]:=r[j+1];                 {line 16}
  end;
until (r[0]<>0) or (r[-1]<0)                                {line 18}
end;

procedure dv(var r:poly; g:poly; y:real);                   {line 21}
var k,m,j:integer;
begin
k:=round(r[-1]); m:=round(g[-1]);                           {line 24}
for j:=0 to m do r[j]:=r[j]-y*g[j];                         {line 25}
trim(r);                                                    {line 26}
end;

begin                                                       {line 29}
write('enter the degree of f '); readln(n); f[-1]:=n;
writeln; writeln('enter, when prompted, the coefficients ');
for i:=0 to n do
  begin
  write('coefficient of x super ',n-i,' '); readln(f[i]);
  end;
writeln;
write('enter the degree of g '); readln(m); g[-1]:=m;
writeln; writeln('enter, when prompted, the coefficients ');
for i:=0 to m do
  begin
  write('coefficient of x super ',m-i,' '); readln(g[i]);
  end;
writeln;
q[-1]:=n-m;                                                 {line 44}
```

```
for i:=0 to round(q[-1]) do q[i]:=0;                    {line 45}
r:=f;                                                   {line 46}
repeat                                                  {line 47}
y:=r[0]/g[0];                                           {line 48}
q[n-round(r[-1])]:=y;                                   {line 49}
dv(r,g,y);                                              {line 50}
until r[-1]<m;                                          {line 51}
for i:=0 to round(q[-1]) do write(q[i]:6:6,' ');        {line 52}
writeln;
if r[-1]<0 then write('0')                              {line 54}
else for i:=0 to round(r[-1]) do write(r[i]:6:6,' ');
end.
```

The procedure trim does just what we said it should do above. In Line 13 it checks whether the leading coefficient of a polynomial $r(x) = a_0x^n + a_1x^{n-1} + \ldots + a_{n-1}x + a_n$ is zero. If it is, Line 15 reduces $\deg(r)$ by 1 and Line 16 redefines (the new) a_0 to be (the old) a_1, (the new) a_1 to be (the old) a_2, etc. (For instance, the old coefficients $a_0 = 0$, $a_1 = -14$, $a_2 = 26$, $a_3 = 0$, $a_4 = 0$, $a_5 = 9$ of the polynomial $0x^5 - 14x^4 + 26x^3 + 9$ get changed to $a_0 = -14$, $a_1 = 26$, $a_2 = 0$, $a_3 = 0$, $a_4 = 9$ so that the new polynomial is $-14x^4 + 26x^3 + 9$.) The loop defined by Lines 12 and 18 repeats the trimming until either the leading coefficient is nonzero or until $f(x)$ gets trimmed all the way down to the zero polynomial (a polynomial is the zero polynomial iff its degree is less than 0).

The second procedure takes as input two polynomials $r(x)$, $g(x)$ and a real number y, and changes $r(x)$ to $r(x) - yx^{\deg(r)-\deg(g)}g(x)$ (it doesn't change $g(x)$ or y; thus the Pascal var designation appears before **r** in Line 21, but not before **g** or **y**). Line 24 defines k to be $\deg(r)$ and m to be $\deg(g)$. Line 25 defines $r(x)$ to be $r(x) - yx^{k-m}g(x)$. This is seen as follows: suppose $r(x) = a_0x^k + a_1x^{k-1} + \ldots + a_{k-1}x + a_k$ and $g(x) = b_0x^m + b_1x^{m-1} + \ldots + b_{m-1}x + b_m$. Then $r(x) - yx^{k-m}g(x)$ will be $(a_0 - yb_0)x^k + (a_1 - yb_1)x^{k-1} + \ldots + (a_m - yb_m)x^{k-m} + a_{m+1}x^{k-m-1} + \ldots + a_{k-1}x + a_k$. Thus for $0 \le j \le m$ the jth coefficient of $r(x) - yx^{k-m}g(x)$ is $a_j - yb_j$, while for $m + 1 \le j \le k$ the jth coefficient is a_j. Line 25 defines — for $0 \le j \le m$ — the new jth coefficient of $r(x)$ to be the old jth coefficient minus y(jth coefficient of $g(x)$). For $m + 1 \le j \le k$ Line 25 does nothing. Finally Line 26 sends $r(x)$ to be trimmed.

The main program starts on Line 29. Lines 30–43 handle the input of $f(x)$ and $g(x)$. Line 44 defines $\deg(q)$ to be $n - m$ and Line 45 sets all coefficients of $q(x)$ equal to 0. Line 46 initially defines $r(x)$ to be $f(x)$. Lines 48 and 50 redefine $r(x)$ to be

$$r(x) - \frac{\text{leading coefficient of } r(x)}{\text{leading coefficient of } g(x)} x^{\deg(r)-m} g(x)$$

(Line 48 computes the fraction and calls it y. Line 50 calls the procedure that changes $r(x)$.) While this is going on, Line 49 sets the coefficient of $x^{\deg(r)-m}$ in the quotient polynomial equal to y. I.e., it sets $q[n-\deg(r)]$ equal to y. (For any term $c_i x^e$ in any polynomial $h(x)$, $i+e=\deg(h)$ so $i=\deg(h)-e$. In our case, $\deg(q)=n-m$ so the index i in the coefficient of $c_i x^{\deg(r)-m}$ is $i=(n-m)-(\deg(r)-m)=n-\deg(r)$.) The loop between Lines 47 and 51 keeps redefining $r(x)$ until $\deg(r)<m$. At this point both $q(x)$ and $r(x)$ have been determined. Line 52 displays $q(x)$. Lines 54, 55 display $r(x)$. (If $r(x)$ is the zero polynomial, i.e., if $\deg(r)<0$, then Line 54 displays it; otherwise, Line 55 does.)

Sample Runs

(1) For the example on Page 20, we get $q(x)=3.5x^3-7x^2+27x-82$ and $r(x)=544x-647$.

(2) For $f(x)=6x^4+4x^3+15x^2-3x+5$ and $g(x)=2x^2+2x+5$, we get $q(x)=3x^2-x+1$ and $r(x)=0$, i.e., $g(x)$ divides $f(x)$.

Program Modifications

(1) The procedure fancy in §Deflation can be added for a fancier on-screen display of the polynomials. Replace Line 52 by "`fancy(q);`" and Line 55 by "`else fancy(r);`".

(2) Arbitrary, changeable parameters: the polynomial degree limit 20 (Line 3); and the real number display parameter 6:6 (Lines 52, 55).

LAGRANGE INTERPOLATION (Section 2.2)

Recall from Page 48 that the problem is to find a polynomial $f(x)$ such that $f(a_i)=b_i$ for $i=1,\ldots,n+1$, where $a_1,\ldots,a_{n+1}$ and $b_1,\ldots,b_{n+1}$ are given numbers and furthermore $a_1,\ldots,a_{n+1}$ are distinct.

The program below will ask the user to input the numbers $a_1,\ldots,a_{n+1}$ (it will refer to these numbers on-screen as a1, a2, etc.; i.e., the subscripts will appear on-line). Next, it will display each a_i on-screen and ask the user to input the corresponding b_i. The program then computes and displays the Lagrangian interpolant polynomial. Here's the program:

```
program lagrange;

type poly = array[-1..20] of real;                    {line 3}
     vector = array[1..21] of real;                   {line 4}

var f,fsubi,g,one:  poly;
    a,b:vector;                                       {line 7}
    i,j,n:integer;
    y:real;
```

```
procedure clearpoly(var p:poly);
var i:integer;
begin
for i:=-1 to 20 do p[i]:=0;                              {line 14}
end;

procedure mult(p,q:poly; var prod:poly);                 {line 17}
var i,j:integer;
    x:real;
begin
prod[-1]:=p[-1]+q[-1];                                   {line 21}
for i:=0 to round(prod[-1]) do                           {line 22}
  begin
  x:=0;
  for j:=0 to i do x:=x+p[j]*q[i-j];
  prod[i]:=x;
  end;
end;

begin                                                    {line 30}
clearpoly(f); clearpoly(fsubi); clearpoly(g); clearpoly(one);
write('enter n '); readln(n); writeln;                   {line 32}
writeln('enter, when prompted, DISTINCT a1,a2,...,a(n+1)');
for i:=1 to n+1 do
  begin
  write('enter a',i,' '); readln(a[i]);
  end;
writeln;
writeln('enter, when prompted, the b numbers');
for i:=1 to n+1 do
  begin
  write('send ',a[i]:6:6,' to '); readln(b[i]);          {line 42}
  end;
writeln;
f[-1]:=n; one[-1]:=0; one[0]:=1;                         {line 45}
for i:=1 to n+1 do
  begin
  clearpoly(fsubi); clearpoly(g);                        {line 48}
  fsubi:=one;                                            {line 49}
  for j:=1 to n+1 do
    begin
    if j=i then g:=one;
    if j<>i then
      begin
```

```
      g[-1]:=1; y:=a[i]-a[j]; g[0]:=1/y; g[1]:=-a[j]/y;
      end;
    mult(fsubi,g,fsubi);                                    {line 57}
    end;
  for j:=0 to n do f[j]:=f[j]+b[i]*fsubi[j];               {line 59}
  end;
  for j:=0 to n do write(f[j]:6:6,' ');                    {line 61}
end.
```

We use 1-dimensional arrays called a and b to hold the numbers $a_1, a_2, \ldots, a_{n+1}$ and $b_1, b_2, \ldots, b_{n+1}$ respectively. These arrays are dimensioned from 1 to 21 (21 being 1 higher than our dimension limit on n). Thus we define the type vector in Line 4 and declare the variables **a, b** to be vectors in Line 7.

Other variables are: **f** for the Lagrangian interpolant polynomial (i.e., the answer); **fsubi** generically for any of the polynomials $f_1, f_2, \ldots, f_{n+1}$ (see the first display on Page 48); **g** for any of the degree-1 polynomials $\frac{x-a_j}{a_i-a_j}$; **one** for the fixed polynomial 1 (it will be convenient to do this); **i** and **j** for various integers where needed; and **y** for various real numbers where needed.

The procedure clearpoly clears (i.e, initializes) a polynomial array by setting all its entries equal to zero.

The procedure mult multiplies the polynomials p and q and calls the result prod. Since prod gets changed by the procedure (changed from being initially undefined to being the product of p and q), the Pascal var designation is used before variable **prod** in Line 17. Line 21 assigns a degree to the polynomial prod; namely the degree $\deg(p) + \deg(q)$. Lines 22–27 compute the ith coefficient of prod as $\sum_j p[j]q[i-j]$ (same definition as on Page 18 except here j replaces $n-k$). Note that in order for this definition to make sense we have to define undefined coefficients of the factor polynomials to be 0. (For instance, suppose $p(x) = a_0x^2 + a_1x + a_2$ and $q(x) = b_0x^2 + b_1x + b_2$. Our definition says that the coefficient c_3 of $p(x)q(x)$ is $c_3 = a_0b_3 + a_1b_2 + a_2b_1 + a_3b_0$, so we'll have to define $a_3 = b_3 = 0$.) This is done by the procedure clearpoly.

The main program begins on Line 30. Line 31 clears all the polynomials. Lines 32–44 handle the input of n and the numbers $a_1, a_2, \ldots, a_{n+1}$ and $b_1, b_2, \ldots, b_{n+1}$. Line 45 defines $\deg(f)$ to be n and also defines the fixed polynomial one. Lines 48–59 compute, for each i, the polynomial f_i. Recall (see the second display on Page 48) that f_i is the product of the degree-1 polynomials $\frac{x-a_1}{a_i-a_1}, \frac{x-a_2}{a_i-a_2}, \ldots$. In other words, we want to do a repeated multiplication. So we initially define fsubi to be one in Line 49. Then for all j Line 57 redefines fsubi as the old value of fsubi times the polynomial g, where g has been defined in the previous 5 lines to be $\frac{x-a_j}{a_i-a_j}$ if $j \neq i$; or to be one if $j = i$.

Once f_i has been computed, the program multiplies it by b_i and then adds that product to the running total it has been keeping for f. I.e., — again refer to the first display on Page 48 — f is computed by initially setting $f = 0$ (this was done by procedure clearpoly) and then, for $1 \leq i \leq n+1$, redefining f to be $f + b_i f_i$ (Line 59; polynomials are added by adding corresponding coefficients). Finally, Line 61 displays f.

Sample Runs

(1) For $n = 4$; $a_1 = 0$, $a_2 = -1$, $a_3 = 1$, $a_4 = -2$, $a_5 = 2$; $b_1 = 1$, $b_2 = 11.65$, $b_3 = -2.25$, $b_4 = 122.1$, and $b_5 = -1.7$, we get $3.7x^4 - 8x^3 + 1.05x + 1$.

(2) For $n = 4$; $a_1 = 1.30277564$, $a_2 = -2.30277564$, $a_3 = 2.61803399$, $a_4 = 0.38196601$, $a_5 = 0$; $b_1 = b_2 = b_3 = b_4 = 0$, $b_5 = -6$, we get $2x^4 - 4x^3 - 10x^2 + 20x - 6$ (see the example on Pages 100, 101).

Program Modifications

(1) The procedure fancy in §Deflation can be added for a fancier on-screen display of the polynomial. Replace Line 61 by "`fancy(f);`".

(2) Arbitrary, changeable parameters: the polynomial degree limit 20 (Lines 3, 14); the array coordinates limit 21 (Line 4); and the real number display parameter 6:6 (Lines 42, 61). The array coordinates limit should be 1 higher than the polynomial degree limit.

EXERCISES

1. Using the procedures mult and clearpoly of program lagrange, write a program to multiply polynomials in $\mathbf{R}[x]$. Try it on $3x^2-6x+2$, $2x^2+5x+7$ (answer: $6x^4 + 3x^3 - 5x^2 - 32x + 14$) and on $2x^4 - 7x^3 + 5x^2 + 2x + 3$, x^3+8x^2-6x+2 (answer: $2x^7+9x^6-63x^5+88x^4-25x^3+22x^2-14x+6$).

2. Write a program to transform the polynomial $f(x) \in \mathbf{R}[x]$ to $g(x) = f(x-b)$. Try it on $f(x) = 2x^3 - 7x^2 + 3x + 2$, $b = 1$ (answer: $g(x) = 2x^3 - 13x^2 + 23x - 10$) and on $f(x) = 3x^6 - x^4 + 7x^3 + 7x + 7$, $b = -2$ (answer: $g(x) = 3x^6 + 36x^5 + 179x^4 + 479x^3 + 738x^2 + 635x + 253$).

3. Modify procedure sr in §Complex Arithmetic to find n-th roots (n an integer) of complex numbers z using De Moivre's theorem. Try it on $n = 3$, $z = 42.382084 + 1.445796i$ (answer: $3.486986 + 0.039637i$) and on $n = 4$, $z = 41 + 840i$ (answer: $5 + 2i$).

4. Using the function bis (of the bisection program), the procedure def, and the quadratic formula, write a program to find all 3 roots of a cubic polynomial. Try it on $x^3 + 25x - 90$ (answer: 2.759486 and $-1.379744 \pm 5.541758i$) and on $2x^3 - 14x^2 + 29x - 18$ (answer: 2 and $(5 \pm \sqrt{7})/2$ ($\doteq 3.822876, 1.177124$)).

5. Write a program to find all 3 roots of a cubic polynomial using Cardan's method. (Use Exercise 3 for complex cube roots.) Try it on the

polynomials given in Exercise 4 and compare results.

6. Let $f(x)$ be a monic cubic polynomial. Let $g(x)$ be the transformed polynomial with missing x^2 term, namely $g(x) = f(x - \frac{f_1}{3}) = x^3 + px + q$, where $p = f_2 - \frac{f_1^2}{3}$ and $q = f_3 - \frac{f_1 f_2}{3} + \frac{2f_1^3}{27}$. Recall that $g(x)$ has a multiple root if and only if the discriminant $\Delta = 4p^3 + 27q^2$ is zero. Moreover it is not hard to show that if $\Delta = 0$, then this multiple root is $-3q/2p$ if $p \neq 0$ and is 0 if $p = 0$. Since all 3 roots sum to -0, the third root then is $3q/p$ if $p \neq 0$ and is 0 if $p = 0$. Transforming back to $f(x)$ (by subtracting $f_1/3$ from roots of $g(x)$) we have: If $\Delta = 0$, then the roots of $f(x)$ are $\frac{-3q}{2p} - \frac{f_1}{3}$, $\frac{-3q}{2p} - \frac{f_1}{3}$, and $\frac{3q}{p} - \frac{f_1}{3}$ if $p \neq 0$ and are $-f_1/3$, $-f_1/3$, and $-f_1/3$ if $p = 0$.

(a) Write a program to check whether or not a cubic polynomial has a multiple root by computing Δ and checking whether or not it is 0 (or, more realistically, whether or not $|\Delta| < 0.000001$). The program should go on to compute the 3 roots (as given above) if $\Delta = 0$ or to describe the nature of the roots if Δ is positive or negative (see the list on Page 84). Try the program on $2x^3 - 22x^2 + 78x - 90$ and on $x^3 - 0.6x^2 + 0.12x - 0.008$. (Sample runs of such a program gave 3.000000, 3.000000, 5.000000 for the first polynomial and 0.200000, 0.200000, 0.200000 for the second.)

(b) Combine the programs in Part (a) and Exercise 4 into a program that will compute Δ for a cubic polynomial; then find all roots by the above method if $\Delta = 0$, or by bisection and deflation if $\Delta \neq 0$.

7. (a) If $f(x) \in \mathbf{R}[x]$ has even degree and if the sum of its even coefficients $(a_0 + a_2 + \ldots + a_n)$ is less than the sum of its odd coefficients, show that $f(-1)$ is negative.

(b) Modify the bisection program to find roots of such polynomials (assuming positive leading coefficients). Try it on $x^4 - 2x^3 - 2x^2 + 8x - 8$ (answer: 2) and on $8x^6 + 72x^5 + 226x^4 + 294x^3 + 152x^2 + 33x + 2.5$ (answer: $(-3 + \sqrt{7})/2 \doteq -0.177124$).

8. Modify the program regulafalsi(real) to use Newton's method rather than regula falsi. Try it on $3x^4 - 12x^3 - 39x^2 + 84x + 126$ and on $2x^3 + 15x^2 - 57x + 30$. (Sample runs of such a program, both using the initial guess 2, produced, for the first polynomial, the same root found by regulafalsi(real), namely 2.645751 and, for the second polynomial, the same root as on Page 173, namely 2.238439.)

9. Combine the function bis, the procedure def, and the Ferrari-method program into a program that will find all 5 roots of a quintic polynomial. Try it on $x^5 - 60x^3 + 195x^2 + 10x - 396$ (answer: $-9, (7 \pm \sqrt{5})/2$ ($\doteq$ 4.618034, 2.381966), and $1 \pm \sqrt{5}$ ($\doteq$ 3.236068, -1.236068)) and on $4x^5 + 4x^4 + 8x^3 + 8x^2 + 8x + 4$ (answer: -0.682328, $-0.5 \pm 0.866025i$, and $0.344164 \pm 1.161541i$).

10. Combine the regulafalsi(complex) program and both deflation procedures into a program that will find a root r of a polynomial by regula falsi, then deflate the polynomial (by r if r is real or by r and r^* if r

is complex), then go back into regula falsi to solve the deflated polynomial, and keep solving and deflating until all roots are found. Try it on $x^6 - 2x^5 - 3x^4 - 7x^3 + 14x^2 + 15x + 18$ (answer: $-0.5 \pm (\sqrt{3}/2)i$, 2, 3, and $-1 \pm \sqrt{2}i$) and on $1.6x^8 - 6.4x^7 + 92x^6 - 144x^5 + 109.7x^4 + 1654x^3 - 5910.4x^2 + 3092x + 450$ (answer: $-1 \pm \frac{\sqrt{13}}{2}$ ($\doteq 0.802776, -2.802776$), $1 \pm \frac{\sqrt{5}}{2}$ ($\doteq 2.118034, -0.118034$), $1 \pm 3i$, and $1 \pm 7i$).

The exercises below involve $Z_p[x]$. Check to see whether or not the mod function on your compiler returns only nonnegative numbers (for instance, for $p = 5$, it should send $1 - 4$ to 2, not to -3). If it does not, first write a function (called, say, altmod) that does.

11. Modify your program of Exercise 1 to multiply polynomials in $Z_p[x]$. Try it on $p = 7$, $3x^5 + 2x^4 + 5x^2 + 3x + 1$, $2x^3 + 6x^2 + x + 4$ (answer: $6x^8 + x^7 + x^6 + 3x^5 + 2x^4 + 4x^3 + x^2 + 6x + 4$) and on $p = 5$, $x^4 + 3x^3 + x^2 + 2x + 3$, $x^4 + 2x^3 + 4x^2 + x + 4$ (answer: $x^8 + x^6 + 2x^5 + 3x^4 + 2x^3 + 3x^2 + x + 2$).

12. Write a program to perform the division algorithm for polynomials in $Z_p[x]$. Try it on $p = 7$, $f(x) = x^6 + 2x^5 + 3x^4 + x + 3$, $g(x) = x^2 + 4x + 1$ (answer: $q(x) = x^4 + 5x^3 + 3x^2 + 4x + 2$ and $r(x) = 3x + 1$) and on $p = 11$, $f(x) = x^7 + x^6 + x^5 + 9x^4 + 2x + 1$, $g(x) = 2x^3 + 5x + 7$ (answer: $q(x) = 6x^4 + 6x^3 + 2x^2 + 7x + 7$ and $r(x) = 6x^2 + 6x + 7$).

13. Modify program lagrange to find Lagrange interpolants in $Z_p[x]$. Try it on $p = 17$, $n = 5$, $a_1 = 0$, $a_2 = 1$, $a_3 = 2$, $a_4 = 3$, $a_5 = 4$, $a_6 = 5$, $b_1 = 4$, $b_2 = 3$, $b_3 = 6$, $b_4 = 14$, $b_5 = 13$, $b_6 = 9$ (answer: $x^5 + 12x^3 + 2x^2 + x + 4$) and on $p = 13$, $n = 5$, $a_1 = 0$, $a_2 = 1$, $a_3 = 2$, $a_4 = 3$, $a_5 = 4$, $a_6 = 5$, $b_1 = 1$, $b_2 = 3$, $b_3 = 1$, $b_4 = 9$, $b_5 = 9$, $b_6 = 1$ (answer: $x5^5 + x^4 + x^2 + 5x + 1$).

14. Write a program to find g.c.d.s of polynomials in $Z_p[x]$ (see Pages 23,24). Try it on $p = 5$, $f(x) = 3x^5 + 4x^4 + 2x^3 + 3x^2 + 4x + 4$, $g(x) = 3x^4 + 3x^3 + 3x^2 + 4x + 2$ (answer: $x^2 + x + 3$ or any scalar multiple) and on $p = 29$, $f(x) = x^7 + 18x^6 + 27x^5 + 23x^4 + 2x^3 + x^2 + 27x + 8$, $g(x) = x^6 + 16x^5 + 23x^4 + 23x^3 + 10x^2 + 3x + 10$ (answer: $x^5 + 12x^4 + 4x^3 + 7x^2 + 11x + 17$ or any scalar multiple).

Note: The analogous problem for polynomials in $\mathbf{R}[x]$ is more difficult owing to a build-up of roundoff errors from the repeated use of the division algorithm.

Index